AF443155

*Engineering Approaches to
High Temperature Design*

Engineering Approaches to High Temperature Design

Volume 2 in the Series
Recent Advances in Creep and Fracture
of Engineering Materials and Structures

Edited by:

B. Wilshire
Department of Metallurgy and Materials Technology,
University College, Swansea

D. R. J. Owen
Department of Civil Engineering,
University College, Swansea

PINERIDGE PRESS

Swansea, U.K.

First Published 1983 by
Pineridge Press Limited
91, West Cross Lane, West Cross, Swansea, U.K.

Copyright © 1983 by Pineridge Press Limited.

Wilshire, B.
Engineering Approaches to High Temperature Design
1. Heat resistant materials
I. Title   II. Owen, D. R. J.
620.1'1217   TA418.26

ISBN 0-906674-29-8

Printed and bound in Great Britain by
Dotesios (Printers) Ltd., Bradford-on-Avon, Wiltshire

PREFACE

The need to base the design of components and structures for high temperature service on the time-dependent properties of the relevant materials has long been recognized. If designers were concerned only with comparatively simple shapes operating under steady conditions of stress and temperature, design procedures would be straightforward. However, for real situations in electricity generating plant, petrochemical installations, aeroengines and other high temperature service applications, components and structures having complicated shapes must perform satisfactorily under complex non-steady stress and temperature conditions, often in hostile environments. Moreover, in order to improve efficiency, the severity of the operating conditions has gradually increased, a trend which has by no means ceased. The safe and economical design of components and structures for high temperature service therefore continues to pose major engineering problems. For this reason, the present text has been devoted to 'Engineering Approaches to High Temperature Design'.

As with the initial volume in this Recent Advances series, authors of international standing have been invited to prepare self-contained articles which combine a general overview of the current position in their chosen speciality with a detailed coverage of their recent activities. Apart from this broad-based directive, no constraints on manuscript length or content were imposed. The structure of the volume was therefore determined primarily by author selection with the overall aim of providing, in a single text, a distinctive new contribution to the literature available on modern concepts in high temperature design.

B. Wilshire

D.R.J. Owen

University College of Swansea,
Singleton Park, Swansea.

May 1983.

CONTENTS

Chapter 1

CRACK GROWTH AT HIGH TEMPERATURE

G.A. Webster

Department of Mechanical Engineering,
Imperial College of Science and Technology,
Exhibition Road, London SW7 2BX, England.

SUMMARY

In this chapter methods of predicting creep crack growth
in engineering components operating at elevated temperatures
are presented. A continuum mechanics approach is adopted.
Initially relevant linear and non-linear fracture mechanics
concepts are reviewed. A creep contour integral termed C^*,
which corresponds to the J-contour integral used in
plasticity, is defined for determining the stress, strain and
strain rate fields ahead of a crack tip in a material
deforming in secondary creep according to the Norton creep
law. Approximate methods for estimating C^*, based on limit
analysis techniques, are outlined and shown to be accurate
to within 25% for most test-piece geometries.

Comparison of the C^* parameter with experimental data on
a range of materials exhibiting ductilities from about 0.1
to 45% reveals that it is capable of correlating creep crack
growth rate $\dot{a}$ by an expression of the form $\dot{a} \propto C^{*\phi}$, where
$\phi = n/(n+1)$ and n is the stress exponent of secondary creep.
This behaviour is explained in terms of exhaustion of the
available creep ductility in a process zone at the crack tip.
It is found that cracking rate is insensitive to the zone
size but inversely proportional to the strain that can be
tolerated at the crack tip at fracture. The model predicts
that the relative resistance of materials to crack extension
at elevated temperatures is not necessarily in proportion
to their creep strengths.

Some evidence is presented to indicate that, when crack
growth under cyclic loading conditions is controlled by
time dependent processes, C^* can be used to correlate the
data.

1.   INTRODUCTION

Several procedures are available for assessing the
useful lives of engineering components which operate in the
creep range.   The particular procedure adopted will depend
upon the application of the component and consequences of
failure.   Important considerations determining lifetimes
include avoidance of excessive creep deformation and
preventing fracture

Two broad categories of structures can be identified;
undefected structures and those containing crack like defects.
Different design methods are needed for each category
although it is possible for an initially unflawed structure
to develop cracks during use so that its category may change.
The chief concern with defected structures is to determine
the influence of the crack on the integrity of the structure.

Life prediction methods for undefected components are
generally based on uni-axial tensile creep deformation and
rupture data.   Creep parameters are often used to rationalize
the data and obtain more reliable extrapolations.   In
addition, established procedures are available for dealing
with complex stress states using equivalent stress criteria
[1-3].   Where appropriate allowance for the development
of damage (usually in the form of voids) resulting from
creep deformation can be incorporated into the predictions
using the concepts of Kachanov[4] and Rabotnov[5,6] or
the void growth models of Cocks and Ashby[7].   Design
methods based on the above procedures are most relevant to
those engineering structures which fail in the same manner
as tensile creep specimens.

Failure in uni-axial tension specimens usually takes
place by the initiation and growth of voids on grain
boundaries perpendicular to the tensile axis[7-10].   It
seldom takes place by crack propagation.   Consequently,
failure lives are governed primarily by the time taken for
the voids to grow to such a size that the tensile strength
of the material is exceeded.

Large engineering components which have been fabricated
by welding, and which contain sites of stress concentration,
are most likely to contain inherent defects or form cracks
during use.   The cracks are most likely to develop in regions
of high stress gradient where sufficient constraint exists
to prevent relaxation of the high local stresses.   In this
circumstance, damage is confined to the immediate crack tip
vicinity and a sharp crack is maintained which can propagate
by creep.   This mode of failure is not characteristic of
the more diffuse development of damage which occurs in
uni-axial tension specimens. It is not apparent therefore,

that uni-axial creep data can be used to provide reliable
estimates of the resistance of a material to creep crack
growth any more than tensile strength can be used as an
accurate indication of fracture toughness. Indeed there is
some evidence to suggest that several alloys that show good
creep rupture strength have poor resistance to creep crack
growth.

Design procedures for predicting creep crack growth in
flawed structures are still being developed. As the use
and accuracy of non-destructive inspection techniques increase
smaller and smaller cracks are being detected and the question
of whether a cracked component can be returned to service,
or must be replaced, is being faced more frequently. There
comes a time when automatic replacement becomes uneconomic.

The primary objective of this chapter is to develop
criteria for assessing the lives of cracked components when
failure is governed by time dependent crack growth. A
continuum mechanics approach will be adopted. Attention will
not be focussed on the individual process (for example
transgranular failure, intergranular failure or void growth)
governing separation at the crack tip although this will
clearly determine the response of a particular material. The
aim will be to establish a parameter to characterise the state
of stress and strain rate local to the crack tip so that the
behaviour of different defected structures may be compared and
predicted. Fracture mechanics will be applied and consideration given to ductile and brittle materials. Some effects of
superimposed cyclic loading will also be mentioned briefly.

2.   REVIEW OF FRACTURE MECHANICS CONCEPTS

The discipline of fracture mechanics has been developed
to assess the integrity of flawed structures. It was applied
first to predict brittle fracture but has since been used to
describe ductile fracture, fatigue crack growth and stress
corrosion cracking [11-16] . It is only relatively recently
that it has been used to characterise creep crack growth,
(see for example two recent reviews [17,18]).

2.1   Principles of Linear Elastic Fracture Mechanics

The principles of linear elastic fracture mechanics
(LEFM) have been dealt with in detail in a number of books
(see for example Knott [14]). Only the essential features
will be covered here. Attention will be restricted to through
thickness cracks and the opening mode (Mode I) of crack
extension since these are the situations which have been
examined most extensively at elevated temperatures. Also
Mode I crack growth is of most practical relevance as it takes
place more easily than the shear modes (Modes II and III).

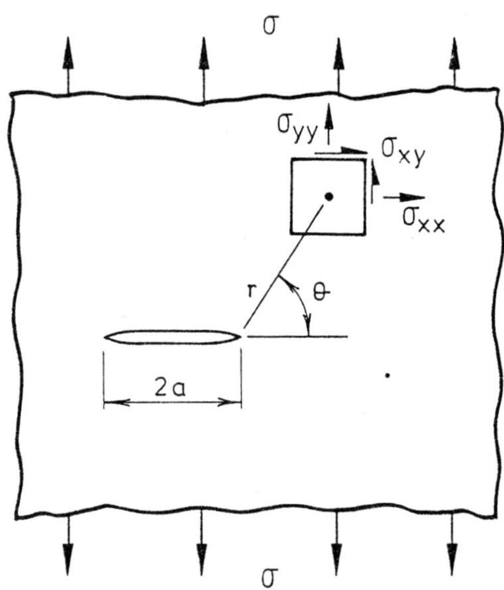

Figure 1. Stress Distribution Ahead of a Crack Tip.

The elastic stresses local to a crack tip can be
determined from stress function solutions. Consider Figure 1.
which shows a through thickness crack of length $2a$ subjected to
a remote stress σ applied perpendicular to the crack plane.
For the coordinate system indicated, it can be shown [19]
that the stresses at coordinates (r,θ) ahead of the crack tip
are given by

$$\sigma_{yy} = \frac{K}{\sqrt{2\pi r}}\; cos\frac{\theta}{2}\left[1\; +\; sin\frac{\theta}{2}\; sin\frac{3\theta}{2}\right] +..... \quad (1a)$$

$$\sigma_{xx} = \frac{K}{\sqrt{2\pi r}}\; cos\frac{\theta}{2}\left[1\; -\; sin\frac{\theta}{2}\; sin\frac{3\theta}{2}\right] +.... \quad (1b)$$

$$\sigma_{xy} = \frac{K}{\sqrt{2\pi r}}\; cos\frac{\theta}{2}\left[sin\frac{\theta}{2}\; cos\frac{3\theta}{2}\right] \quad\quad +.... \quad (1c)$$

Equations (1) apply for any geometry and hold for stresses
local to the crack tip. For stresses further away additional
terms involving $r^{\frac{1}{2}}$, $r^{3/2}$ ...etc. must be included.
However, close to the crack tip it is clear that the singular
term in $r^{-\frac{1}{2}}$ dominates. The paramter K, which defines the
magnitude of the singularity at the crack tip for all the
stress components, is called **the stress** intensity factor.

It is normally written as K_I for Mode I loading. However, as
this is the only mode being considered here the subscript
will be omitted. For an infinite plate, K is given by

$$K = \sigma\sqrt{\pi a} \qquad (2)$$

For any other geometry it can be written as

$$K = Y\,\sigma\sqrt{a} \qquad (3)$$

where Y is a non-dimensional factor which in general will
depend upon crack size and component dimensions. Values of
Y for a selection of common test-piece geometries are included
in the Appendix. Other solutions can be obtained from
reference handbooks [20,21] .

An alternative characterisation of the crack tip region
is possible in terms of energy release rate G, where G is the
energy expended in extending a crack over unit area.

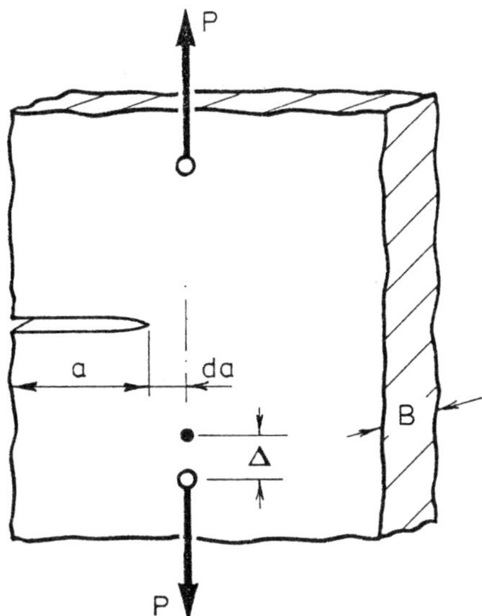

Figure 2. Loading of a Cracked Body.

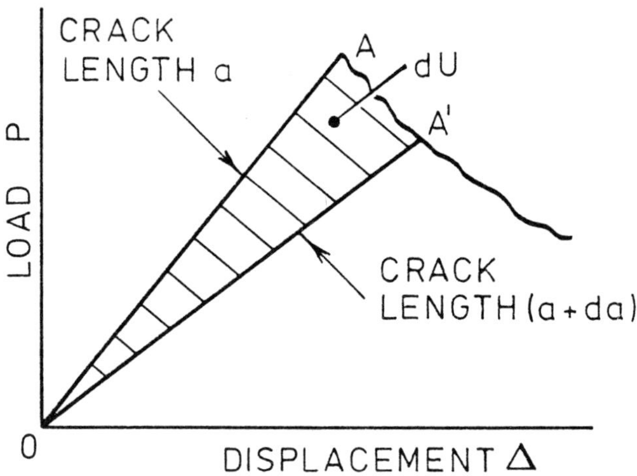

Figure 3. Elastic Load P - Displacement Δ Lines for
Different Crack Lengths.

Consider Figure 2 which shows a body of thickness B and
arbitrary shape containing an edge crack of length a. Its
corresponding elastic load P versus load point displacement
Δ diagram will be represented by the line OA in Figure 3. At
A the crack is allowed to extend da along the path AA'. The
line OA' will then correspond to the elastic loading line of
the same body having a crack length $(a+da)$. The shaded area
dU is therefore the energy required to extend the crack, da
and G becomes

$$G = \frac{dU}{B\,da} \qquad (4)$$

which can be expressed in terms of specimen compliance
$c = \Delta/P$ as

$$G = \frac{P^2}{2B}\,\frac{dc}{da} \qquad (5)$$

Equation (5) is valid irrespective of the path AA'. For crack
extension at constant load it has the special form

$$G = \frac{P}{2B}\,\frac{d\Delta}{da} \qquad (6)$$

and for crack extension at constant displacement it becomes

$$G = \frac{-\Lambda}{2B} \frac{dP}{da} \qquad (7)$$

Equations (5) to (7) are convenient expressions for evaluating G experimentally or by computation for any cracked body. The conventional Griffith fracture theory provides a link between the stress intensity and energy characterisations of fracture. For an infinite plate loaded as indicated in Figure 1, it can be shown [14] that for plane stress conditions

$$G = \frac{\sigma^2 \pi a}{E} \qquad (8)$$

and for plane strain

$$G = \sigma^2 (1 - \nu^2) \frac{\pi a}{E} \qquad (9)$$

Comparison of these equations with equation (2) gives

$$K^2 = EG \qquad (10)$$

for plane stress and

$$K^2 = \frac{EG}{(1 - \nu^2)} \qquad (11)$$

for plane strain. In these equations E is the elastic modulus and ν Poisson's ratio. Relations (10) and (11) apply for all cracked bodies irrespective of shape. Consequently the state of stress around a crack tip can be characterised in terms of K or G. Both approaches will be applied to the creep situation. Clearly K and G can both be evaluated as a function of load and crack length for any geometry. In the Griffith theory it is postulated that unstable fracture takes place when K and G reach their corresponding critical values K_c and G_c respectively. Data will be presented to show that time dependent crack growth can take place at elevated temperatures before these values are attained.

2.2   Non-Linear Fracture Mechanics

In practice because of plasticity and/or creep the stresses local to a crack tip will be modified from those calculated assuming   linear elastic behaviour. We will first consider the case in the absence of creep deformation. The extent of the modification will depend on the yield stress of

8

the material and its work hardening characteristics. For
substantial plastic deformation K (and G) and equations (1)
will cease to describe adequately the stress state local to
the crack tip. In these circumstances it is necessary to
resort to non-linear fracture mechanics concepts. This can
be achieved by making reference to Figure 4 which shows an
arbitrary contour Γ surrounding a crack tip in a two
dimensional body of unit thickness. The contour starts on
the lower face of the crack and finishes on the upper surface.

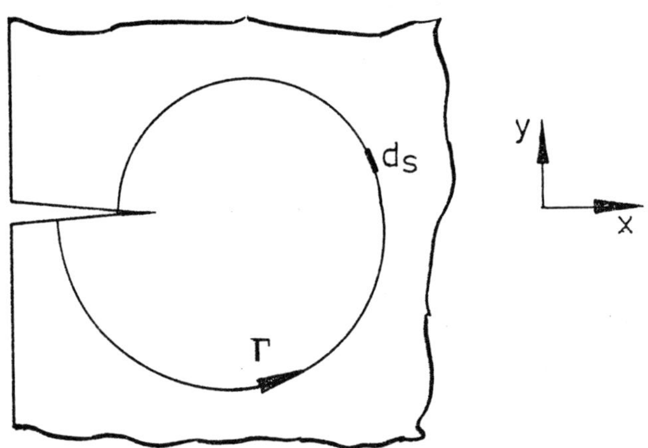

Figure 4. Arbitrary Contour Around a Crack Tip.

It is possible [20] to define a line integral J around
Γ such that

$$J = \int_{\Gamma} W_s \, dy - T_i \frac{\partial u_i}{\partial x} \, ds \qquad (12)$$

where W_s is strain energy density given by

$$W_s = \int \sigma_{ij} \, d\varepsilon_{ij} \qquad (13)$$

with σ_{ij} and ε_{ij} stress and strain tensors respectively, T_i
and n_i components of the traction and displacement vectors,
x and y the coordinates and s arc length along Γ. For non-
linear elastic situations or when 'deformation' plasticity
theory is valid J is path independent (hence its common
name, path independent J contour integral) and can be used
to characterise the stress state ahead of a crack tip in the
same way as G for linear elasticity. Experimentation and

computation [16] have also shown, despite lack of proof, that the J concept can be extended to situations involving incremental plasticity behaviour.

The actual stress distribution ahead of a crack tip depends on the work-hardening characteristics of the material. For a power-law hardening material deforming in uni-axial tension according to the expression

$$\frac{\varepsilon}{\varepsilon_Y} = \left(\frac{\sigma}{\sigma_Y}\right)^N \tag{14}$$

where σ_Y and ε_Y are yield stress and strain respectively and N is a constant, the corresponding multi-axial stress equation is

$$\frac{\varepsilon}{\varepsilon_Y} = \frac{3}{2}\left(\frac{\sigma_e}{\sigma_Y}\right)^{(N-1)} \frac{\sigma'_{ij}}{\sigma_Y} \tag{15}$$

where σ'_{ij} is the deviatoric stress tensor and σ_e is the equivalent stress which may be defined in terms of the Tresca or Von-Mises equivalent stress criteria. Using equation (15), the stress field at coordinates (r,θ) ahead of the crack tip is given by [21,22]

$$\frac{\sigma_{ij}}{\sigma_Y} = \left[\frac{J}{I_N \sigma_Y \varepsilon_Y r}\right]^{1/(N+1)} \tilde{\sigma}_{ij}(\theta,N) \tag{16}$$

where $\tilde{\sigma}_{ij}(\theta,N)$ is a non-dimensional function of N and θ chosen so that $\tilde{\sigma}_e(\theta,N)$ has a maximum value of unity. The factor I_N has been tabulated by Hutchinson [21] and is given approximately by McClintock [23] as,

$$I_N = 10.3(0.13 + \frac{1}{N})^{\frac{1}{2}} - \frac{4.8}{N}$$

The corresponding expression for the strain field ahead of the crack tip becomes by substitution of equation (16) into equation (15)

$$\frac{\varepsilon_{ij}}{\varepsilon_Y} = \left[\frac{J}{I_N \sigma_Y \varepsilon_Y r}\right]^{N/(N+1)} \tilde{\varepsilon}_{ij}(\theta,N) \tag{17}$$

where $\tilde{\varepsilon}_{ij}(\theta,N)$ is another non-dimensional function of θ and N.

10

Equation (16) characterises the stress distribution ahead of a crack in a power-law hardening material in the same way that equations (1) do for a linear elastic material. When $N = 1$ and $\sigma_Y/\varepsilon_Y = E$ in equation (14), $J = G$ and equation (16) reduces to equations (1) as required for linear elastic behaviour. Application of the J contour is not restricted to power-law hardening materials, but this form of behaviour has been considered because of the similarity of the stress dependence in equation (14) to that in the Norton creep law.

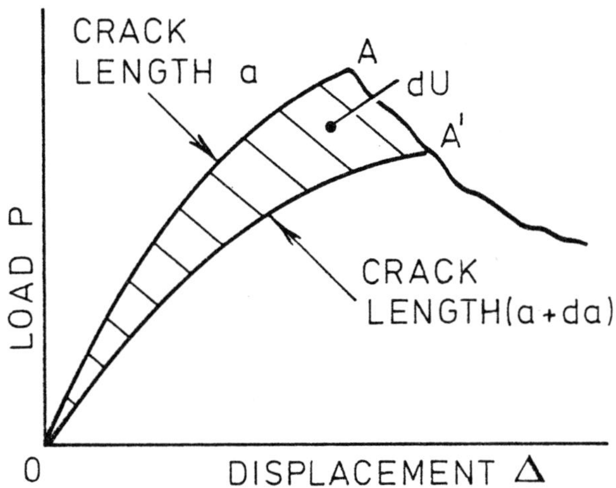

Figure 5. Load-Displacement Diagram for a Power-Law
Hardening Material

Rice [26] has shown that J can be obtained from a load-displacement diagram of a cracked body in the same way as G. Following the arguments of section 2.1 J becomes

$$J = \frac{dU}{B\,da} \qquad (18)$$

where dU is the shaded area in Figure 5. For power-law hardening behaviour the deformation and stress histories at all points in the body will be proportional and it can be shown from Ilyushin's theorem [27] that

$$\Delta = C_N P^N \qquad (19)$$

where C_N can be regarded as the plastic compliance of the cracked body. In general C_N will depend upon crack length. Application of equation (19) gives,

$$dU = (Pd\Delta - N\Delta dP)/(N+1)$$

$$= \frac{P^{(N+1)}}{(N+1)} \, dC_N \tag{20}$$

Hence substitution in equation (18) gives J for a power-law hardening material as,

$$J = \frac{P^{(N+1)}}{(N+1)B} \, \frac{dC_N}{da} \tag{21}$$

As with the corresponding linear elastic equation (5), this expression is valid for any path AA'. For crack extension at constant load it reduces to,

$$J = \frac{P}{(N+1)B} \cdot \frac{d\Delta}{da} \tag{22}$$

and for crack extension at constant displacement it becomes,

$$J = \frac{-N\,\Delta}{(N+1)B} \cdot \frac{dP}{da} \tag{23}$$

Equations (21) to (23) are convenient expressions for determining J as a function of crack length, either experimentally or by computation, for any cracked body exhibiting power-law hardening behaviour. When $N = 1$ these equations reduce to the corresponding elastic relations (equations (5) to (7)) as required.

It will be demonstrated how this approach can be extended to creeping materials.

3. APPLICATION OF FRACTURE MECHANICS CONCEPTS TO CREEP

In this approach it is argued that creep crack growth is controlled by the state of stress and strain rate local to the crack tip. Immediately on loading the stress distribution at the crack tip will be elastic (in the absence of plastic deformation) and described by K. For 'creep brittle' circumstances where little creep deformation accompanies fracture, the stress distribution will remain virtually unaltered by creep and K will continue to characterise the stresses around the crack. However, when substantial creep deformation accompanies fracture stress redistribution will occur at the crack tip and K will no longer adequately describe the local state of stress.

12

Consider a material which deforms in secondary creep according to the Norton uni-axial creep law;

$$\frac{\dot{\varepsilon}}{\dot{\varepsilon}_o} = \left(\frac{\sigma}{\sigma_o}\right)^n \tag{24}$$

where $\dot{\varepsilon}$ is creep strain rate and $\dot{\varepsilon}_o$, σ_o and n are constants at constant temperature. Note the similarity between this equation and the power-law hardening relation (equation (14)). Equilibrium and compatibility conditions dictate that if $N = n$ the stress distribution at a crack tip in two identically loaded components, one deforming plastically according to equation (14) and the other creeping according to equation (24), will be the same. (This can be visualized more clearly, for example, by comparing the stress distributions in plastic and creeping beams subjected to bending). The only difference will be that a load P applied to the former will result in a displacement Δ whereas it will cause a displacement rate $\dot{\Delta}$ in the latter. By analogy with the work-hardening case, therefore, it is possible to define a contour C^* (sometimes called $\dot{J}$ or J^*) [18,28-31] which characterizes the state of stress ahead of a crack tip in a creeping material in the same way that J does for the work-hardening case. The definition of C^* for a contour Γ (such as shown in Figure 4) surrounding a crack tip in a two dimensional body of unit thickness becomes by comparison with equation (12)

$$C^* = \int_\Gamma W_s^* \, dy - T_i \, \frac{\partial \dot{u}_i}{\partial x} \, ds \tag{25}$$

where W_s^* is strain energy rate density and is given by

$$W_s^* = \int \sigma_{ij} \, d\dot{\varepsilon}_{ij} \tag{26}$$

and $\dot{u}_i$ is the displacement rate vector. When secondary creep dominates all other strain rate terms will be negligible. Consequently, for a given crack length and loading conditions, C^* will be independent of time and constant creep stress and strain rate fields will be obtained at the crack tip corresponding to the work-hardening situation. The C^* integral is equivalent to the Rice [22] J-contour integral (equation (12)) but with strain replaced by strain rate in equations (25) and (26). C^* will be subject to the same restrictions with respect to path independence as J. When path independence is valid, C^* can be used to characterise the stress and strain rate fields around a crack tip in a material in which creep deformation dominates. The stress field at coordinates (r, θ)

ahead of a crack tip is given, by comparison with equation (16), as

$$\frac{\sigma_{ij}}{\sigma_o} = \left[\frac{C^*}{I_n \sigma_o \dot{\varepsilon}_o r}\right]^{1/(n+1)} \tilde{\sigma}_{ij}(\theta, n) \qquad (27)$$

where $\tilde{\sigma}_{ij}(\theta, n)$ is a non-dimensional function of n and θ chosen so that $\tilde{\sigma}_e(\theta, n)$ has a maximum value of unity in the same manner as for the plasticity case. The factor I_n is the same as I_N with $N = n$ and is

$$I_n = 10.3(0.13 + \frac{1}{n})^{\frac{1}{2}} - \frac{4.8}{n} \qquad (28)$$

The corresponding equation for the strain rate field ahead of the crack tip is,

$$\frac{\dot{\varepsilon}_{ij}}{\dot{\varepsilon}_o} = \left[\frac{C^*}{I_n \sigma_o \dot{\varepsilon}_o r}\right]^{n/(n+1)} \tilde{\dot{\varepsilon}}_{ij}(\theta, n) \qquad (29)$$

where $\tilde{\dot{\varepsilon}}_{ij}(\theta, n)$ is another non-dimensional function of θ and n. The non-dimensional factors in these equations can be obtained from the previous work-hardening solutions [23,24] by substituting $N = n$.

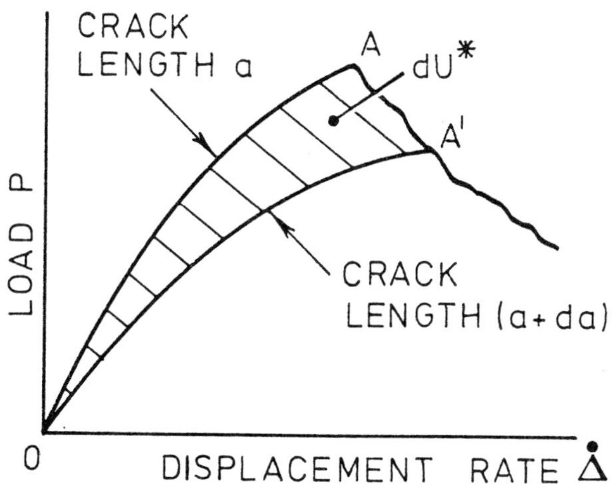

Figure 6. Load-Displacement Rate Diagram for a Material Deforming in Secondary Creep.

Just as J can be determined from the area under the load-deflection diagram for a work-hardening material, C^* can be obtained from the area under a load-displacement rate $\dot{\Delta}$ diagram for a creeping material [29,30,32].

Consider Figure 6 which shows creep displacement rate plotted as a function of load for a body containing a crack of length a and then a crack of length $(a+da)$. The value of C^*, which can be regarded as the rate of change of creep energy dissipation rate with respect to crack length, can be obtained from the shaded area dU^* as

$$C^* = \frac{dU^*}{Bda} \qquad (30)$$

for a body of thickness B. Expression (30) can be evaluated in the same way as equation (18) by making reference to the Norton creep law. When equation (24) dominates, $\dot{\Delta}$ can be written as

$$\dot{\Delta} = C_n P^n \qquad (31)$$

where C_n is a measure of the creep compliance of the body. Substituting equation (31) in equation (30) gives

$$C^* = \frac{P^{(n+1)}}{(n+1)B} \cdot \frac{dC_n}{da} \qquad (32)$$

which is the creep equivalent of equation (21). The equation can be rewritten as

$$C^* = \frac{P\dot{\Delta}}{BW} \left[\frac{W}{(n+1)} \frac{1}{C_n} \cdot \frac{dC_n}{da} \right] \qquad (33)$$

where W is the width of the body and the term in square brackets is a non-dimensional factor. Expressing this factor as F gives

$$C^* = \frac{P\dot{\Delta}}{BW} \cdot F \qquad (34)$$

Equation (34) is a convenient expression for C^*. It is valid along any path AA' of crack extension in Figure 6. For crack extension at constant load,

$$C^* = \frac{P\dot{\Delta}}{BW} \left[\frac{W}{(n+1)} \cdot \frac{1}{\dot{\Delta}} \cdot \frac{d\dot{\Delta}}{da} \right] \qquad (35)$$

and for crack extension of constant displacement rate

$$C^* = \frac{P\dot{\Delta}}{BW} \left[\frac{-n}{(n+1)} \cdot \frac{W}{P} \cdot \frac{dP}{da} \right] \qquad (36)$$

Any of the terms in square brackets in equations (33), (35) and (36) can be used to evaluate the factor F.

Analytical solutions to these equations are only possible in very restricted circumstances [30,32]. In most cases approximate procedures or numerical computations have to be employed. These will be discussed in more detail in the next section since estimates of C^* are needed if this parameter is to be used to characterise creep crack growth.

4.   APPROXIMATE ESTIMATES OF THE PARAMETER C^*

Reliable numerical computations of C^* using equations (32) to (36) depend upon the accuracy with which the secondary creep equation (24) describes the material behaviour and how well the component and loading conditions can be modelled using Finite Element or Boundary Integral Equation methods [33,34]. When these methods are not available approximate procedures must be adopted.

The approximate procedures for C^* are based on dimensional analysis arguments and limit analysis techniques. The latter are justified on the basis that for most metals $n \gg 1$ in equation (24). In this case limit analysis techniques can be applied to express P as a proportion of the plastic collapse load of the cracked body P_{LC} [35]. Provided the proportionality factor is only weakly dependent on crack length it is possible to write [36],

$$\frac{1}{P} \cdot \frac{dP}{da} \approx \frac{1}{P_{LC}} \cdot \frac{dP_{LC}}{da} \qquad (37)$$

so that from equation (36),

$$F \approx \left[\frac{-n}{n+1} \cdot \frac{W}{P_{LC}} \cdot \frac{dP_{LC}}{da} \right] \qquad (38)$$

16

Substitution of equation (38) into equation (34) allows
approximate estimates of C^* to be obtained for a range of
values of n and crack configurations provided appropriate
limit load solutions are available. Strictly, the limit
analysis should be based on a 'yield stress' of σ_o from
equation (24), and not the actual yield stress σ_Y but as the
factor F is non-dimensional the yield stress cancels and
plastic collapse solutions can be used directly. Similarly
it is not necessary to know the equivalent stress criterion
obeyed by the material, and whether plane stress or plane
strain conditions apply, provided dP_{LC}/da is not affected.
These conditions will automatically be taken into account by
the $\dot{\Delta}$ term in equation (34). The solutions will increase in
accuracy with increase in n and will be correct for $n = \infty$

Values of F as a function of crack length and n are given
in the Appendix for the most common test-piece geometries
used. For example, for a single edge notch tension (SENT)
specimen

$$F = \left[\frac{n}{n+1} \cdot \frac{1}{(1-a/W)} \right] \qquad (39)$$

The corresponding dependence of F on crack length, for the
standard compact tension (CT) test-piece [37] , using the limit
analysis estimates of Merkle and Corten [38] , is shown in
Figure 7 for $n = 3, 7$ and 20. Similar trends can be obtained
for the other geometries

The limit analysis method of estimating F is in direct
agreement with the procedure of Harper and Ellison [31] and
the dimensional analysis arguments of Nikbin et al [39] .
A further approximate solution for F can be obtained by
considering the approach of Rice, Paris and Merkle [40] .
Applied to the SENT geometry, expressing their equations in a
form appropriate to a material creeping according to equation
(24) leads to,

$$\dot{\Delta} \ \alpha \ (W - a) \ \left(\frac{P}{P_{LC}}\right)^n \qquad (40)$$

Substitution in equation (33) yields

$$F_{RPM} = \left[\frac{n-1}{n+1} \ \frac{1}{(1-a/W)} \right] \qquad (41)$$

where the subscript has been added to identify the method of
solution. Comparison of this equation with equation (39)
indicates that they only differ by the ratio $(n-1)/n$

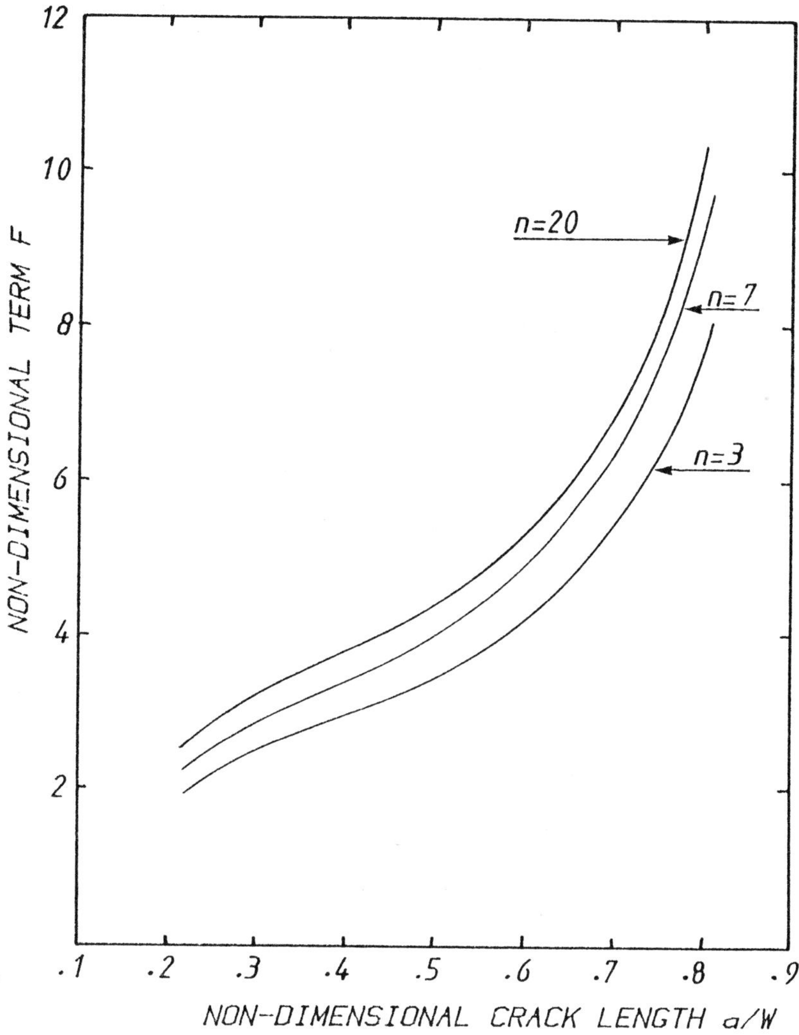

Figure 7.   Limit Analysis Estimates of F for the CT
Geometry.

This difference applies to all predominantly tensile loading
situations. For bending circumstances the Rice, Paris, Merkle
method leads to identical estimates of F to those obtained from
equation (38). For $n \gg 1$ both methods predict essentially the
same values of F for all geometries.

A comparison of the approximate determinations of F with
some numerical calculations by Kumar et al [33] and Hutchinson
et al [41] has been made by Smith and Webster [37].

18

The greatest discrepancies were found at short crack lengths
and low values of n. Generally, for $a/W > 0.375$ and $n > 7$,
a difference of less than 25% was obtained for mainly tensile
loading and less than 10% for bending conditions. Examination of the expressions in the Appendix for F (and F_{RPM})
reveals them to be relatively insensitive to n for $n > 7$.
Consequently an accurate knowledge of n is not needed in this
region. Values of $n > 7$ are often experienced so that this
is of practical relevance.

The form of equation (34) is particularly suitable for
experimental determinations of C^* since it enables C^* to be
obtained from experimental measurements of the displacement
rates at the loading points. In addition, very little
information need be known about the creep properties of the
material as it has already been shown that F is insensitive
to the value of n or equivalent stress criterion adopted.
This is not true of numerical estimates of C^* which involve
calculated determinations of the displacement rate. These
are particularly sensitive to the values assumed for $\dot{\varepsilon}_o$, σ_o
and n and to the equivalent stress definition assumed.
Furthermore, side grooves are often introduced into test-
pieces in practice to promote flat fractures and control the
direction of crack growth. Accurate estimates of C^* by
numerical means will, therefore, need to be able to allow for
the influence of the side grooves on creep displacement rates.
Additional complications may also arise with those materials
which do not exhibit a constant stress index n since the
numerical expressions for C^* are especially sensitive to this
term. For these reasons most estimates of C^* have been
obtained from experimentally determined displacement rates.

5.   EXPERIMENTAL MEASUREMENTS OF CREEP CRACK GROWTH

Methods of measuring crack growth at elevated temperatures
have been discussed by a number of authors [42,43] . Broadly
the procedures adopted are similar to those used to determine
uni-axial creep data. A typical arrangement is shown in
Figure 8. Usually the specimen is inserted in a three zone
furnace capable of maintaining a uniform and constant temperature. An extensometer system is employed to obtain displacements and often windows are provided in the furnace to enable
crack extension to be monitored visually.

The ease with which a crack can be viewed at elevated
temperatures depends upon the intensity of illumination it is
possible to provide at the crack tip, the amount of surface
oxidation and the extent of deformation which accompanies the
cracking. Some of these features are illustrated in Figure 9.
The figure gives an indication of the range of creep deform-
tion which can accompany creep crack growth. In Figure 9a the
Double Cantilever Beam (DCB) test-piece arms have remained

Figure 8. Typical Experimental System for Measuring Crack
Growth at Elevated Temperatures.

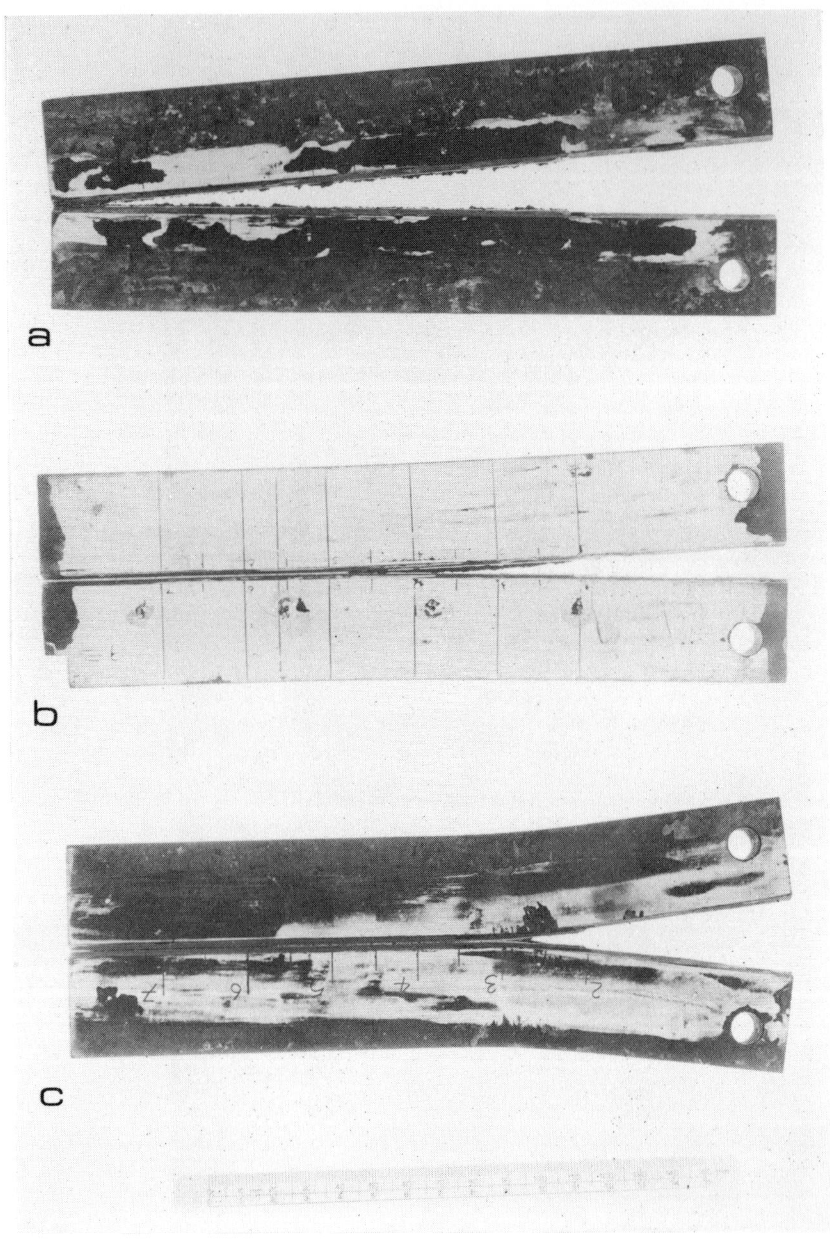

Figure 9. Extent of Creep Deformation Associated with Crack Growth in DCB Test-Pieces a) $\frac{1}{2}\%$CrMoV Steel, at 565°C, b) 1% CrMoV Steel at 565°C, c) $2\frac{1}{4}\%$ CrMoV Steel at 538°C.

straight throughout and there is little evidence of permanent deformation having accompanied the cracking process. In contrast Figure 9c shows considerable arm bending and indicates that substantial creep deformation has been associated with the crack growth. An intermediate situation is shown in Figure 9b. When creep crack propagation takes place in the absence of much creep deformation the crack tip remains sharp and there is little opening. Also there is often evidence of discontinuous cracking ahead of the main crack tip and consequently some ambiguity in precisely locating the crack tip. When extensive creep deformation accompanies cracking, crack blunting takes place and there is usually little difficulty in identifying the position of the crack tip.

Another complication arises with optical methods when surface oxidation occurs as the crack may become obscured against the dull background of the oxidised surface (Figure 9a). Crack extension is most easy to detect when the background remains relatively reflecting. A way of overcoming this problem is to paint the sides of the specimen adjacent to the crack with a white high temperature ceramic coating as shown in Figure 9b. This method has been found to be successful by Nikbin, Webster and Turner [30,39] but care has to be taken in the selection of paint and method of application otherwise the paint may not crack as the specimen cracks. If the paint is too brittle it may flake ahead of the crack tip whereas if it is too ductile it could bridge the crack and indicate too short a crack length.

Optical methods only give the crack length on the surface and a reliable indication of average crack length will only be obtained if the crack front is relatively straight through the thickness of the specimen. A way of encouraging straight fronted cracks is to increase through thickness constraint by introducing side grooves as shown in Figure 9. If these are not provided, and extensive creep deformation takes place, crack tunnelling may occur which could result in surface readings seriously underestimating average crack length. It is possible that some previous reports of long incubation times prior to the outset of cracking may have been overestimated due to the time needed for crack growth which had begun at the interior of the specimen to become apparent at the surface. Bowed cracks may not only cause unreliable estimates of crack propagation rates but could also lead to inaccurate estimates of K or C^*.

Alternative methods of monitoring crack growth include the compliance technique and electrical potential measurements [43]. With each an appropriate calibration curve is needed.

The routine for obtaining crack growth by the elastic compliance technique is to make periodic partial unloadings and reloadings of the sample at suitable time intervals to obtain the change in compliance from which the crack extension may be estimated. This procedure may be automated. Since the displacements are measured remote from the crack tip, an indication of overall specimen response will be obtained and approximately average crack lengths determined.

The compliance method of determining crack growth is most satisfactory for those circumstances where load changes are required during a test, such as in thermal fatigue or creep/fatigue investigations, when one of the normal load changes can be used to obtain an instantaneous estimate of compliance. For constant load studies a periodic partial unload may cause a transient response [44] due to the regeneration of an elastic stress distribution at the crack tip. If this happens the subsequent crack propagation rate may not be characteristic of that for steady loading as stress redistribution by creep may be prevented.

Several electrical potential methods are available which can use either a direct current (DC) or an alternating current (AC) supply. In each case a constant current is applied to the specimen and a potential difference measured across two points either side of the crack plane. This potential difference is converted into crack length with the aid of a calibration curve [45,46]. If a DC system is used a uniform current density is obtained through the test-piece thickness and an average crack length determined. With an AC system the current is concentrated mainly in the surface due to a 'skin' effect and consequently surface crack lengths are measured. The AC system does, however, allow lower current densities to be employed.

Provided it is plotted in non-dimensional form an electrical potential calibration curve obtained at room temperature should be valid at elevated temperature. However, experimental data [47] have shown reduced sensitivity under creep conditions. This has been attributed to crack shorting due to surface roughness (particularly when the fracture is intergranular), uncracked ligaments left behind the crack front, voiding around the crack tip and oxide bridging across the crack faces. Because the extent of these features may be rate sensitive, different calibration curves may be needed for different loading conditions, even for the same material, and it is recommended that periodic checks of crack length are made visually. If this is not possible the test should be stopped prior to final fracture so that the final crack length can be measured and the calibration curve made to pass through the known end points.

Most of the test-piece geometries referred to in the Appendix have been used for making measurements of crack growth at elevated temperatures. It is common practice to use side-grooves to develop straight fronted cracks when required. If side-grooves are introduced it is necessary to replace gross thickness B in the expressions for K, G and C^* by the net thickness between the side-grooves B_n so that, for example, equation (34) becomes

$$C^* = \frac{P\dot{\Delta}}{B_n W} \cdot F. \qquad (42)$$

6. CHARACTERISATIONS OF CREEP CRACK GROWTH RATE

There is still much discussion in the literature over the most appropriate parameter to use for characterising creep crack growth rate $\dot{a}$. The most commonly used parameters are stress intensity factor K, the contour integral C^* and the net section or reference stress σ_{ref} across the uncracked ligament [48,49]

$$\sigma_{ref} = \sigma_y \frac{P}{P_{LC}} \qquad (43)$$

Relations that have been produced are usually of the form

$$\dot{a} = AK^m \qquad (44)$$

$$\dot{a} = D\sigma_{ref}^{\,p} \qquad (45)$$

and

$$\dot{a} = HC^{*\phi} \qquad (46)$$

where A,D,H,m,p and ϕ are material constants. Typically it is found that $m \sim p \sim n$ and ϕ is slightly less than unity with in some instances a tendency towards a threshold below which creep crack growth does not take place.

Generally, descriptions in terms of stress intensity factor are restricted to 'creep brittle' conditions [50-52] whereas reference stress characterisations are more appropriate to ductile situations and $n \gg 1$ [53-55]. The C^* parameter has probably had most success over the widest range of conditions [17,18]. However, despite these correlations in

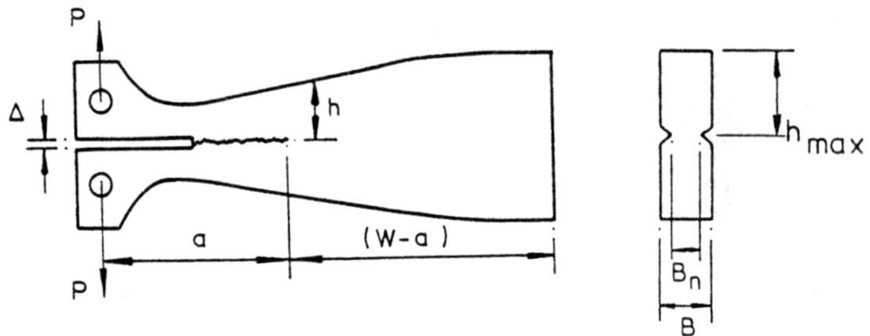

Figure 10. Constant K contoured DCB Test-Piece.

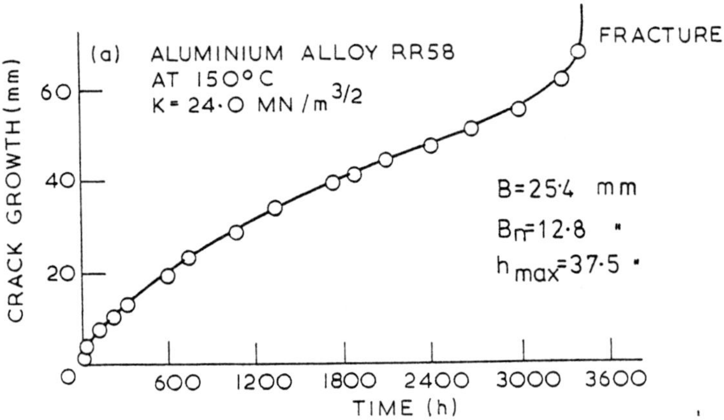

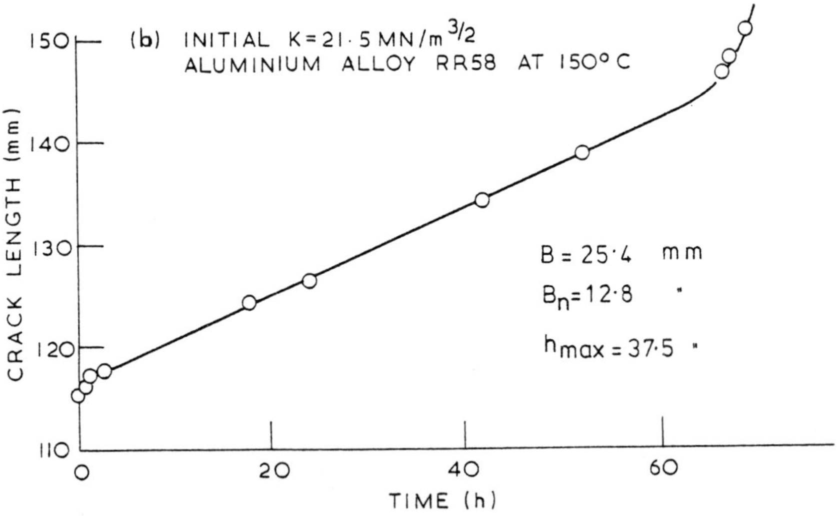

Figure 11. Typical Creep Crack Growth Curves for RR58
a) at Constant K and b) at Constant C^*.

particular circumstances, widespread acceptance of the range
of applicability of these parameters has not been achieved.
This can be illustrated by considering the results of an
extensive series of investigations by Webster and co-workers
[30,36,39,44,56] some of which were carried out in
collaboration with Bristol University [57]. These studies
were performed on a selection of alloys exhibiting creep
ductilities from < 1% to approximately 50%.

To examine the relevance of LEFM one series of experiments [30] was made on two alloys (one a precipitation
hardenable $2\frac{1}{2}\%Cu$ $1\frac{1}{2}\%Mg$ aluminium alloy designated RR58 and
the other a $\frac{1}{2}$ CrMoV steel heat-treated to simulate heat-
affect zone material) of limited creep ductility on constant
K contoured DCB test-pieces (shown in Figure 10). Examples
of the results obtained are presented in Figure 11. Figure
11a) indicates that a constant cracking rate is not obtained
at constant K whereas Figure 11b) shows that it is if load
is altered to maintain C^* constant. This is further
illustrated by a wider spread of data shown in Figure 12.
Although results for the RR58 only have been presented,
similar features were demonstrated by the $\frac{1}{2}$CrMoV steel. It
is clear, even for these brittle materials that C^* gives a
better correlation in this instance than K.

Additional results for a wider range of test-piece
geometries on a 1CrMoV steel of appreciable creep ductility
are depicted in Figure 13. Here C^* gives a better description
of the cracking rate than σ_{ref} despite the ductility
exhibited by the material.

Further satisfactory correlations of creep crack growth
rate with C^* are shown in Figure 14 for a more diverse range
of materials. These results serve to illustrate that C^* can
characterise creep crack propagation over a wide spread of
materials and test conditions.

In principle K would only be expected to correlate the
results if crack extension takes place in a predominantly
elastic stress field before stress redistribution by creep
is possible. It would seem, from the tests on RR58 and
$\frac{1}{2}$CrMoV steel, that only a small amount of creep is required
to invalidate predictions in terms of stress intensity factor.
Similarly from the 1CrMoV steel data it would appear that
very extensive creep deformation, sufficient to cause
appreciable crack tip blunting, is needed for reference stress
descriptions of cracking to be appropriate. With crack tip
blunting, damage will become more distributed around the crack
and failure by net section rupture will be more likely than
by crack propagation. In the experiments reported here, the
crack remained relatively sharp and crack growth was described
most satisfactorily by C^*. It is apparent, therefore, that

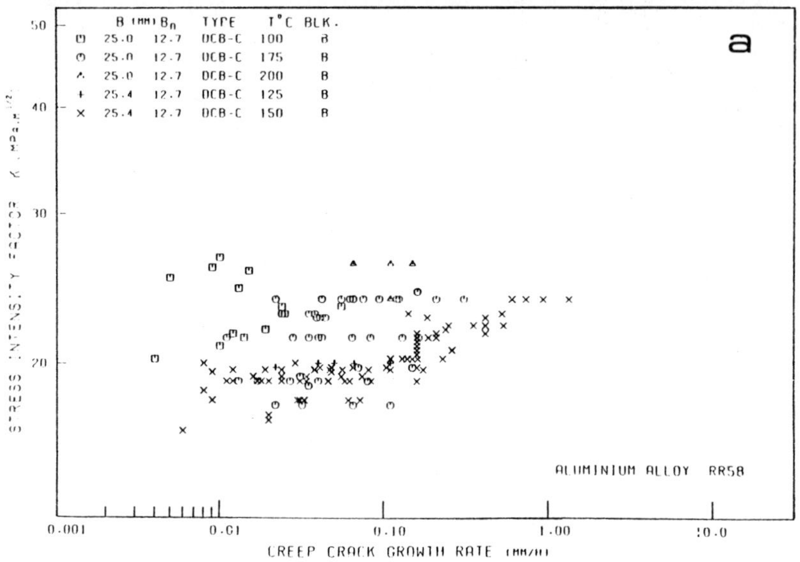

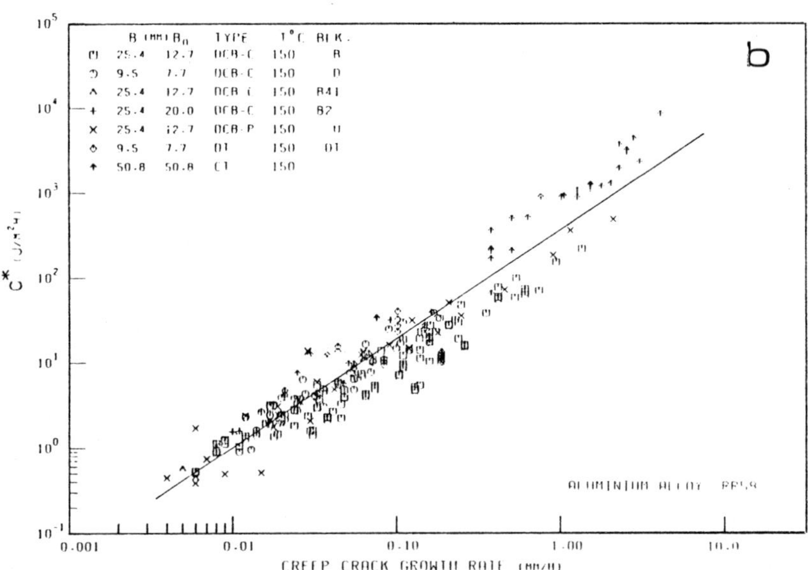

Figure 12. Correlations of Creep Crack Growth in RR58 at $150^{\circ}C$ in terms of a)K and b)C^*.[39]

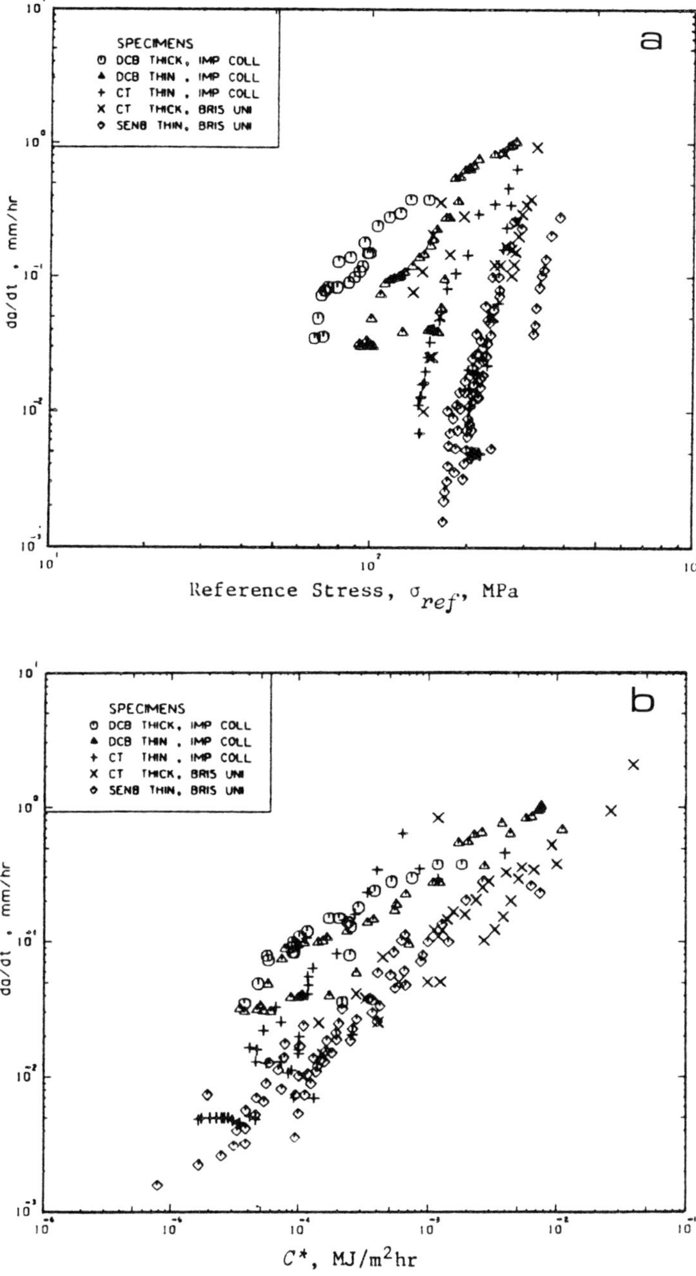

Figure 13. Dependence of Creep Crack Growth Rate on a 1CrMoV steel at 565°C on a)Reference Stress and b)C^*. [57]

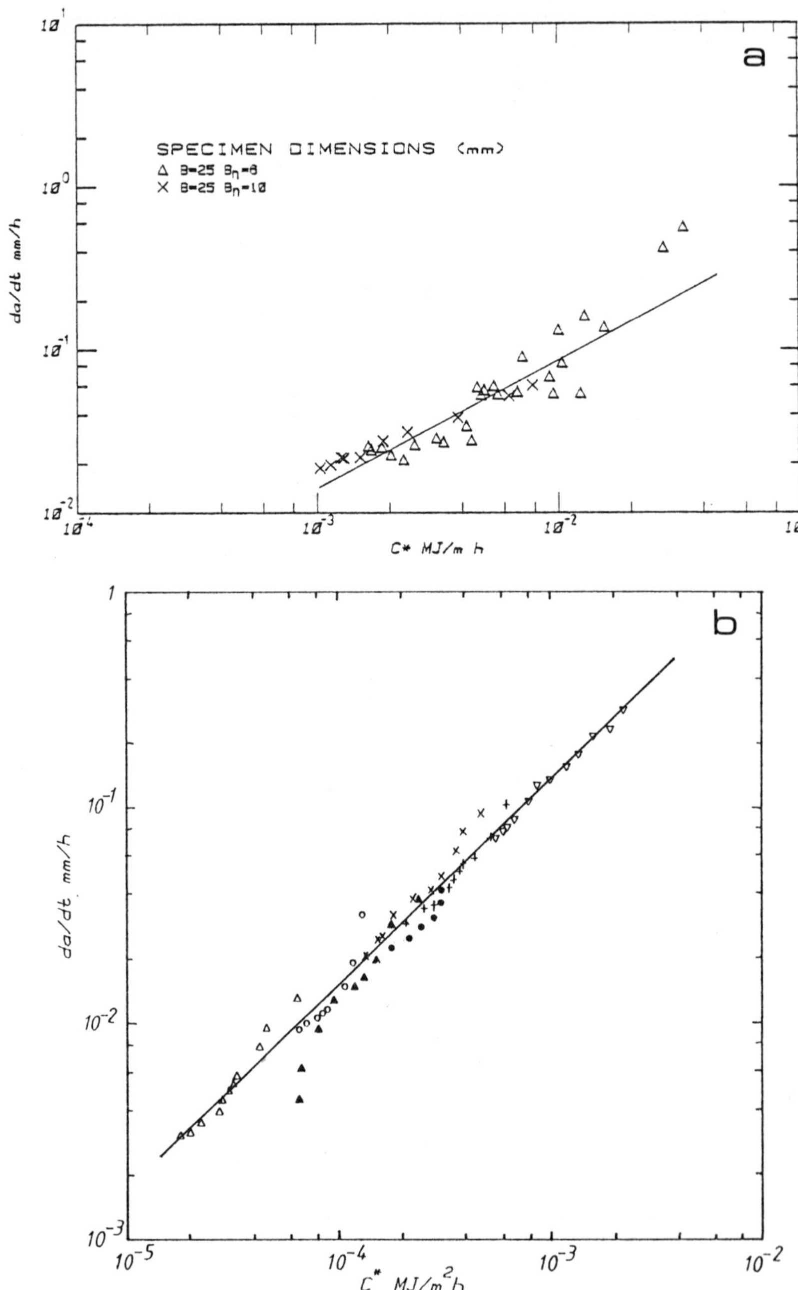

Figure 14. Characterisations of Creep Crack Growth Rate for a) $2\frac{1}{4}$ CrMo Steel at 538°C [56] b) Lead at Room Temperature [58].

for the majority of materials and test conditions examined, C^* provides a better characterization of the local stress distribution ahead of a creeping crack than either K or σ_{ref} and hence a more satisfactory correlation of cracking rates.

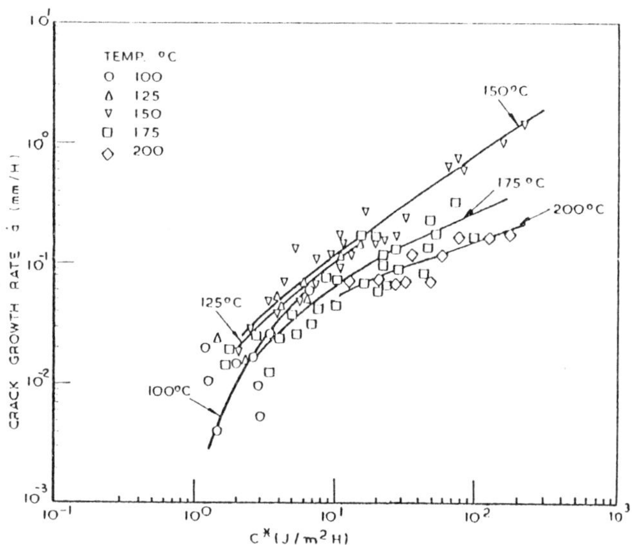

Figure 15.   Dependence of Creep Crack Growth of RR58 on Temperature [59].

Figure 15 shows the dependence of creep crack growth rate on temperature for the aluminium alloy RR58. Cracking rate initially increases at a given C^* with increase in temperature to 150°C and then decreases. Consequently creep crack propagation rate does not increase with temperature in the same manner as creep strain rate. It was found [59] that the maximum cracking rate corresponded to the temperature (150°C) where minimum creep deformation was measured consistent with cracking rate increasing with decrease in material creep ductility.

Constraint is another important factor influencing creep crack growth. This was studied by Nikbin, Smith and Webster [56] who introduced side-grooves of different depths into CT and DCB test-pieces. A selection of the results obtained is shown in Figure 16. They indicate that cracking rate increases with increase in constraint (decrease in the ratio Bn/B) demonstrating that when creep deformation is inhibited the propensity to cracking is increased.

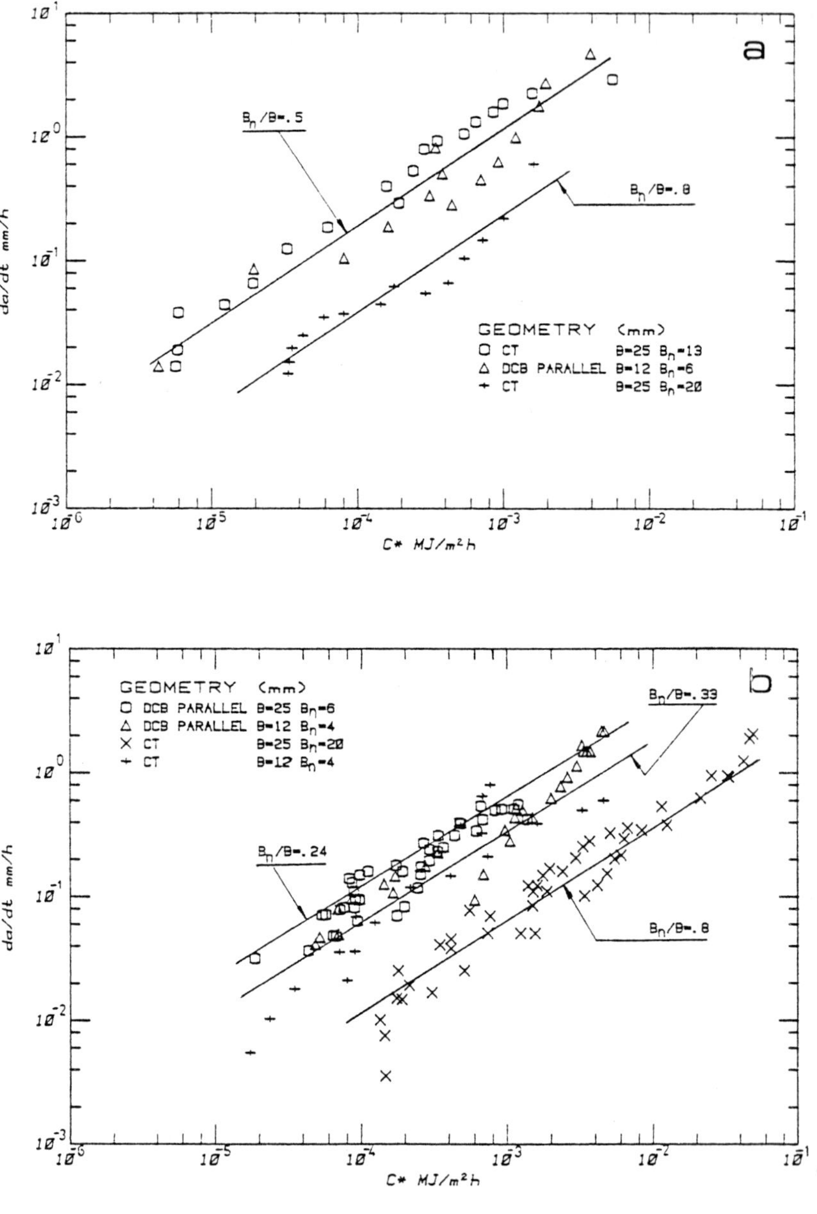

Figure 16. Effect of Constraint on Creep Crack Growth Rate at 565°C of a)½% CrMoV Steel and b)1% CrMoV Steel

7. MICROSTRUCTURAL ASPECTS OF CRACK GROWTH

It is possible to assess the sensitivity of a material to cracking at elevated temperatures by considering the stress and strain rate distributions local to the crack tip [56,59]. When creep strain dominates they will be given by equations (27) and (29) respectively. Consider Figure 17 which shows a creep process zone ahead of a crack which is assummed to be extending at constant speed by creep. It may be postulated that this zone encompasses the region over which creep damage is accumulating ahead of the crack tip. Material will first experience damage when it enters the creep zone at $r = r_c$ and will have accumulated strain ε_{ij} by the time it is at a distance r from the crack tip such that

$$\varepsilon_{ij} = \int_{r=r_c}^{r} \dot{\varepsilon}_{ij} \, dt \qquad (47)$$

Substituting equation (29) into equation (47) therefore gives

$$\varepsilon_{ij} = \int_{r=r_c}^{r} \dot{\varepsilon}_o \left[\frac{C^*}{I_n \sigma_o \dot{\varepsilon}_o r} \right]^{n/(n+1)} \tilde{\varepsilon}_{ij}(\theta,n) \, \frac{dt}{dr} \cdot dr$$

Integrating for a constant crack growth rate by substituting $dr/dt = -\dot{a}$ results in

$$\varepsilon_{ij} = (n+1)\dot{\varepsilon}_o \left[\frac{C^*}{I_n \sigma_o \dot{\varepsilon}_o} \right]^{n/(n+1)} \frac{\tilde{\varepsilon}_{ij}(\theta,n)}{\dot{a}} \left[r_c^{\frac{1}{n+1}} - r^{\frac{1}{n+1}} \right] \qquad (48)$$

It is possible to propose that fracture will occur at the crack tip when the available creep ductility ε_f^* of the material is exhausted. Hence substituting $\varepsilon_{ij} = \varepsilon_f^*$ at $r = o$ in equation (48) gives

$$\dot{a} = \frac{(n+1)\dot{\varepsilon}_o}{\varepsilon_f^*} \left[\frac{C^*}{I_n \sigma_o \dot{\varepsilon}_o} \right]^{n/(n+1)} \tilde{\varepsilon}_{ij}(\theta,n) \, r_c^{1/(n+1)} \qquad (49)$$

The equation has the same form as equation (46) which was previously determined experimentally. The model implies that in the experimental expression $\emptyset = n/(n+1)$. For large values of n, $\emptyset$ will be a fraction close to unity in agreement with

32

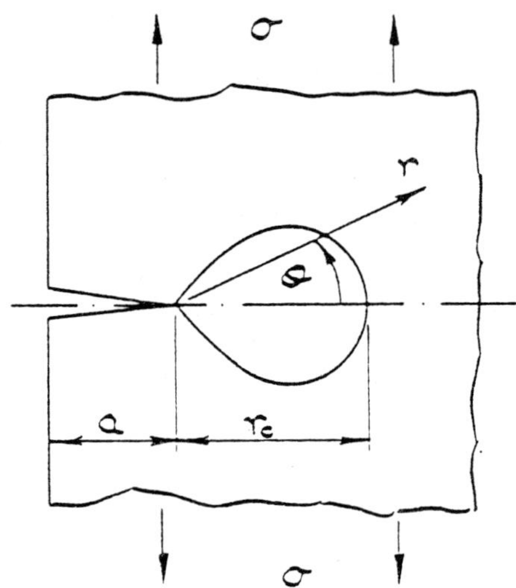

Figure 17. Zone Ahead of a Creeping Crack in Which Creep
Damage Takes Place.

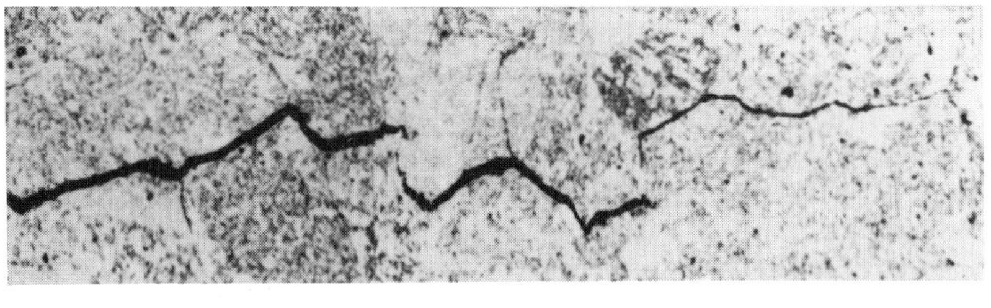

Figure 18. Discontinuous Cracking Ahead of Main Crack-Tip
in $\frac{1}{2}$CrMoV Steel at 565°C.

most practical observations.

An argument for a creep zone has also been proposed by
Reidel [60]. The available experimental evidence [59,61]
suggests that voiding and microcracking can extend up to
several grains ahead of the main crack and that crack advance
takes place in a series of small steps in quasi-static
fashion by the progressive linking of the main crack to this
damage (see Figure 18). However, since for $n \gg 1$, r_c is
raised to a small fractional power in equation (49) and
relative insensitivity to the size of the creep damage zone
is predicted. In contrast the equation indicates that crack
growth rate should be inversely proportional to the material
creep ductility. It should be noted however, that the
ductility referred to, ε_f^* , is that appropriate to the state
of stress at the crack tip and not necessarily the uni-axial
creep ductility ε_f . This is of practical importance since it
has been shown that hydrostatic tension can reduce failure
strain significantly in creep [7,62,63] and plasticity
experiments [64,65].

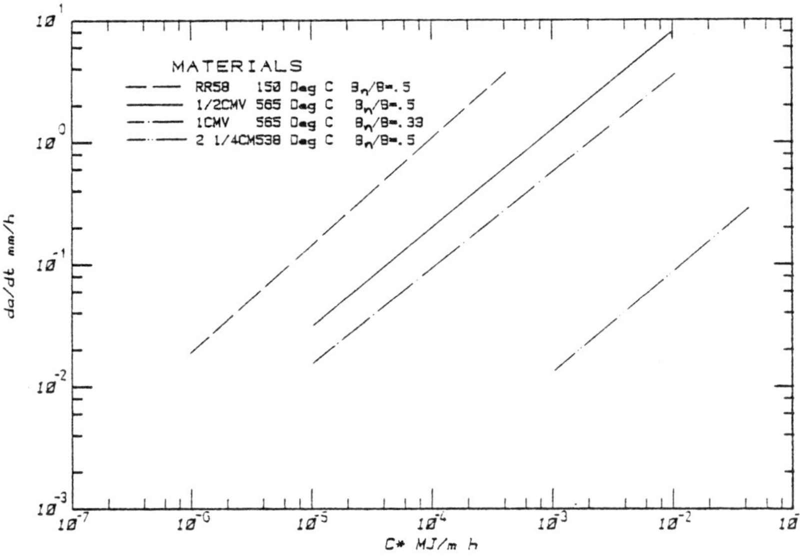

Figure 19. Comparison of Creep Crack Growth Rates for RR58
and Low-Alloy Steels.

This model will now be used to explain the experimental
cracking data in the low alloy steels and aluminium alloy RR58
reported previously in Figures 12 to 16 and which are compared
in Figure 19. The corresponding uni-axial creep properties of

these materials are given in Figure 20. Average straight lines have been drawn through the secondary creep rate data of Figure 20a) to obtain the constants in the Norton creep law, equation (24). These are listed for each material in Table 1 together with the material grain sizes and uni-axial creep ductilities ε_f. A pronounced decrease in ductility with decrease in stress was observed in some instances. The overall range in ductilities obtained was between 0.1% and 45%.

TABLE 1.

UNI-AXIAL CREEP PROPERTIES

MATERIAL	TEMP ^oC	GRAIN SIZE (μm)	n	$^+\sigma_0$ (MPa)	DUCTILITY	
					ε_f %	ε_f^* %
$\tfrac{1}{2}$CMV	565	250	15	570	2 - 0.1	0.6
1CMV	565	40	9	780	20 - 5	1
$2\tfrac{1}{4}$CMV	538	20	9	365	45	55
RR58	150	150	35	370	7 - 2	0.25
RR58	175	150	24	350	9 - 3	0.7
RR58	200	150	15	350	11 - 3	2

$^+$ σ_0 has been calculated with $\dot{\varepsilon}_0 = 1/hr$.

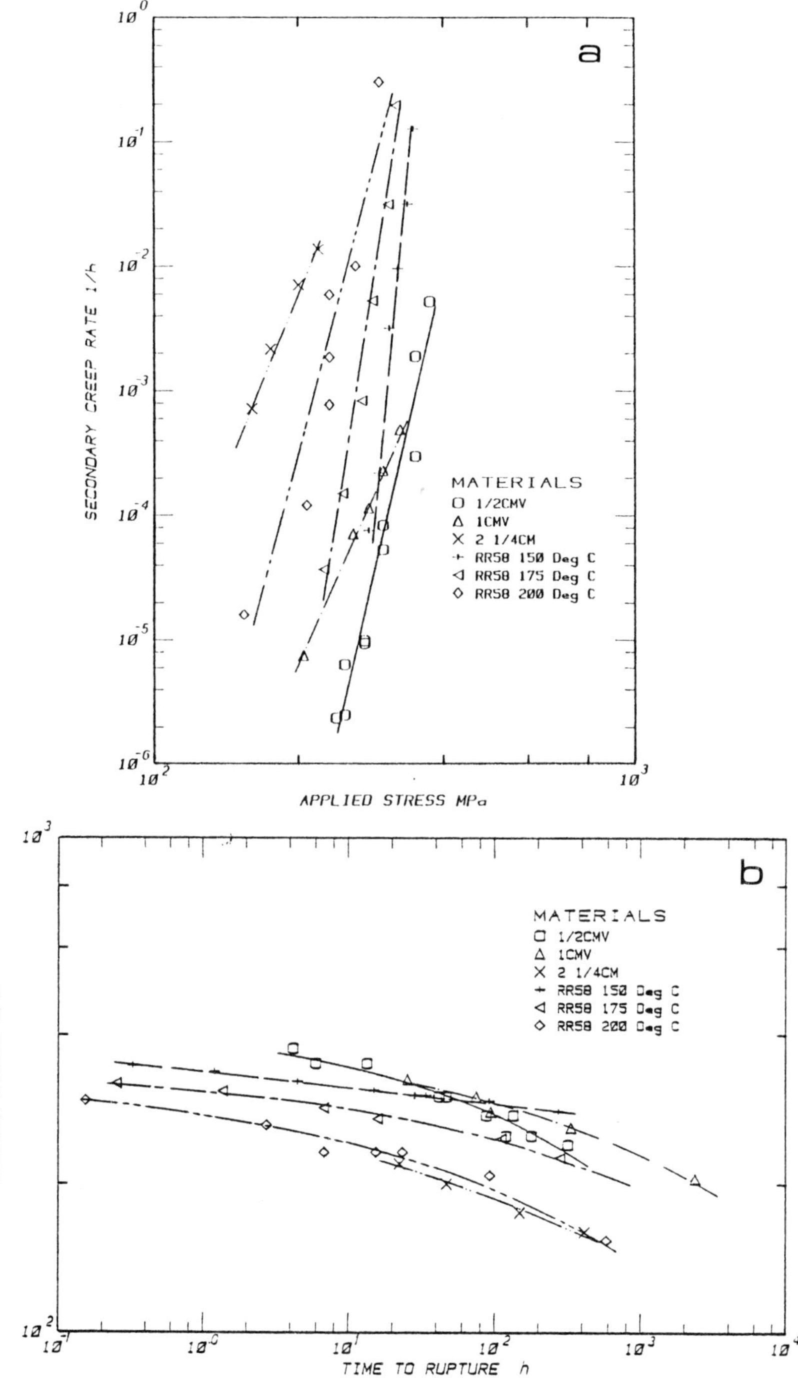

Figure 20. Uni-Axial Creep Properties of RR58 and Low
Alloy Steels a)Secondary Creep Rate Data,
b)Rupture Data.

For the test conditions investigated it is clear, from Figures 12 to 16 (except in the region of a threshold), that the experimental $\dot{a}$ versus C^* relation can be represented by a straight line of slope $\emptyset$ slightly less than one. Although broad agreement of $\emptyset$ with the value of $n/(n+1)$ predicted in equation (49) is obtained, strict comparison is difficult because of the nature of the uni-axial creep data. Figure 20a) shows that although an average straight line has been drawn through each set of results, in most instances there is a progressive decrease in slope with decrease in stress implying a change in exponent n in equation (24). The most appropriate value of n for the region ahead of the crack will depend upon the stress range over which most of the damage takes place. Nevertheless, the range of $\emptyset$ measured from 0.8 to 0.95 is consistent with the values of n listed in Table 1.

A comparison of Figures 15, 19 and 20 reveals that the resistance to creep crack growth of the materials does not correspond with their respective creep strengths. For the aluminium alloy, creep strength decreases progressively with increase in temperature, whereas the resistance to crack growth increases at constant C^* with increase in temperature above 150°C. For the steels an inverse trend is obtained. The alloy with the highest creep strength ($\frac{1}{2}$%CrMoV steel) cracks fastest whilst the $2\frac{1}{4}$%CrMo steel with the lowest strength cracks slowest. There is, however, a trend with the uni-axial ductilities shown in Table 1. A progressive increase in resistance to crack growth is indicated with increase in material ductility. This behaviour can be explained by considering equation (49) in more detail.

For the range in values of n of interest in Table 1., $I_n \sim 4$. Similarly $\dot{\varepsilon}_0 = 1/hr$ by choice and using the maximum value of the effective non-dimensional term $\tilde{\varepsilon}_e(\theta,n) = 1$ equation (49) becomes

$$\dot{a} = \left[\frac{C^*}{4\sigma_0}\right]^{n/(n+1)} \frac{(n+1)}{\varepsilon_f^*} \cdot r_c^{1/(n+1)} \tag{50}$$

It has already been established [59,61] that the creep damage zone r_c is of the order of the grain size. Substituting appropriate values in equation (50) from Table 1 confirms that crack growth rate is influenced primarily by the term ε_f^*. The actual values of ε_f^* needed to correlate the data are listed in the Table. For the $2\frac{1}{4}$ CrMo steel ε_f^* is close to the material uni-axial creep ductility ε_f. For the other materials it can be up to an order of magnitude less than ε_f. Such a decrease has been reported by MacKenzie, Hancock and Brown [64] for the plastic deformation of notched bars. The extent to which hydrostatic tension is likely to affect creep

ductility will depend on the mechanisms governing deformation
and void growth. A different response may be anticipated
depending upon whether void growth is deformation or diffusion
controlled [7,62]. It is apparent from the experimental
results on the side-grooved specimens (Figure 16) that an
increase in constraint can cause an increase in crack growth
rate by as much as a decade.

When the appropriate values from Table 1 are inserted
into equation (50) all the data presented can be described
approximately (to within a factor of about 2) by the relation

$$\dot{a} = \frac{300}{\varepsilon_f^*} (C^*)^{0.85} \tag{51}$$

where $\dot{a}$ is in mm/h with ε_f^* as a percentage and C^* in units
of MJ/m^2h. It is clear therefore that available creep
ductility ε_f^* is the most important parameter governing the
susceptibility of a material to creep crack growth.
Consequently heat-treatments and fabrication procedures
leading to reduced ductilities should be avoided if the risk
of failure by creep crack propagation is to be minimised.
Similarly, as an increase in constraint reduces ε_f^* , crack
growth is most likely to take place from sites of stress
concentration and other regions of high constraint. For
plane stress situations the experimental results imply that
uni-axial creep ductility can be used in equation (51) to
indicate creep crack growth rates.

8.   INFLUENCE OF CYCLIC LOADING

Many components operating at elevated temperatures are
subjected to combined steady and cyclic loading which can
lead to creep/fatigue interaction and environmental effects
[18,66,67]. The extent of these effects for a given
material is likely to depend upon the temperature, period
of cycling and ratio of the load amplitude to mean load.
Under steady loading the initial elastic stress distribution
(in the absence of plasticity) becomes modified by creep until
it is described by equation (27). Any load change will upset
the approach to steady state and may affect crack growth rate.
An illustration of the transient effects that can be caused
is shown in Figure 21. An enhanced cracking rate and
displacement rate were obtained immediately after a load
increase. When the load was reduced a prolonged incubation
period was observed before measurable crack extension and
displacement were detected.

Transient effects can be explained in terms of the

38

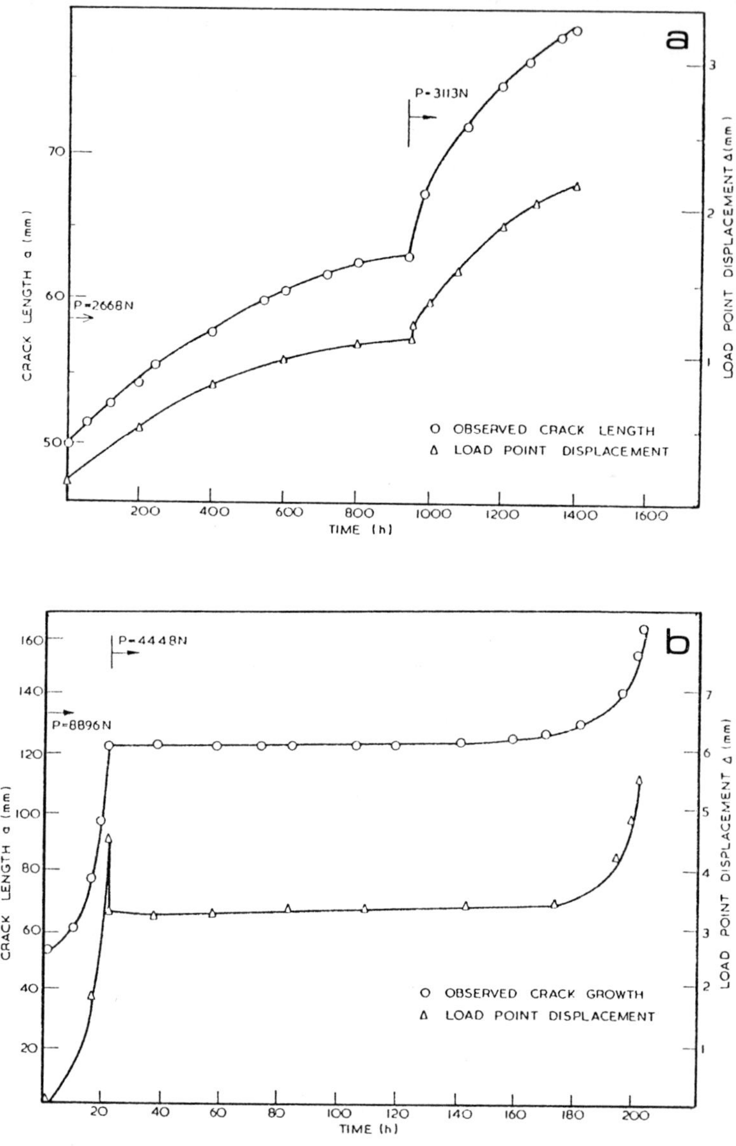

Figure 21. Transient Response of $\frac{1}{2}$% CrMoV Steel at 565°C to a)a Load Increase and b)a Load Reduction [44].

influence of the load change on the local stress state at the
crack tip. Any load change will cause an instantaneous
elastic stress distribution to be superimposed and time will
be needed for stress redistribution to occur and a new steady
state to be developed. Immediately after a load increase,
the stresses closest to the crack tip will be higher than
their equilibrium values and an enhanced cracking rate will
be expected whilst they are relaxing. Conversely for a load
reduction, the stresses local to the crack tip will have to
increase to reach their steady-state values. Whilst this
redistribution is occurring a reduced cracking rate, and
possibly an incubation period if the load reduction is large
enough, will be anticipated.

With cyclic loading the above process will be repeated.
The stress redistribution that is achieved at the crack tip
will depend upon the amount of stress redistribution that
can take place by creep. Major factors will include the
frequency and amplitude of the cyclic loading and the
operating temperature.

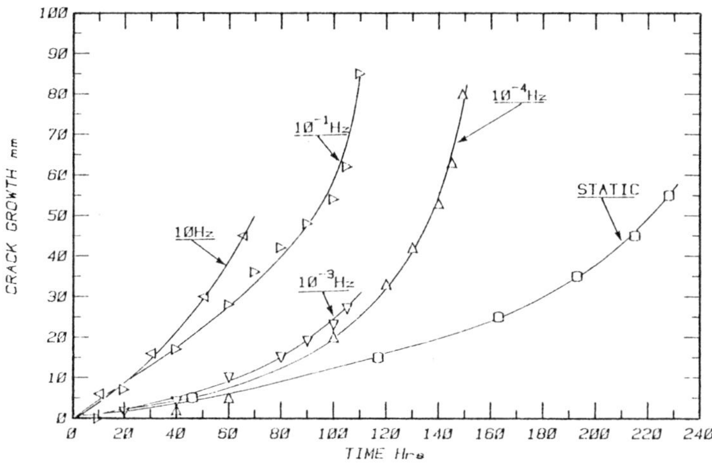

Figure 22. Effect of Frequency on Crack Growth of $\frac{1}{2}$%CrMoV
Steel at 565°C and $R = 0.8$.

40

An indication of how creep crack growth rate can be
accelerated with increase in frequency at a constant R-ratio
(minimum to maximum load ratio) is shown in Figure 22 for a
$\frac{1}{2}$% CrMoV steel [68]. The frequency effect is clearly
demonstrated when the data are plotted as crack growth/cycle,
da/dN, against stress intensity factor range ΔK in Figure 23.
A similar effect has been observed (Figure 24) in tests
carried out on a nickel base superalloy at 700°C. For both
materials as frequency is decreased, the crack growth per
cycle increases indicating significant time dependent
crack growth.

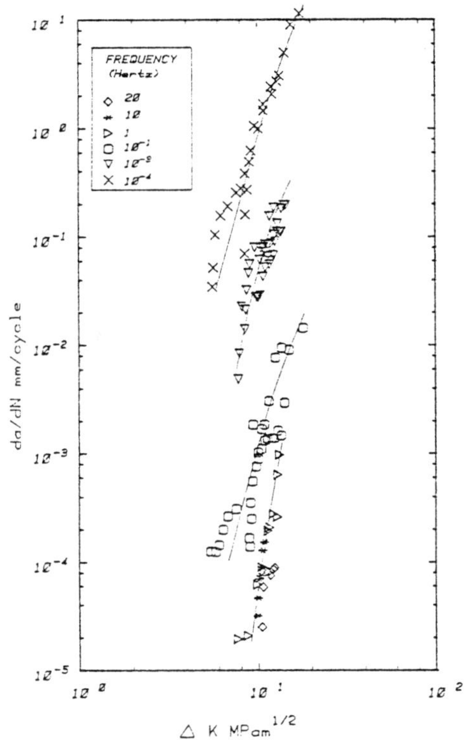

Figure 23. Dependence of Crack Growth/Cycle on Frequency
for $\frac{1}{2}$% CrMoV Steel at 565°C and $R = 0.8$ [68].

Because of the time dependent features observed the data
of Figures 23 and 24 are shown plotted against C^* in Figure 25.
Also included in this figure are some steady load crack growth
results. It is apparent that the cyclic cracking rates are
correlated satisfactorily by the static data. This
demonstrates that the cyclic crack growth took place mainly by
a creep process although some slight acceleration due to
fatigue is evident. In general situations involving high
temperatures and high R-ratios would be expected to favour
crack growth by creep. In contrast low temperatures and low
R-ratios should result in fatigue dominated crack extension.

It is important in practice to establish the mechanism controlling crack propagation in order that the correct characterising parameter is used. When cyclic crack growth is predominantly time dependent C^* should be employed and when it is cycle dependent ΔK.

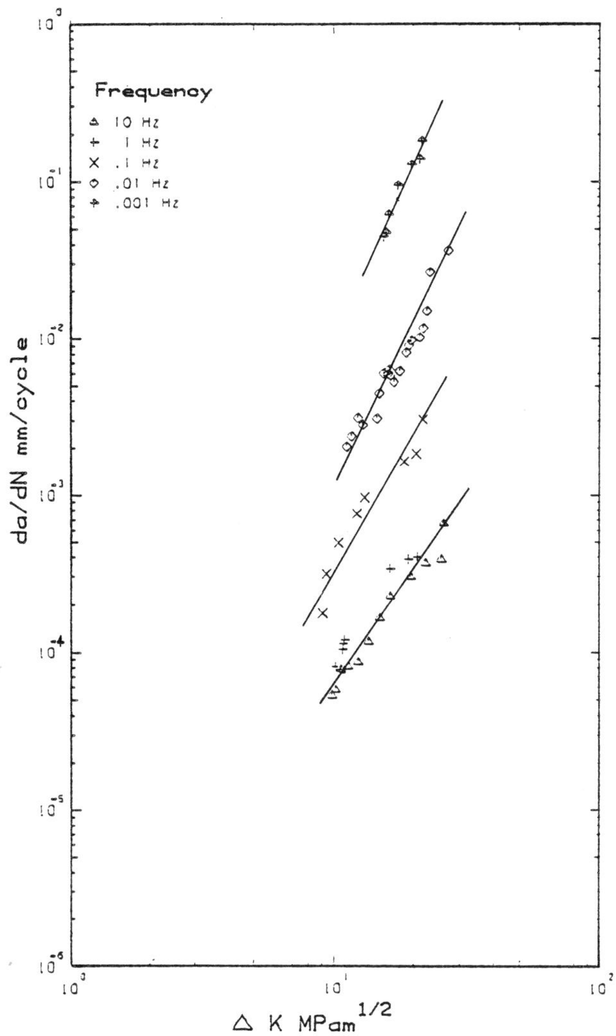

Figure 24. Dependence of Crack Growth/Cycle on Frequency for Nickel-Base Alloy AP1 at 700°C and $R = 0.7$

9.   PRACTICAL APPLICATIONS

Before the life of a cracked component operating at elevated temperatures can be determined it is necessary to decide which of the characterising parameters C^*, K or σ_{ref}

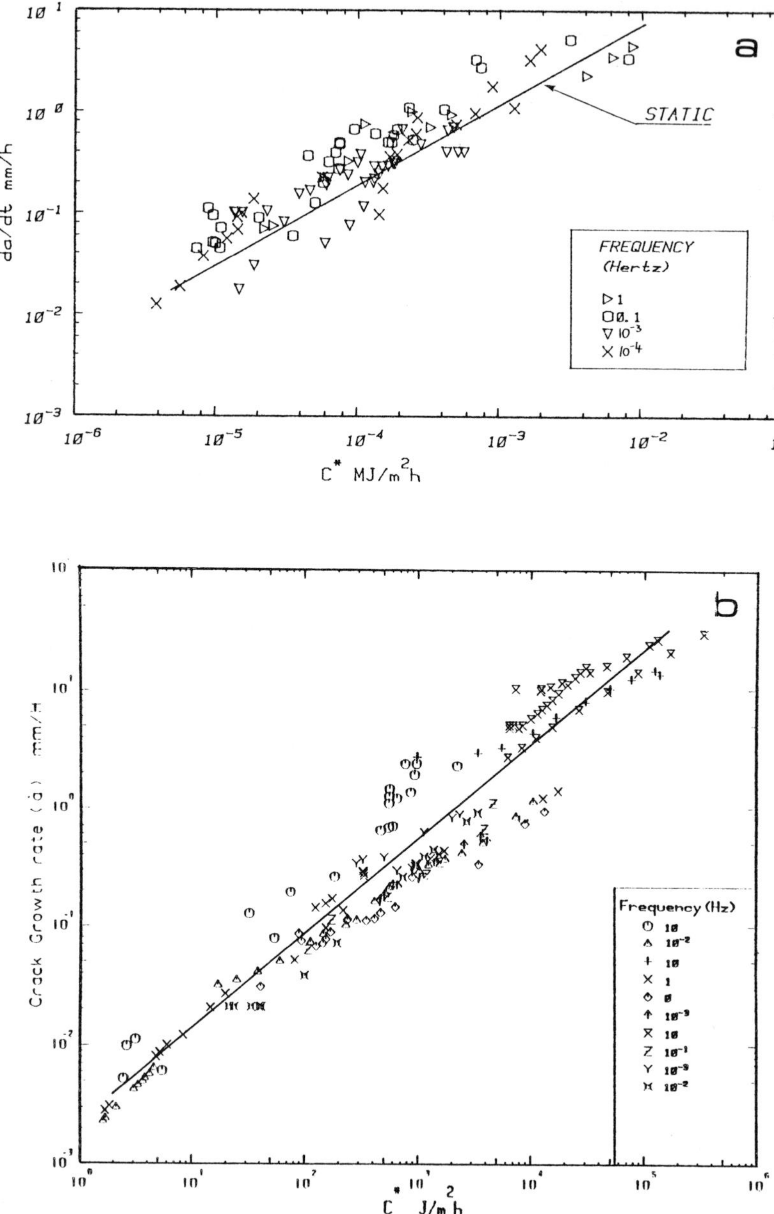

Figure 25. Dependence of Crack Growth Rate under Cylic
Loading Conditions on C^* for
a) $\frac{1}{2}$% CrMoV Steel at 565°C and b)AP1 alloy
at 700°C.

is most appropriate for describing the crack growth. Despite
some of the early correlations in special circumstances with
K and σ_{ref}, the available experimental evidence strongly
suggests that C^* has the greatest universal applicability.
It is recommended that this term is used in general.

A complication with applying the C^* parameter is the
difficulty of calculating it for practical components. The
experimental procedure combined with the limit analysis
approach [36,37] cannot be used unless the cracked body can
be monitored. In general numerical integration will be
needed.

The DCB geometry is an example where an analytical
solution is possible when secondary creep predominates. For
the dimensions given in the Appendix it can be shown that
[30,32],

$$C^* = Xa^{(n+1)} \qquad (52)$$

where
$$X = \frac{2\dot{\varepsilon}_o}{(n+1)B}\left[\frac{2n+1}{2Bn\sigma_o}\right]^n \frac{P^{(n+1)}}{(h/2)^{(2n+1)}} \qquad (53)$$

Substituting this in equation (46) and integrating for constant
load and $\emptyset = n/(n+1)$ gives the time taken for the crack to
extend from a_o to a as,

$$t = \frac{1}{HX^{\emptyset}(n-1)} \cdot \left[\frac{1}{a_o^{(n-1)}} - \frac{1}{a^{(n-1)}}\right] \qquad (54)$$

It should be noted that the analysis presented in this
chapter is strictly only applicable when secondary creep
dominates. Further work is needed to extend the approach to
include situations involving significant elastic strains and
also primary and tertiary creep deformation.

10.   CONCLUSIONS

A continuum mechanics approach to describing creep crack
growth has been presented which is consistent with microstructural observations of fracture. Fracture mechanics concepts for characterising the stress, strain and strain rate
fields ahead of the crack have been developed. A creep
contour integral termed C^*, which corresponds to the J-contour
integral used in plasticity, has been defined and applied to
materials deforming according to the Norton Creep law.

44

Powerful approximate methods of estimating C^* for
cracked bodies based on limit analysis techniques have been
presented and compared with numerical calculations.
Generally agreement to within 25% is obtained for mainly
tensile loading and to within 10% for bending situations.
This level of accuracy is regarded as acceptable when the
normal scatter in experimental creep data is taken into
account.

Experimental creep crack growth results have been
provided for a range of materials exhibiting uni-axial creep
ductilities from about 0.1% to 45%. It is shown that the
C^* parameter describes these data more satisfactorily than
stress intensity factor K or reference stress σ_{ref}
irrespective of material ductility. It is also found that
the resistance to crack growth of these materials does not
correspond with their relative uni-axial creep strengths.

A model of creep crack growth has been presented to
explain the experimental results. The model postulates a
process zone at the crack tip in which damage is accumulating.
It predicts an expression for creep crack growth rate $\dot{a}$ of the
form $\dot{a} \propto (C^*)^{\emptyset}$, where $\emptyset = n/(n+1)$ and n is the stress
exponent of secondary creep in agreement with the experimental
data. The proportionality factor in the relation is found to
be insensitive to the process zone size but to be inversely
proportional to the material creep ductility ε_f^* appropriate
to the state of stress at the crack tip. An approximate
expression has been presented in terms of only ε_f^* and C^*
with $\emptyset = 0.85$ for determining propagation rates to within
a factor of about two for the materials examined. Crack
growth rate for plane stress conditions can be predicted
approximately by making ε_f^* equal to the uni-axial creep
ductility of the material. As constraint is increased,
faster cracking rates are obtained and a reduced value of
ε_f^* is needed.

Some experimental results determined under cyclic loading
conditions have been included. These demonstrate that, when
crack growth is controlled by time dependent processes, the
results can be correlated by C^* and agree with the
corresponding static load data.

The expressions developed for C^* have been determined
on the assumption that secondary creep predominates.
Extension of the approach is needed to include elastic strains
and also primary and tertiary creep deformation when these
are significant.

11.   ACKNOWLEDGEMENTS

The author is indebted to Dr. K.M. Nikbin and D.J. Smith who undertook most of the investigations reported. Grateful thanks for financial support over the years are due to; Science and Engineering Research Council, Ministry of Defence and the General Electric Company.

12.   APPENDIX

12.1   <u>Stress Intensity Factor and C^* Solutions for a
Selection of Common Test-Piece Geometries.</u>

The relations for K are given in a form suitable for expressing as

$$K = Y\sigma \sqrt{a} \qquad (3)$$

In all instances C^* is given by

$$C^* = \frac{P\dot{\Delta}}{BW} \cdot F \qquad (34)$$

where F has been obtained by limit analysis methods or the Rice, Paris, Merkle approach. When the latter has been adopted the term F_{RPM} is used for identification purposes.

(i)   <u>Single Edge Notch Tension (SENT)</u>

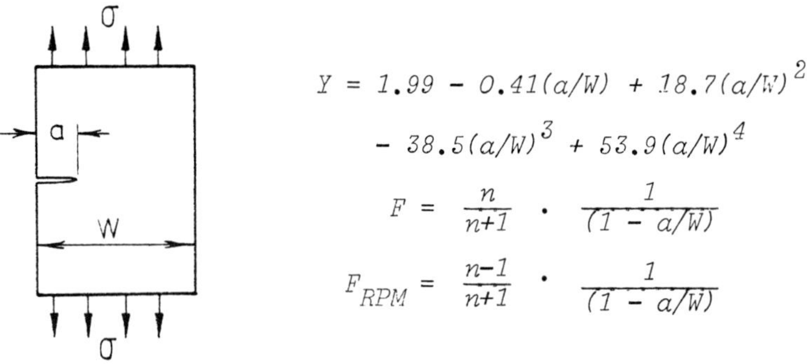

$$Y = 1.99 - 0.41(a/W) + 18.7(a/W)^2$$
$$- 38.5(a/W)^3 + 53.9(a/W)^4$$
$$F = \frac{n}{n+1} \cdot \frac{1}{(1 - a/W)}$$
$$F_{RPM} = \frac{n-1}{n+1} \cdot \frac{1}{(1 - a/W)}$$

(ii)   <u>Centre Cracked Plate (CCP)</u>

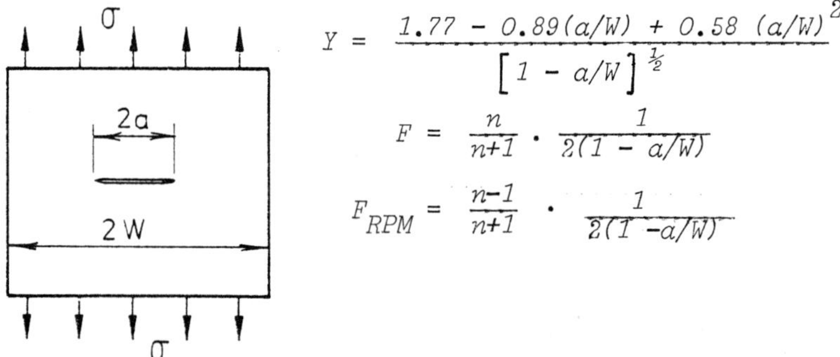

$$Y = \frac{1.77 - 0.89(a/W) + 0.58 (a/W)^2}{\left[1 - a/W\right]^{\frac{1}{2}}}$$
$$F = \frac{n}{n+1} \cdot \frac{1}{2(1 - a/W)}$$
$$F_{RPM} = \frac{n-1}{n+1} \cdot \frac{1}{2(1 - a/W)}$$

(iii)   <u>Double Edge Notch Tension (DENT)</u>

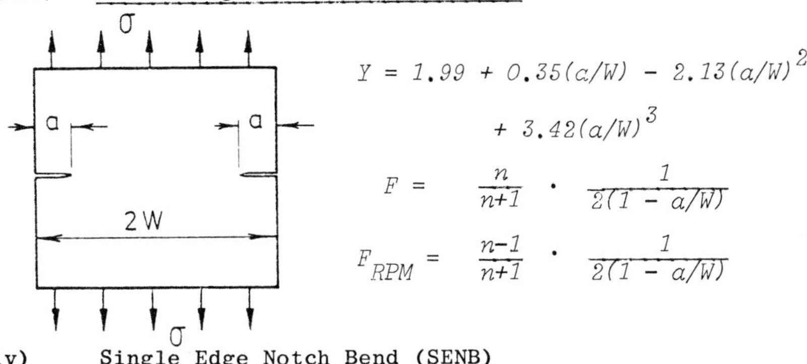

$$Y = 1.99 + 0.35(a/W) - 2.13(a/W)^2$$
$$+ 3.42(a/W)^3$$

$$F = \frac{n}{n+1} \cdot \frac{1}{2(1 - a/W)}$$

$$F_{RPM} = \frac{n-1}{n+1} \cdot \frac{1}{2(1 - a/W)}$$

(iv)   <u>Single Edge Notch Bend (SENB)</u>

a) Pure Bend

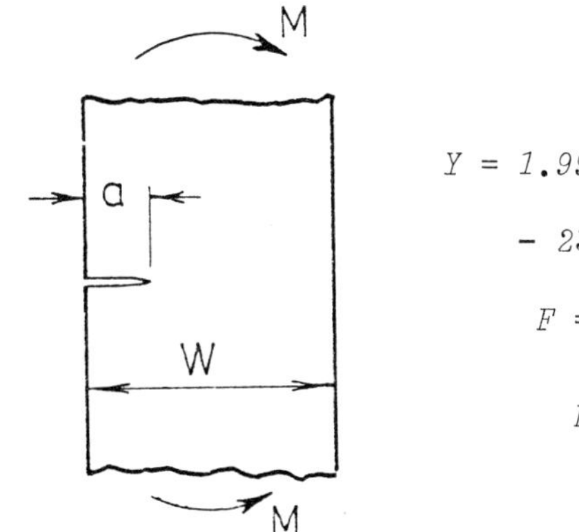

$$\sigma = \frac{6M}{BW^2}$$

$$Y = 1.99 - 2.46(a/W) + 13.0(a/W)^2$$
$$- 23.2(a/W)^3 + 24.8(a/W)^4$$

$$F = \frac{2n}{n+1} \cdot \frac{1}{(1 - a/W)}$$

$$F_{RPM} = F.$$

b) Three Point Bend

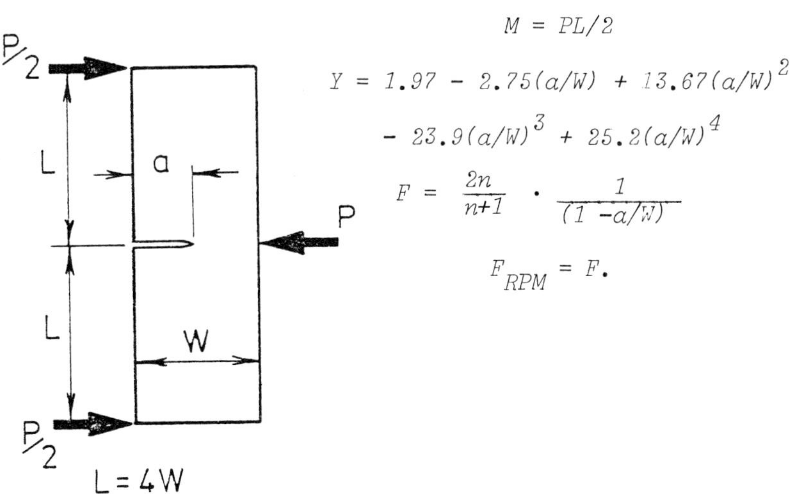

$$M = PL/2$$

$$Y = 1.97 - 2.75(a/W) + 13.67(a/W)^2$$
$$- 23.9(a/W)^3 + 25.2(a/W)^4$$

$$F = \frac{2n}{n+1} \cdot \frac{1}{(1 - a/W)}$$

$$F_{RPM} = F.$$

(v)   <u>Double Cantilever Beam (DCB)</u>

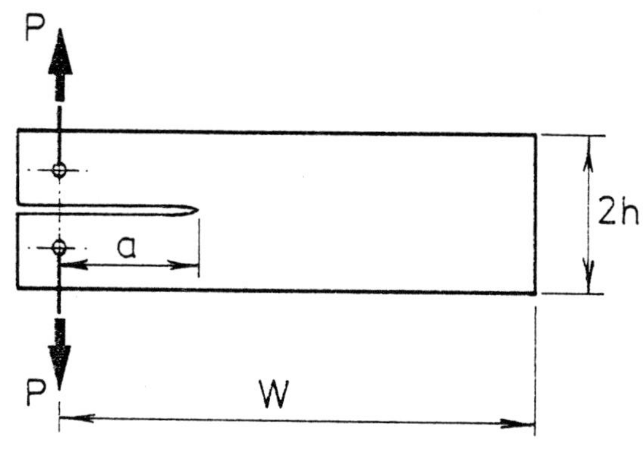

$$K = \frac{2P}{B}\left[\frac{3a^2}{h^3} + \frac{1}{h}\right]^{\frac{1}{2}}$$

$$F = \frac{n}{(n+1)} \cdot \frac{1}{a/W}$$

$$F_{RPM} = \frac{1}{a/W}$$

(vi)   <u>Compact Tension (CT)</u>

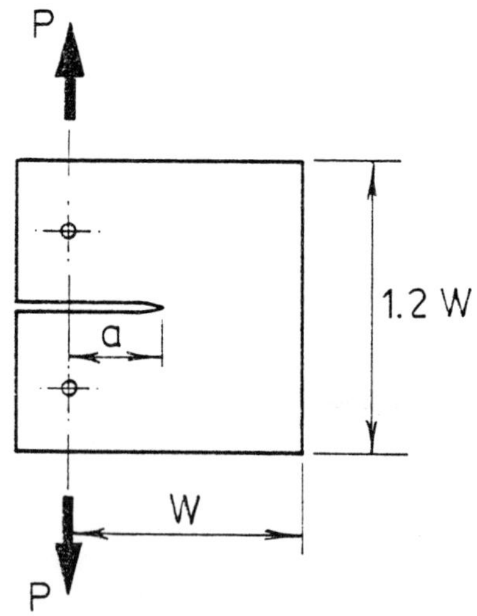

$$K = \frac{P}{BW^{\frac{1}{2}}} \left[29.6(a/W)^{\frac{1}{2}} - 185.5(a/W)^{3/2} \right.$$

$$+ 655.7(a/W)^{5/2} - 1017(a/W)^{7/2}$$

$$\left. + 638.9(a/W)^{9/2} \right]$$

$$F = \frac{n}{n+1} \cdot X$$

a/W	0.25	0.375	0.5	0.625	0.75
X	1.96	3.13	4.70	6.04	7.90

13. REFERENCES

1. FINNIE, I. and HELLER, W.R. - 'Creep of Engineering Materials', McGraw-Hill, New York ,1959.

2. ODQVIST, F.K.G. - 'Mathematical Theory of Creep and Creep Rupture', Oxford University Press, London, 1966.

3. PENNY, R.K. and MARRIOTT, D.L. - 'Design for Creep', McGraw-Hill, London, 1971.

4. KACHANOV, L.M. -'The Theory of Creep (in Russian), Gos. Izdat.Fiz-Mat Lit, Moscow, 1960.

5. RABOTNOV, Yu. M. - 'Creep Problems in Structural Members' Ed. Leckie F.A., North-Holland, Amsterdam, 1969.

6. RABOTNOV, Yu. M. - 'Advances in Creep Design', Ed. Smith, A.I. and Nicholson, A.M., Applied Science, London, 1971.

7. COCKS, A.C.F. and ASHBY, M.F. - 'On Creep Fracture by Void Growth', Prog. Mater. Sci, 1982, 27, 189.

8. GAROFALO, F. - 'Fundamentals of Creep and Creep Rupture in Metals', MacMillan, New York, 1965.

9. GITTUS, J. - 'Creep Viscoelasticity and Creep Fracture in Solids', Applied Science, London, 1975.

10. GITTUS, J. - 'Cavities and Cracks in Creep and Fracture', Applied Science, London, 1981.

11. TETELMAN, A.S. and McEVILY, A.J. - 'Fracture of Structural Materials' Wiley, New York, 1968.

12. LIEBOWITZ, H. (Ed) - 'Fracture' Vol. 1. Academic Press, New York, 1968.

13. YOKOBORI, T. - 'An Interdisciplinary Approach to Fracture and Strength of Solids' Wolters-Noordhoff, Groningen, 1968.

14. KNOTT, J.F. - 'Fundamentals of Fracture Mechanics' Butterworths, London, 1973.

15. RITCHIE, R.O. - 'Near Threshold Fatigue-Crack Propagation in Steels' Int. Met. Reviews No. 5 and 6, 1979, p.205.

16. LATZKO, D.G.H. (Ed) - 'Post-Yield Fracture Mechanics' Applied Science, London, 1979.

17. ELLISON, E.G. and HARPER, M.P. - 'Creep Behaviour of Components Containing Cracks - a Critical Review '

17. (Continued).
    J. Strain Anal., $\underline{13}$, 1978, p.35.

18. SADANANDA, K. and SHAHINIAN, P. - 'Creep-Fatigue Crack
    Growth' Cavities and Cracks in Creep and Fatigue,
    Ed Gittus, J. Applied Science, London, 1981.

19. IRWIN, G.R. - 'Analysis of Stresses and Strains near the
    End of a Crack Traversing a Plate' Trans. ASME J. App.
    Mech. $\underline{24}$, 1957, p.361.

20. ROOKE D.P. and CARTWRIGHT D.J. - 'Compendium of Stress
    Intensity Factors' HMSO, London, 1976.

21. TADA. H, PARIS P.C., IRWIN G.R. - 'The Stress Analysis
    of Cracks Handbook' Del Research, Hellertown, Pa., 1973.

22. RICE, J.R. 'A Path Independent Integral and the Approxi-
    mate Analysis of Strain Concentration by Notches and
    Cracks' Trans. ASME, J. App. Mech., $\underline{35}$ 1968, p.379.

23. HUTCHINSON, J.W. - 'Singular Behaviour at the End
    of a Crack in a Hardening Material' J.Mech. Phys. Solids,
    $\underline{16}$,1968, p.13.

24. RICE, J.R. and ROSENGREN, G.F. - 'Plane Strain Deformation
    near a Crack Tip in a Power-Law Hardening Material'
    J.Mech. Phys. Solids, $\underline{16}$, 1968, p.1.

25. McCLINTOCK, F.A. - 'Mechanics in Alloy Design',
    Fundamental Aspects of Structural Alloy Design. Eds.
    Jaffee,R.I. and Wilcox,B.A. Plenum, New York, 1977.

26. RICE,J.R. - 'Mathematical Analysis in the Mechanics
    of Fracture', Treatise on Fracture, vol.2, Ed. Liebowitz
    H. vol.2, Academic Press, New York, 1968.

27. ILYUSHIN, A.A. - 'The Theory of Small Elastic-Plastic
    Deformations' Prikadnaia Matematika i Mekhanka PMM,
    $\underline{10}$, 1946, p.347.

28. TURNER, C.E. and WEBSTER, G.A. 'Application of
    Fracture Mechanics to Creep Crack Growth' Int. J. Fracture
    1974, p.455.

29. LANDES, J.D. and BEGLEY, J.A. - 'A Fracture Mechanics
    Approach to Creep Crack Growth', Westinghouse Research
    Lab Report 74-1E7-FESGT-P1, Dec. 1974.

30. NIKBIN, K.M., WEBSTER, G.A. and TURNER, C.E. - 'Relevance
    of Non-Linear Fracture Mechanics to Creep Cracking',
    Cracks and Fracture, ASTM STP 601, 1976, p.47.

52

31. HARPER, M.P. and ELLISON, E.G. - 'Use of the C^*
    Parameter in Predicting Crack Propagation Rates'
    J. Strain Anal., 12, 1977, p.167.

32. WEBSTER, G.A. - 'The Application of Fracture Mechanics
    to Creep Cracking' Conf. on Mechanics and Physics of
    Fracture, Inst. of Physics, Cambridge, 1975, Paper 18.

33. KUMAR, V., GERMAN, M.D. and SHIH, C.F. - 'Estimation
    Technique for the Prediction of Elastic-Plastic Fracture
    of Structural Components of Nuclear Systems' Report
    no. SRD-80-094, EPRI, June 1981.

34. BANTHIA, V. and MUKHERJEE, S. - 'Boundary Element
    Analysis of Stresses in a Creeping Plate with a Crack'
    Presented at Int. Conf. Elastic-Plastic Fracture
    Mechanics, Philadelphia, 1981, to be published in ASTM
    STP.

35. HAIGH, J.R. and RICHARDS, C.E. - 'Yield Point Loads and
    Compliance Functions of Fracture Mechanics', Central
    Electricity Generating Board, CERL Memo No. RD/L/M/461,
    1974.

36. SMITH, D.J. and WEBSTER, G.A., - 'Characterizations of
    Creep Crack Growth in 1% CrMoV Steel', J. Strain Anal.,
    16, 1981, p.137.

37. SMITH, D.J. and WEBSTER, G.A. - 'Estimates of the C^*
    Parameter for Crack Growth in Creeping Materials'
    Presented at Int. Conf. Elastic-Plastic Fracture
    Mechanics, Philadelphia, 1981, to be published in
    ASTM STP.

38. MERKLE, J.G. and CORTEN, H.T. - 'A J-Integral Analysis for
    the Compact Specimen Considering Axial as well as
    Bending Effects' J. Press. Vessel Tech., Trans. ASME,
    94, 1974, p.286.

39. NIKBIN, K.M., WEBSTER, G.A. and TURNER, C.E. 'A Comparison
    of Methods of Correlating Creep Crack Growth' in Fracture
    1977, 2, ICF4, Waterloo, Canada, June 1977, p.627.

40. RICE, J.R., PARIS, P.C. and MERKLE, J.G. - 'Some Further
    Results of J-Integral and Analysis Estimates' Progress
    in Flaw Growth and Fracture Toughness Testing, ASTM STP
    536, 1973, p.231.

41. HUTCHINSON, J.W., NEEDLEMAN, A. and SHIH, C.F. - 'Fully
    Plastic Crack Problems in Bending and Tension', Proc.
    O.N.R. Int. Sym. on Fracture Mechanics, Washington, D.C.
    Sept, 1978.

53

42. KENYON,J.L., WEBSTER,G.A., RADON,J.C. and TURNER,C.E., -
'An Investigation of the Application of Fracture Mechanics
to Creep Cracking' Int. Conf. on Creep and Fracture
in Elevated Temperature Applications' I. Mech. E,
Sept., 1973.

43. WEBSTER, G.A.-'Methods of Measuring Crack Growth at
Elevated Temperatures' in Measurement of High
Temperature Mechanical Properties of Materials Ed.
Loveday, M.S., Day, M.F. and Dyson, B.F., HMSO, London,
1982, p.255.

44. WEBSTER,G.A. and NIKBIN,K.M. 'History of Loading Effects
on Creep Crack Growth in $\frac{1}{2}$% CrMoV Steel' in Creep
in Structures Ed. Ponter,A.R.S. and Hayhurst,D.R.
Springer-Verlag, Berlin, 1981, p.576-581.

45. KLINTWORTH,G.C. and WEBSTER,G.A. 'Optimisation of
Electrical Potential Methods of Measuring Crack Growth'
J. Strain Anal. <u>14</u>, 1979, p.187.

46. ARONSON, G.H. and RITCHIE, R.O. 'Optimisation of the
Electrical Potential Technique for Crack Growth
Monitoring in Compact Test-Pieces Using Finite Element
Analysis' J. Testing and Evaluation 7, 1979, p.208.

47. FREEMAN, B.L. and NEATE,G.J. 'Limitations of the
DC Potential Drop Technique of Crack Length Measurement
at Elevated Tempertures' CEGB. Report SSD/MID/R24/78,
1978.

48. LECKIE, F.A. 'Some Structural Theorems of Creep and
their Implications' in Advances in Creep Design, Ed.
Smith A.I. and Nicholson A.M., Applied Science, London,
1971.

49. MARRIOTT, D.L. 'A Review of Reference Stress Methods
for Estimating Creep Deformation' Sym. on Creep in
Structures, Ed. Hult J., Gothenburg <u>2</u>, 1970, p.137.

50. SIVERNS, M.J. and PRICE, A.T. 'Crack Propagation under
Creep Conditions in a Quenched $2\frac{1}{4}$CrMoV Steel,
Int. J. Fracture <u>9</u>, 1973, p.199.

51. NEATE, G.J. and SIVERNS, M.J.-'The Application of
Fracture Mechanics to Creep Crack Growth' Int. Conf.
on Creep and Fracture in Elevated Temperature
Applications, I. Mech. E,, Sept., 1973.

52. FLOREEN, S. - 'The Creep Fracture of Wrought Nickel Base
Alloys by a Fracture Mechanics Approach' Met. Trans.
A, <u>6A</u>, 1975, p.1741.

54

53. HARRISON, C.B. and SANDOR, G.N. 'High Temperature
    Crack Growth in Low Cycle Fatigue' Eng. Fract. Mech.,
    3, 1971, p.403.

54. LECKIE, F.A. and HAYHURST, D.R. - 'Creep Rupture of
    Structures' Proc. Roy. Soc., A340, 1974, p.323.

55. NICHOLSON, R.D. and FORMBY, C.L.-'The Validity of
    Various Fracture Mechanics Methods at Creep Temperatures'
    Int. J. Fracture, 11, 1975, p.595.

56. NIKBIN, K.M., SMITH, D.J. and WEBSTER, G.A.-'Influence
    of Creep Ductility and State of Stress on Creep Crack
    Growth' Int. Conf. Advances in Life Prediction Methods
    at Elevated Temperatures, Albany, April 1983, ASME,
    to be published.

57. CHRISTIAN, E.M., SMITH, D.J., WEBSTER, G.A. and
    ELLISON, E.G. - 'Critical Examination of Parameters for
    Predicting Creep Crack Growth' in Advances in Fracture
    Research, Ed. Francois, D. Pergammon Oxford, 1980, p.1295.

58. HYDE, T.H., LOW, K.C. and WEBSTER, J.J.- private
    communication.

59. NIKBIN, K.M. and WEBSTER, G.A. - 'Temperature Dependence
    of Creep Crack Growth in Aluminium Alloy RR58' in
    Micro and Macro Mechanics of Crack Growth Ed,
    Sadananda, K., Rath, B.B. and Michel, D.J., Met. Soc.
    AIME, 1982, p.137.

60. REIDEL, H.-'Cracks Loaded in Anti-Plane Shear under
    Creep Conditions' 2 Metallkunde, 69, 1978, p.755.

61. NIKBIN, K.M.-'Relevance of Fracture Mechanics to Creep
    Crack Growth' Ph.D. Thesis, University of London, 1977.

62. DYSON, B.F., LOVEDAY, M.S. and ROGERS, M.J. - 'Grain
    Boundary Cavitation under Various States of Applied
    Stress' Proc. Roy. Soc., A349, 1976, p.245-259.

63. AL-FADDAGH, K.D., WEBSTER, G.A. and DYSON, B.F.
    - 'Influence of State of Stress on Creep Fracture of
    $2\frac{1}{4}$% Cr 1%Mo Steel' to be presented at Fourth
    Int. Conf. on Mechanical Behaviour of Materials,
    Stockholm, 1983,to be published.

64. MACKENZIE, A.C., HANCOCK, J.W. and BROWN, D.K.-'On the
    Influence of State of Stress on Ductile Failure Initiation
    in High Strength Steels' Eng. Fract. Mech., 9, 1977,
    p.167-188.

65. HANCOCK, J.W. and COWLING, M.J. - 'Role of State of
Stress on Crack-Tip Fracture Processes' Met. Sci.,
Aug.- Sept., 1980, p.293-304.

66. SADANANDA, K. and SHAHINIAN, P. - 'Creep Crack Growth
Behaviour and Theoretical Modelling' Met. Sci., 15,
1981, p.425-432.

67. SADANANDA, K. and SHAHINIAN, P. - 'High Temperature
Crack Growth Behaviour in Nimonic PE16 and Alloy 718'
Met. Tech., Jan., 1982, p.18-25.

68. SMITH, D.J. and WEBSTER, G.A. and 'Influence of Cyclic
Loading on Crack Growth of a $\frac{1}{2}$% CrMoV Steel' to
be presented at Fourth Int. Conf. on Mechanical
Behaviour of Materials, Stockholm, 1983, to be
published.

Chapter 2

THE POTENTIAL THEORY OF CREEP

D.W.A. REES
Department of Mechanical & Manufacturing
Engineering,
Trinity College,
Dublin 2

SUMMARY

The potential theory of creep is assessed from available experimental data for a wide range of polycrystalline materials. It is shown that at test temperatures $< {}^{T}m/_2$ a flow rule can realistically represent experimental observations under a combined tension-torsion stress state, provided that the creep potential is either formulated from both the second and third stress deviator invariants or taken to be of anisotropic form.

A homogenous potential function is further shown to correlate satisfactorily creep curves between tension and torsion for 0.18% C-steel at 400°C and between tension, compression and internal pressure for aluminium tubes at 250°C on the basis of an associated normalised equivalent strain.

Modifications are discussed which can account for creep strain history through successive translation in the dissipation potential during deformation, the accumulation of damage preceding fracture, the effects of hydrostatic stress and incompressibility.

1.   NOTATION

A,b,c,d,g	material constants in constitutive equations
k,n,p,q,s	for creep
A',B,C,D',M,Q	creep rate equation constants
$C_{ij}, C_{ijkl}, C_{ijklmn}$	tensors of initial anisotropy
f	creep potential function

F	time function
J_1, J_2, J_3	absolute stress invariants
J_2', J_3'	deviatoric stress invariants ($J_1' = 0$)
K	hardening constant
m	time exponent
λ	stress ratio
t	time
T	absolute temperature
$\dot{W}$	dissipation creep work rate
Y	tensile yield stress
δ_{ij}	Kronecker delta
$\dot{\varepsilon}_{ij},\ \dot{\varepsilon}_{ij}'$	absolute and deviatoric creep strain rate tensors
$\bar{\varepsilon},\ \dot{\bar{\varepsilon}}$	equivalent creep strain and strain rate
$\varepsilon,\ \gamma$	normal and shear strain creep components
$\dot{\varepsilon},\ \dot{\gamma}$	normal and shear strain creep rate components
$\sigma_{ij}, \sigma_{ij}'$	absolute and deviatoric stress tensors
$\bar{\sigma}$	equivalent stress
$\sigma,\ \tau$	normal and shear stress components
θ	normalised equivalent creep strain parameter
$\dot{\phi}$	creep flow rule scalar multiplier
t_{ij}'	the deviator of the square of σ_{ij}'
r, θ, z	subscripts for cylindrical co-ordinates
$1, 2, 3,$	subscripts denoting principal stress and strain
ω	damage state variable

2. INTRODUCTION

In the theory of creep, which provides strain rates under any combination of applied stress, it is assumed that the strain rate vector lies in the position of the outward normal to a surface which represents the dissipation potential. This potential has been traditionally identified with the von Mises second invariant of deviatoric stress function for which the associated stress-creep rate relations, supplied by the flow rule, were originally derived by Marin (1) and Soderberg (2). This theory assumes incompressibility and neglects the effects of the third stress deviator invariant and anisotropy through previous creep or plastic strain history.

From the present examination of results reported from laboratory experiments, the simple theory is shown to provide a first approximation to low-medium temperature primary and secondary creep deformation under biaxial stress, following an initial loading path in which the stress components were increased proportionately. In seeking a better representation of the many features observed for a crystalline solid, isotropic dissipation potentials, which are functions of both stress deviator invariants, are examined.

When the creep behaviour of a material is known to
invalidate isotropic theory alternative anisotropic potential
functions are examined that describe the form of anisotropy
which is inherent in the grain texture or that which is
induced by creep deformation.

Modifications are made to each class of multiaxial creep
theory that account for the growth of microscopic damage
that is associated with the increase in strain rates during
tertiary creep.

3. POTENTIAL THEORY

For the foundation of the creep potential we follow
Ziegler (3) who generalised Onsager's extremum principal for
a linear dependence between flux and force to the non linear
case. With creep, in particular, the rate of dissipation
work is given by

$$W = \sigma_{ij}\,\dot{\varepsilon}_{ij} = (\sigma'_{ij} + \frac{1}{3}\delta_{ij}\sigma_{kk})\,\dot{\varepsilon}_{ij}$$

$$= \sigma'_{ij}\,\dot{\varepsilon}_{ij} + \frac{1}{3}\sigma_{kk}\dot{\varepsilon}_{ii} = \sigma'_{ij}\dot{\varepsilon}_{ij}, \ (\dot{\varepsilon}_{ii} = 0). \tag{1}$$

According to the principal, if a strain rate (flux) vector $\dot{\varepsilon}_{ij}$
is prescribed the stress (force) vector σ_{ij} maximises
equation (1), subject to the side condition in which a scalar
dissipation function f is given by

$$f(\sigma_{ij},T) = \sigma_{ij}\dot{\varepsilon}_{ij} = \sigma'_{ij}\,\dot{\varepsilon}_{ij}$$

where, for a given temperature, $f(\sigma_{ij},T) = $ constant are convex
surfaces. In geometric terms, the flux vector $\dot{\varepsilon}_{ij}$ corresponding to a given force vector σ_{ij} lies in the exterior
normal to f and passes through the end point of σ_{ij} (inset
Fig. 1). It is then possible to express normality through
the flow rule

$$\dot{\varepsilon}_{ij} = \dot{\phi}\,\frac{\partial f(\sigma_{ij})}{\partial \sigma_{ij}} \quad , \quad \dot{\phi} = f\left\{\frac{\partial f}{\partial \sigma_{kl}}\,\sigma_{kl}\right\}^{-1}$$

$$\tag{2}$$

which has been further founded in thermodynamics by Besseling
(4) and from dislocation and continuum slip in the microstructure by Rice (5) as outlined in Section 9.

3.1. Isotropy

The assumption of material isotropy and incompressibility
limits the formulation for $f(\sigma_{ij}, T)$ to a function of the two
non zero stress deviator invariants $J'_2 = \frac{1}{2}\sigma'_{ij}\sigma'_{ij}$, $J'_3 = \frac{1}{3}\sigma'_{ij}\sigma'_{jk}\sigma'_{ki}$
when the temperature is constant.

60

Equation (2) then provides,

$$\dot{\varepsilon}_{ij} = \dot{\phi} \, \frac{\partial f(J_2', J_3')}{\partial \sigma_{ij}}$$

$$= \dot{\phi} \left\{ \frac{\partial f}{\partial J_2'} \, \sigma_{ij}' + \frac{\partial f}{\partial J_3'} \, t_{ij}' \right\} \tag{3}$$

where $\quad t_{ij}' = \dfrac{\partial J_3'}{\partial \sigma_{ij}'} = \sigma_{ik}' \sigma_{kj}' - \dfrac{2}{3} J_2' \delta_{ij}$.

If the alternative expression $\dot{W} = \bar{\sigma} \dot{\bar{\varepsilon}}$ is used in equation (1) for the rate of dissipation work, equation (2) is written

$$\dot{\bar{\varepsilon}} = \dot{\phi} \, \frac{\partial f(\bar{\sigma})}{\partial \bar{\sigma}} \tag{4}$$

from which $\dot{\phi} = \dot{\bar{\varepsilon}} / (\partial f / \partial \bar{\sigma})$ in equation(3) can be conveniently established from known tensile creep properties because, then, by definition $\dot{\bar{\varepsilon}} = \dot{\varepsilon}$ and $\bar{\sigma} = \sigma$. When, for example, the third invariant is ignored, as in the simplified Marin-Soderberg theory, then $f = J_2' = \bar{\sigma}^2/3$, equation (4) becomes $\dot{\bar{\varepsilon}} = 2\dot{\phi}\bar{\sigma}/3$ and substitution into equation (3) gives

$$\dot{\varepsilon}_{ij} = \frac{3\dot{\bar{\varepsilon}}}{2\bar{\sigma}} \, \sigma_{ij}' \tag{5}$$

where, in general $\bar{\sigma} = \sqrt{\dfrac{3}{2}(\sigma_{ij}'\sigma_{ij}')}^{\frac{1}{2}}, \; \dot{\bar{\varepsilon}} = \sqrt{\dfrac{2}{3}(\dot{\varepsilon}_{ij}\dot{\varepsilon}_{ij})}^{\frac{1}{2}}$. However, if the dependence of equivalent secondary creep rate on equivalent stress is established from tensile creep tests to be given by the Norton law $\dot{\bar{\varepsilon}} = A\bar{\sigma}^n$, then equation (5) simplifies to

$$\dot{\varepsilon}_{ij} = \frac{3A}{2} \, \bar{\sigma}^{n-1} \, \sigma_{ij}' \, . \tag{6}$$

A direct comparison between the theory and laboratory experiments is possible on the basis of a strain rate ratio for the particular case of creep under tension (σ) combined with torsion (τ) where $\lambda = \tau/\sigma$. Table I gives the corresponding constitutive relation and creep rate ratio, derived from equation (3), corresponding to each of five creep potential functions. These are, respectively, the Mises potential, three homogenous functions and one non-homogenous function, in which Y is the tensile yield stress at the test temperature.

3.2.   Anisotropy

The two anisotropic potentials given in Table I are taken form the general series function ;

$$\bar{\varepsilon} = C_{ij} \, \sigma_{ij}' + \tfrac{1}{2} C_{ijkl} \sigma_{ij}' \sigma_{kl}' + \tfrac{1}{3} C_{ijklmn} \sigma_{ij}' \sigma_{kl}' \sigma_{mn}' \cdots$$

The flow rule, equation (2), supplies the following expression
for the creep rate tensor in which incompressibility is
implied

$$\dot{\varepsilon}_{ij} = \dot{\phi}\,[C_{\{ij\}} + \tfrac{1}{2}C_{\{ij\}kl}\,\sigma'_{kl} + \tfrac{1}{3}C_{\{ij\}klmn}\,\sigma'_{kl}\sigma'_{mn}] \tag{7}$$

where

$$C_{\{ij\}} = C_{ij} - \tfrac{1}{3}\,\delta_{ij}C_{kk}$$

$$C_{\{ij\}kl} = C_{ijkl} - \tfrac{1}{3}\delta_{ij}C_{ppkl}$$

$$C_{\{ij\}klmn} = C_{ijklmn} - \tfrac{1}{3}\delta_{ij}C_{ppklmn}$$

The quadratic function has been discussed by Rabotnov (6).
The determination of the anisotropy parameters $C_{\{ij\}kl}$ was
outlined though no comparison with experimental results was
made.   The constant of proportionality K, appearing in
Table I, between the strain rate ratio and the stress ratio
was given as (3-2 β) for tests in tension combined with
torsion.   Hence an examination of the value of β , or K,
for a material would identify the nature of its creep deformation.   When a material is known to be initially anisotropic
and K is found to be constant, for example, then the anisotropy is preserved during creep.   By the addition of the
linear term in Table 1 the flow rule provides the two strain-
rate ratio expressions given which are not proportionally
dependent upon stress ratio.   The more simplified anisotropic ratio given in the first expression is found from the
second by assuming no difference in the ratio, M, between
uniaxial and shear yield stresses for positive and negative
directions in the material.   Though creep potentials involving the cubic term are not considered here they have
found great use in providing a representation of the distortion observed in subsequent yield loci for time independent
plasticity (7).

4.   FUNDAMENTAL RESEARCH

The literature reveals that results from creep tests on
thin-walled cylinders under combined tension-torsion surpass
those for other biaxial stress systems.   These tests were
performed in specially designed test rigs,almost all of
which involve incremental loading to the creep test stresses
followed by the measurement of axial and shear strains for
the duration of creep.   Some authors have reported the
creep strain rates for a particular time in the primary region,
while others, observing a secondary stage, report the minimum
axial and shear creep rates corresponding to the particular
ratio between axial and shear stresses.   An alternative
presentation of the present author's test results is given
in Fig. 1(8). Four thin-walled cylinders of commercially
pure annealed aluminium were loaded proportionately by
increments to the test stresses and allowed to creep for
50-100 hr while axial and shear strains were recorded continuously.   The figure shows that, in the absence of

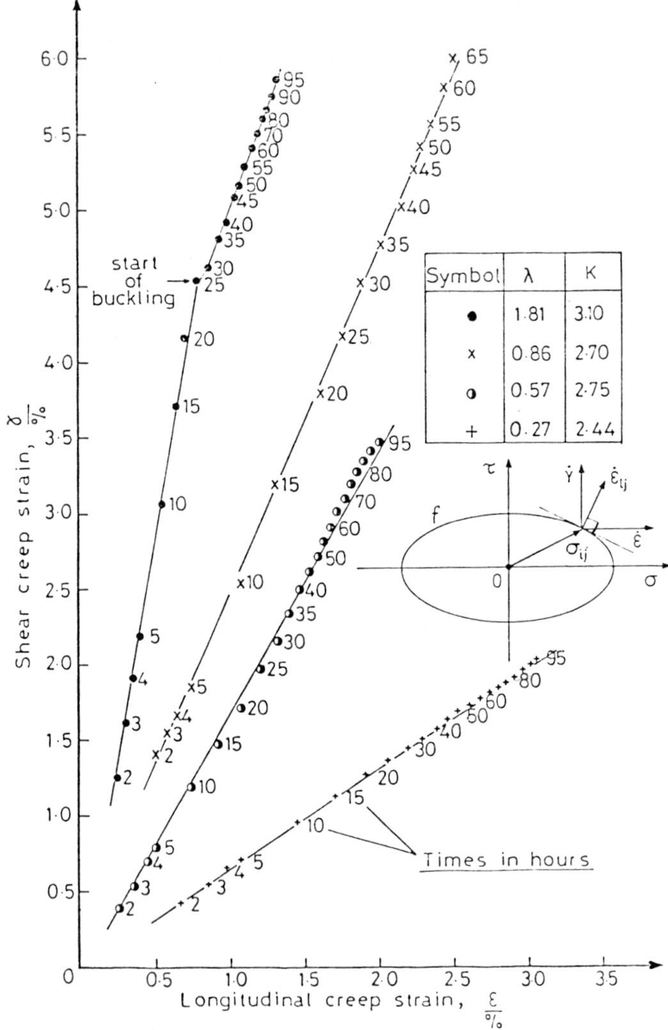

Fig.1 Component σ vs ε creep strain paths for combined
tension-torsion tests on thin-walled cylinders of
commercially pure annealed aluminium (ElA) under the
indicated stress ratios ($\lambda = \tau/\sigma$) at 21°C(8).
Strain rate vector $\dot{\varepsilon}_{ij}$ is normal to creep potential
surface f at the stress point σ_{ij}.

torsional buckling, the component creep strain plot corres-
ponding to each stress ratio is linear. Thus the ratio $\dot{\gamma}/\dot{\varepsilon}$
is constant and, in accordance with the theory, is independent
of the time in primary or secondary creep. The corresponding
creep potential f, shown inset in Fig. 1, appears as an
ellipse in $\sigma - \tau$ space containing the stress vector σ_{ij}.
The direction of the outward normal vector, that is $\dot{\varepsilon}_{ij}$,
at the stress point defines the ratio $\dot{\gamma}/\dot{\varepsilon}$. The results
show that only the magnitude of the creep strain rate vector

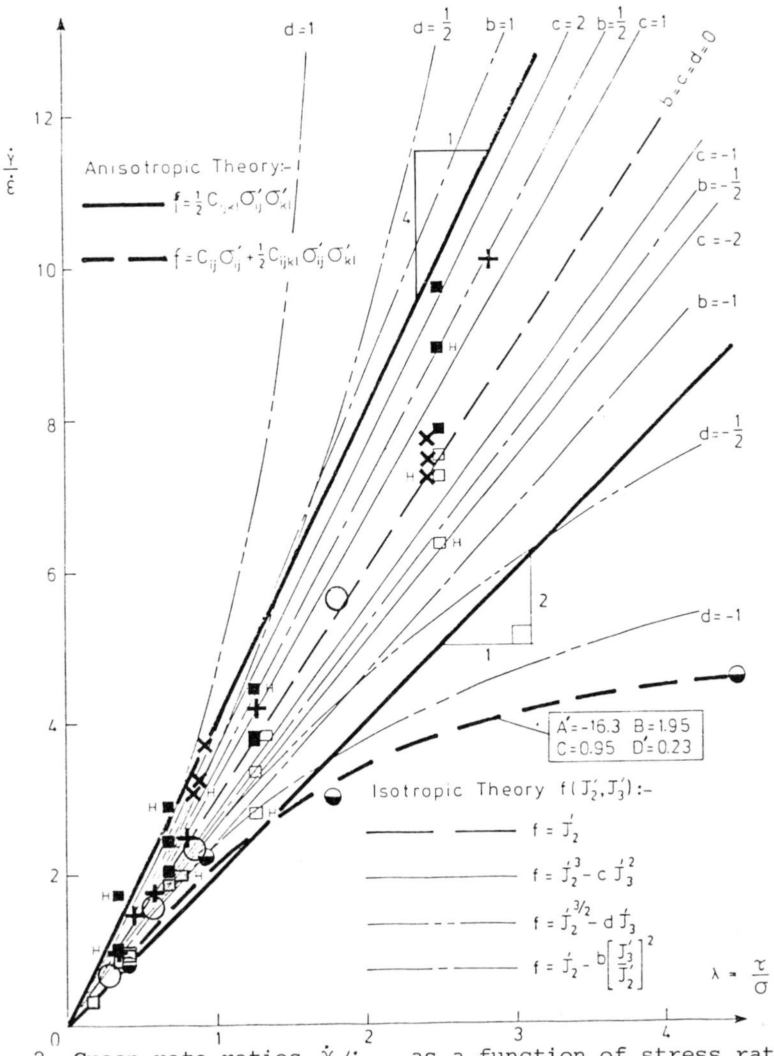

Fig.2 Creep rate ratios $\dot{\gamma}/\dot{\varepsilon}$ as a function of stress ratio
$\lambda=\tau/\sigma$ from both isotropic and anisotropic theory
(Table I) and from experiment. Data shown applies to
thin-walled cylinders under combined tension and
torsion for the following test conditions:

○ E1A A1 (commercially pure) at 21°C and 100 hr.(8).
◑ "   "   "   "   "   "   "   "   "(8)as
                                              extruded).

■ Mg-alloy at 20°C and 150 hr.(9).
□ "   "   " 50°C "   "   "   "
✕ 2S-0 Al (commercially pure) at room temperature and
                                              100 hr. (13).
✚ 6061-T6 Al-alloy at 400°F under steady state
                                  secondary creep (16).
All materials were heat treated prior to test except
where otherwise stated. H indicates the highest stress
levels where more than one test was performed under the
same λ.

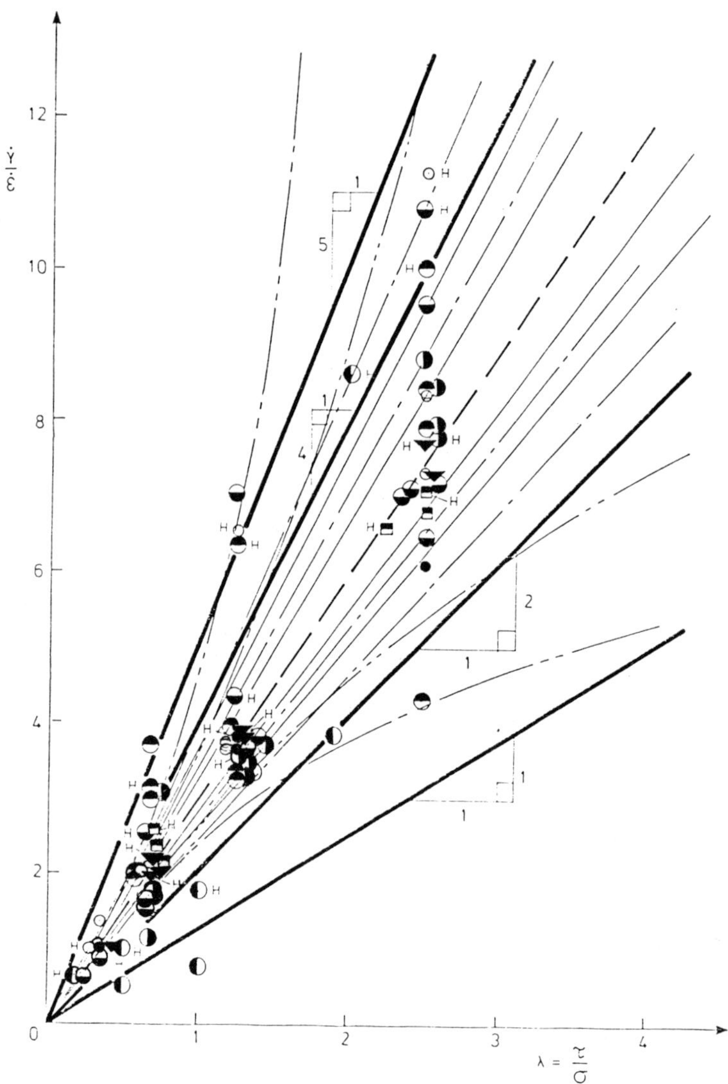

Fig.3. Continuation of test results in Fig.2 for :
● RR59 Al-alloy at 150°C and 150 hr. (10).
○   "   "   "   " 200°C "   "   "   "
◐ 0.17% C-steel at 455°C and 150 hr (11).
◒   "   "   " 350°C "   "   " (12).
◓   "   "   " 450°C "   "   "   "
◑   "   "   " 550°C "   "   "   "
▤ 17-7PH stainless steel at 972°F and 45 min. (14).
▼ Nimonic 75 at 650°C and 150 hr. (15).

is affected by creep time.   This might not be so if previous
loading from the virgin state for the material were not proportional (8), i.e. if the creep test stresses were achieved
following plastic prestraining or by a stepped stress path then
the resulting strain history can also influence the direction
of $\dot{\varepsilon}_{ij}$.

4.1   Isotropic Theory

Experimental hardening constants, calculated from $K = \dot{\gamma}/\lambda\dot{\varepsilon}$
and given in Fig.1, allow direct comparison with the simpli-
fied J_2' theory, where, from Table 1, K = 3.   Reasonable
agreement with a Mises creep potential is found for this
material.   Figs. 2 and 3 present this comparison graphically
with further results taken from creep tests for which the test
stresses were achieved by proportional loading.   The predictions corresponding to each of the first four isotropic
creep potentials in Table I are shown in each figure for the
values of the constants indicated.   While the plot of the
experimental ratios for aluminium from Fig.1 lie close to the
line representing the Mises potential ($f = J_2'$), it appears
that the other materials shown in Fig.2 and those for Fig.3
are better represented by the more general $f(J_2'$, $J_3')$ theory.
The behaviour of the pure aluminium in the as-extruded
condition is included with the results in Fig.2 to show that
the effect of annealing this material at 400°C for 3 hr. was
to turn it into a form whose creep behaviour closely conformed
to isotropic theory.   All other materials in Figs. 2 and 3
were heat treated and we therefore might reasonably expect
agreement with isotropic theory.   The pioneering tests
performed by Johnson show the effects of test temperature
and the magnitude of the stress levels in the ratio $\dot{\gamma}/\dot{\varepsilon}$
for Mg-alloy (9) in Fig.2 and Al-alloy (RR59) (10) and 0.17%
C-steel (11,12) in Fig.3.   When taken with further results
on the effect of stress level alone for aluminium 2S0 (13) in
Fig.2, stainless steel (14) and nimonic 75 (15) in Fig.3,
there would appear to be no marked effect on $\dot{\gamma}/\dot{\varepsilon}$ between
materials in raising the test temperature or the stress
levels corresponding to a particular stress ratio.   This
provides the supporting experimental evidence for a homogenous
creep potential function.   The fact that an accumulation of
points lies on or close to the line from the Mises potential
($\dot{\gamma}/\dot{\varepsilon}$ = 3λ) in both figures gives some support to a simplified creep theory in which J_3' is disregarded. However,
the many points lying away from this line cannot be ignored
and the figures indicate that the creep observed in typical
engineering materials under combined service stresses might
be better represented by any of the three homogenous functions
given in Table I with a suitable material constant e.g. Al-
alloy (16) in Fig.2 is well represented for c = 1 during
secondary creep.   The indicated high stress points (H) for
steel in Fig. 3 would appear to have raised the ratio $\dot{\gamma}/\dot{\varepsilon}$

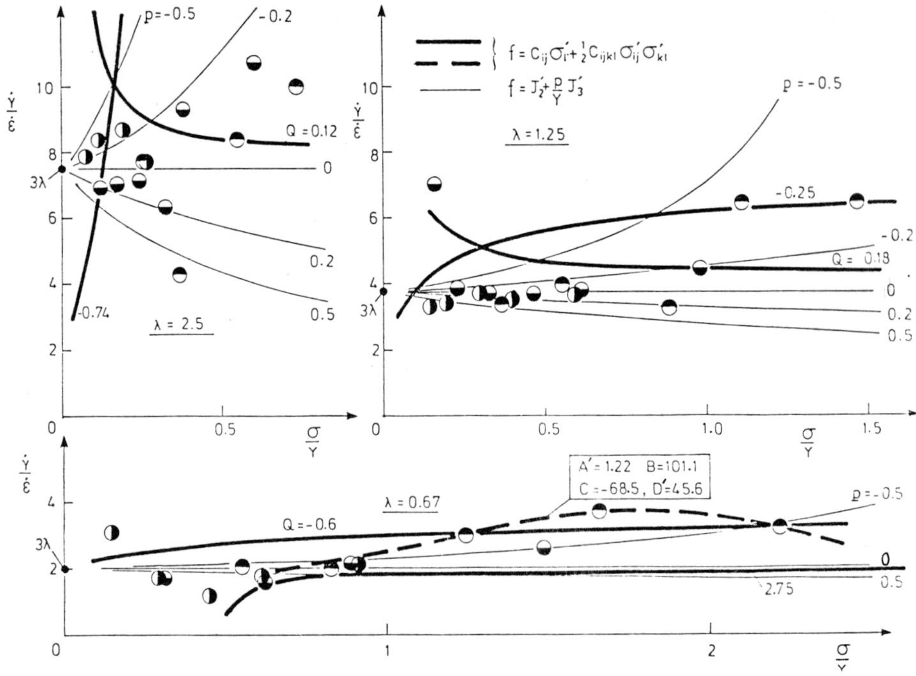

Fig. 4. Variation in the ratio $\dot{\gamma}/\dot{\epsilon}$ with normalised
tensile creep stress σ/Y for three combined
tension-torsion stress ratios (λ) in 0.17%
C-steel (12) at ● 350°C, ◐ 450°C and ◑ 550°C.
Comparisons are made with predictions from an
inhomogenous creep potential $f = J'_2 + PJ'_3/Y$
for the   P values indicated and the aniso-
tropic potential $f = C_{ij}\sigma'_{ij} + \tfrac{1}{2}C_{ijkl}\sigma'_{ij}\sigma'_{kl}$.

corresponding to a particular stress ratio and temperature.
Fig.4 examines this in greater detail.   The plots of $\dot{\gamma}/\dot{\epsilon}$
against normalised axial stress component σ/Y for each
of three stress ratios show that at high stresses the ratio
of creep rates departs slightly from the constant value,
according to a homogenous potential function.   Predictions
shown from the non-homogenous potential given in Table I
will account for this effect through a suitable choice of
the material constant P, although possible scatter in the
test results may be misleading in this respect.

TABLE I : Constitutive Relations and Strain-Rate Ratio
Predictions from Seven Creep Potentials

Creep Potential $f =$	Constitutive Relation $\dot{\varepsilon}_{ij} = \dot{\phi}\,\partial f/\partial\sigma'_{ij} =$	Strain-Rate Ratio $\dot{\gamma}/\dot{\varepsilon} = 2\dot{\varepsilon}_{12}/\dot{\varepsilon}_{11} =$
J'_2	$\dot{\phi}\,\sigma'_{ij}$	3λ
$J'^3_2 - c\,J'^2_3$	$\dot{\phi}\left(3 J'^2_2 \sigma'_{ij} - 2c\,J'_3\,t'_{ij}\right)$	$\dfrac{3\lambda(\lambda^2 + \frac{1}{3})^2 - \frac{2\lambda c}{9}(\lambda^2 + \frac{2}{9})}{(\lambda^2 + \frac{1}{3})^2 - \frac{c}{27}(\lambda^2 + \frac{2}{9})(2 + 3\lambda^2)}$
$J'_2 - b\left(\dfrac{J'_3}{J'_2}\right)^2$	$\dot{\phi}\left\{\sigma'_{ij} - \dfrac{2b}{J'^4_2}\left(J'^2_2 J'_3 t'_{ij} - J'_2 J'^2_3 \sigma'_{ij}\right)\right\}$	$\dfrac{81\lambda(\lambda^2 + \frac{1}{3})^3 - 2\lambda b(\lambda^2 + \frac{2}{9})}{27(\lambda^2 + \frac{1}{3})^3 - 9b(\lambda^2 + \frac{2}{9})\left\{(\lambda^2 + \frac{1}{3})(\lambda^2 + \frac{2}{3}) - \frac{2}{3}(\lambda^2 + \frac{2}{9})\right\}}$
$J'^{3/2}_2 - d\,J'_3$	$\dot{\phi}\left(\frac{3}{2} J'^{1/2}_2 \sigma'_{ij} - d\,t'_{ij}\right)$	$\dfrac{9\lambda(\lambda^2 + \frac{1}{3})^{1/2} - 2d\lambda}{3(\lambda^2 + \frac{1}{3})^{1/2} - d(\lambda^2 + \frac{2}{3})}$
$J'_2 + \dfrac{P}{Y} J'_3$	$\dot{\phi}\left(\sigma'_{ij} + \dfrac{P}{Y} t'_{ij}\right)$	$\dfrac{\lambda\left(3 + P\,\sigma/Y\right)}{1 + \frac{P\sigma}{2Y}(\lambda^2 + \frac{2}{3})}$
$\frac{1}{2} C_{ijkl}\,\sigma'_{ij}\,\sigma'_{kl}$	$\frac{1}{2}\dot{\phi}\,C_{\{ij\}kl}\,\sigma'_{kl}$	$K\lambda$
$C_{ij}\sigma'_{ij} + \frac{1}{2} C_{ijkl}\,\sigma'_{ij}\,\sigma'_{kl}$	$\dot{\phi}\left(C_{\{ij\}} + \frac{1}{2} C_{\{ij\}kl}\,\sigma'_{kl}\right)$	$\dfrac{M^2\lambda\frac{\sigma}{Y} + Q\left(\frac{\sigma}{Y}-1\right)}{\frac{\sigma}{Y} + \frac{Q}{M}\left(\lambda M\frac{\sigma}{Y}-1\right)}$ OR $\dfrac{A' + B\lambda\frac{\sigma}{Y} + C\frac{\sigma}{Y}}{1 + D'\frac{\sigma}{Y} + C\lambda\frac{\sigma}{Y}}$

4.2 Anisotropic Theory

An alternative representation of the test results in
Figs. 2 and 3 is presented from the quadratic potential
(Table I) for K values of 1,2,4 and 5. Clearly, some
materials are well represented by the linear predictions
given in each figure but the decision as to whether a material
is properly represented by an isotropic or anisotropic
potential rests with the results of preliminary studies that
were not, in general, made by the experimenters. Most bil-
lets or rods from which tubes are machined would exhibit
anisotropic creep properties unless they are carefully homo-
genised in order to remove texture. Some proof of the
initial conditions of the material is therefore desirable
by comparing, say, uniaxial tensile creep behaviour for ortho-
gonal directions (17) or, in a comparison between tensile and
compressive creep behaviour (18) with an associated study of

68

the grain structure.   Alternatively, by analogy with the
choice of yield functions for time independent plasticity ,
clarification in the choice of either isotropic or anisotropic theory for creep can be shown from the measured ratio
between transverse and axial strains under uniaxial stress
(19).   Unfortunately transverse strains are rarely measured
in uniaxial creep tests.   In tension, for example, the compressive transverse strain may be taken as one half of the
axial tensile strain provided creep deformation is controlled
by an isotropic potential.   This ratio is then independent
of the function $f(J_2', J_3')$.   On the basis of the author's
preliminary isotropy studies for annealed aluminium and its
combined stress test results in Fig.2 ,which conform reasonably closely to $f = J_2'$ control, it is likely that the texture of some materials in Figs. 3 and 4 has resulted in the
departure from $f = J_2'$.

The stress dependence of $\dot{\gamma}/\dot{\epsilon}$ for steel is further
compared in Fig.4 with predictions from the second potential
(Table I) that is composed of the sum of linear and quadratic
terms.   Employing the simpler expression in two constants,
values of Q are given in Fig.4 for an M value of 2 corresponding to each of the three stress ratios (λ).   The marked
difference between the predictions from isotropic and anisotropic theory emphasises that the use of an appropriate
potential, which represents the stress dependence of the
strain-rate ratio, is provisional upon convincing evidence
which reveals the condition of the material initially.   A
further prediction from the more complex strain rate ratio
in four constants, corresponding to the same creep potential,
is given in Fig. 4 for $\lambda = 0.67$.   The indicated constants
A', B, C and D' provide a description of the creep deformation
observed for 0.17%C - steel at 350°C.   Further use of this
anisotropic ratio and the physical significance of the constants is illustrated from Fig. 5 which gives an equivalent
plot to Fig. 1 but for the aluminium in an initially anisotropic as-extruded condition.   While the plots of the instantaneous loading strains reveal initial rotations in their
paths under low stress ratios, the strain paths for the
duration of creep are linear from which the indicated K values
have been found.   The corresponding points in Fig.2 clearly
do not conform to a linear prediction as supplied by the
quadratic potential.   However, the prediction shown for the
indicated constants, A' ,B,C and D' well represents the data
points and further reflect initial anisotropy in the difference that was observed between tensile and compressive
yield stresses (8) , for the material in this condition.

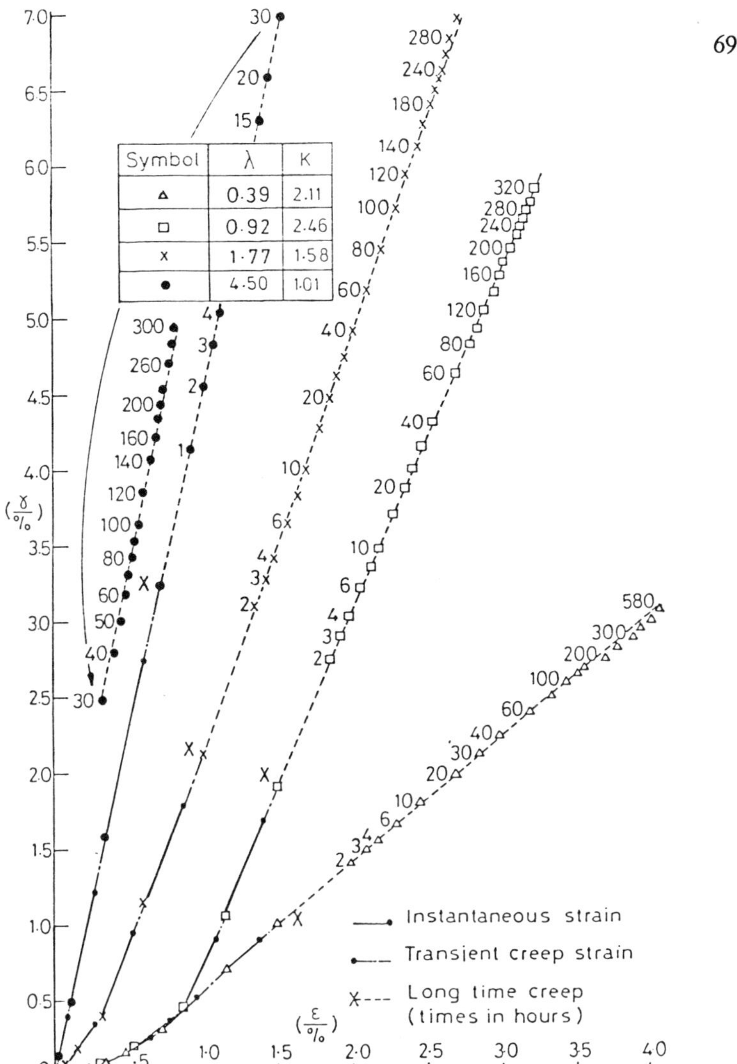

Fig. 5 : Component γ vs ε strain paths for combined tension-torsion tests on thin walled cylinders of commercially pure as-extruded aluminium (E1A) under indicated stress ratios (λ) at 21°C (8).

5. CREEP CURVE CORRELATIONS

The effect of the third stress deviator invariant may further be examined from a correlation between the creep curves obtained in tension and in torsion. For this purpose test results obtained by Crossland, Patton and Skelton (20), for a 0.18% carbon steel at 400°C, are employed. Their results from tensile creep tests gave a primary time hardening creep law of the form

$$\varepsilon = A\sigma^n t^m$$

for $0 < t < 3000$ hr. A normalised equivalent strain is defined from the law as

$$\frac{1}{A}\frac{\bar{\varepsilon}}{\bar{\sigma}^n} = t^m$$

(8)

in which any particular definition of equivalent stress $\bar{\sigma}$ and equivalent strain $\bar{\varepsilon}$ for the combined stress case must degenerate to σ and ε respectively for the tension test. For a Mises potential it follows from equations (5) and (8) that the primary creep rates are given by

$$\dot{\varepsilon}_{ij} = \frac{3mA}{2}\bar{\sigma}^{n-1}t^{m-1}\sigma'_{ij}$$

The time index m was identified with the slope of $\log \varepsilon$ vs $\log t$ plots (not shown) for each of six tensile creep stresses in the range $185 < \sigma < 263$ N/mm^2. These plots were approximately parallel and from them an average value $m = 0.273$ was found. Eight isochronous plots of $\log \sigma$ vs $\log \varepsilon$ were also of approximately constant slope and this was identified with an average stress index $n = 4.3$. When a constant A has been found from the average of the intercept values for the chosen times, equation (8) becomes

$$\left\{\frac{1752}{\bar{\sigma}}\right\}^{4.3}\bar{\varepsilon} = t^{0.273} .$$

(9)

In Fig. 6 a comparison is made between equation (9) and the experimental tensile creep data. The solid line shown represents the right-hand side of equation (9), while the experimental points given are calculated from the left-hand side for the indicated stresses and representative measured creep strains at eight times taken from the six creep tests. Equation (9) clearly provides a good correlation of primary tensile creep data. The representation of torsion or other multiaxial creep data by equation (9) is made by assuming the same time and stress dependence (the respective indices found by Crossland et al. for torsion were a comparable 0.313 and 4.7, a fact which further confirms that the time and stress depdendence of creep strain can be separated), while the definition of $\bar{\sigma}$ and $\bar{\varepsilon}$ depends upon the creep potential function. By von Mises $\bar{\sigma} = \sqrt{3}\tau, \bar{\varepsilon} = \gamma/\sqrt{3}$ for torsion and therefore equation (9) becomes

$$\left\{\frac{890.2}{\tau}\right\}^{4.3}\gamma = t^{0.273} .$$

(10)

A representative sample of shear creep stresses and strains, taken at the same times from nine tests, are used in the calculation of the left-hand side of equation (10). The corresponding points shown in Fig. 6 are identified according to the nominal shear stresses indicated. While there is now a greater spread in the data there is evidently no correlation for this material between creep tests performed in torsion and those performed in tension on the basis of $f = J'_2$.

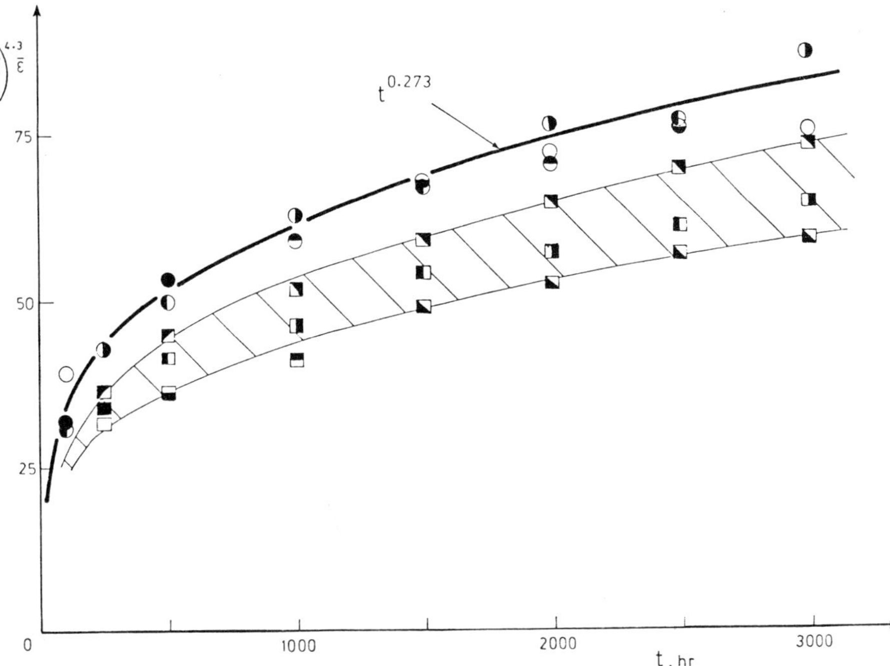

Fig. 6 :   Correlation between long-term creep test
results for thin-walled cylinders of 0.18%
C-steel at 400°C under axial tension and
torsion (20). The normalised strain ordinate
is established from tensile creep tests for
true stresses and logarithmic strains.
Equivalent stress $\bar{\sigma}$ and equivalent creep
strain $\bar{\varepsilon}$ are defined according to the Mises
creep potential function $f = J_2{}'$. Points
shown correspond to the following stresses
(N/mm²) in separate tests :
Tension - ○ 185.4, ◖ 200.85, ◑ 216.3, ◐ 231.75,
◑ 247.2, ● 262.7.
Torsion □ 102.06, ◪ 117.62, ◪ 125.41 ◪ 133.18,
◪ 140.35, ◪ 149.06 ◪ 156.62, ◪ 164.4, ■ 173.

If the second function in Table I is employed as a creep
potential, for tension $J'_2 = \bar{\sigma}^2/3$, $J'_3 = 2\bar{\sigma}^3/27$ and thus at
completion of loading to the creep test stress :

$$f = \bar{\sigma}^6(1/3^3 - 4c/9^3). \qquad (11)$$

For torsion $J'_2 = \tau^2$, $J'_3 = 0$ and hence $f = \tau^6$, where,
from equation (11) for the same potential, the equivalent
stress is

$$\bar{\sigma} = \tau/(1/3^3 - 4c/9^3)^{1/6}. \qquad (12)$$

Assuming that the dissipation work of creep in torsion is a
fundamental measure of the amount of deformation

$$\dot{W} = \sigma_{ij}\dot{\varepsilon}_{ij} = \bar{\sigma}\,\dot{\bar{\varepsilon}} = \tau\,\dot{\gamma}.$$

Substituting from equation (12) and integrating with respect
to time, the equivalent creep strain for torsion is expressed

72

by

$$\bar{\varepsilon} = (1/3^3 - 4c/9^3)\gamma . \tag{13}$$

Substituting equations (12) and (13) into equation (9), the normalised strain becomes

$$(1/3^3 - {}^{4c}/_{9^3})^{0.883}\left\{\frac{1752}{\tau}\right\}^{4.3}\gamma = t^{0.273} . \tag{14}$$

Fig. 7 shows the much improved correlation between tension and torsion for the same test results as in Fig. 6 when for torsion, c = -2.14 in equation (14). The scatter band indicates the magnitude of the likely error if just one test in torsion was made to establish the constant c for the material.

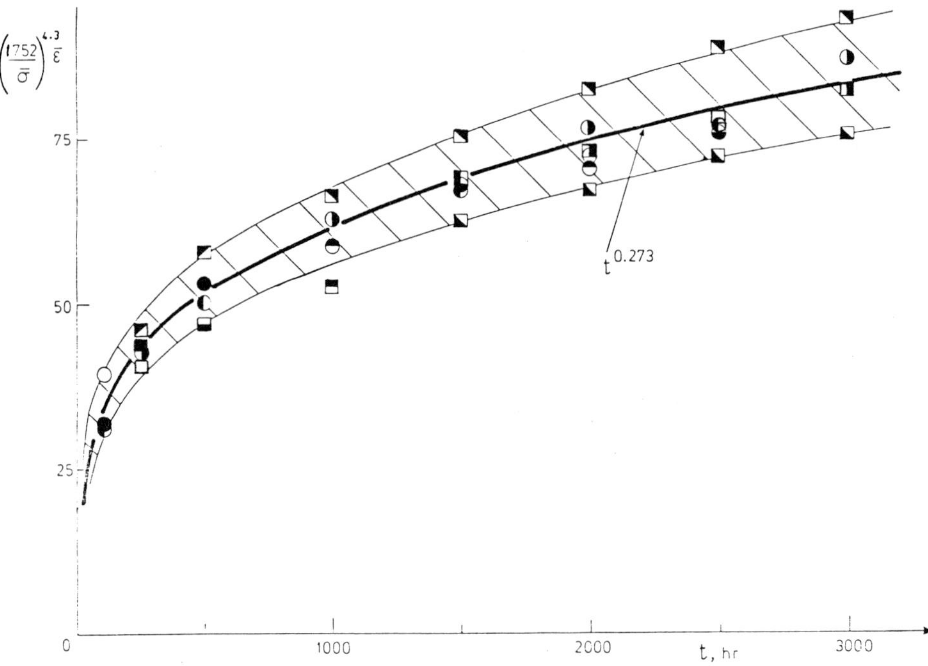

Fig. 7   Improved correlation between tension and torsion
         creep tests for the same test results as Fig.6 but
         with $\bar{\sigma}$ and $\bar{\varepsilon}$ in the ordinate now defined from the
         homogenous creep potential $f = J_2'^3 - cJ_3'^2$
         for c = -2.14.

The scalar multiplier $\dot{\phi}$ in the respective constitutive relation (Table I) is found from the corresponding relation for tension when $\dot{\varepsilon}_{11} = \dot{\bar{\varepsilon}}$, $\sigma'_{11} = 2\bar{\sigma}/3$,

$$t'_{11} = 2\bar{\sigma}^2/9 , \quad J'_2 = \bar{\sigma}^2/3 \quad \text{and} \quad J'_3 = 2\bar{\sigma}^3/27 . \quad \text{Then}$$

$$\dot{\phi} = \dot{\bar{\varepsilon}}\Big/\frac{2\bar{\sigma}^5}{9}(1 - \frac{4c}{27}) . \tag{15}$$

Combining equations (8) and (15), the constitutive relation

becomes, in general form

$$\dot{\varepsilon}_{ij} = 9Am\bar{\sigma}^{-n-5}(3J_2'^2\sigma_{ij}' - 2cJ_3't_{ij}')t^{m-1} / 2(1-4c/27)$$

and for the particular material in question (A $=$ $1752^{-4;3}$
n = 4.3, m = 0.273 and c = -2.14)

$$\dot{\varepsilon}_{ij} = 105.34 \times 10^{-12}\bar{\sigma}^{-0.7}(3J_2'^2\sigma_{ij}' + 4.28J'_3t_{ij}')t^{-0.727}.$$

The influence of J_3' in this correlation is apparently quite
strong.   This influence is now further examined from a
correlation based on different stress systems employing the
test results of Finnie (21), who obtained short-time creep
data for aluminium cylinders at 250°C under circumferential
tension (σ_θ), axial compression (σ_z) and internal pressure.

A stress exponent of n $\cong$ 6 was established from isochronous plots in each of the three tests.   Fig. 8 shows
that the left-hand side of the normalised strain function,

$$\left\{\frac{3460}{\bar{\sigma}}\right\}^6 \bar{\varepsilon} = F(t) , \tag{16}$$

adequately correlates uniaxial creep data for a single test
made under tension and compression.   If $\bar{\sigma}$ and $\bar{\varepsilon}$ are
defined from the Mises potential, for tests under internal
pressure (p)

$$\bar{\sigma}^2 = \sigma_z^2 - \sigma_z\sigma_\theta + \sigma_\theta^2 = 3f . \tag{17}$$

As $\sigma_\theta/\sigma_z = 2$ for $\sigma_\theta = p^D/2h$, in which D and h are the
cylinder mean diameter and thickness respectively from
equation (17)

$$\bar{\sigma} = \frac{\sqrt{3}}{2}\sigma_\theta . \tag{18}$$

According to the potential (equation (17)), the axial strain
$\varepsilon_z = \phi \, \delta f/\delta\sigma_z = 0$, which was confirmed by Finnie's experi-
ments.   The work hypothesis then simplifies to $\dot{W} = \sigma_\theta\dot{\varepsilon}_\theta = \bar{\sigma}\dot{\bar{\varepsilon}}$
from which integration gives

$$\bar{\varepsilon} = \frac{2}{\sqrt{3}}\varepsilon_\theta . \tag{19}$$

Substituting equations (18) and (19) into equation (16), the
normalised equivalent creep strain under internal pressure
is given by

$$\left\{\frac{4092.34}{\sigma_\theta}\right\}^6 \varepsilon_\theta = F(t) . \tag{20}$$

Fig. 8 shows that equation (20) will not provide a satisfactory correlation between internal pressure and tension/ compression creep data.   Finnie thought that because $J_3' = 0$
for internal pressure, the discrepancy was due either to the
development of anisotropy during creep or to the influence
of hydrostatic stress at the homologous test temperature,
i.e. $T/T_m = 0.56$.   The success of the creep potential

found for Fig. 1-6, where $^T/_{T_m} < 0.5$, lies in the assumption that the strain hardening mechanisms of creep deformation are equivalent to those for lower temperature time-independent plasticity, where the plastic potential is well established.

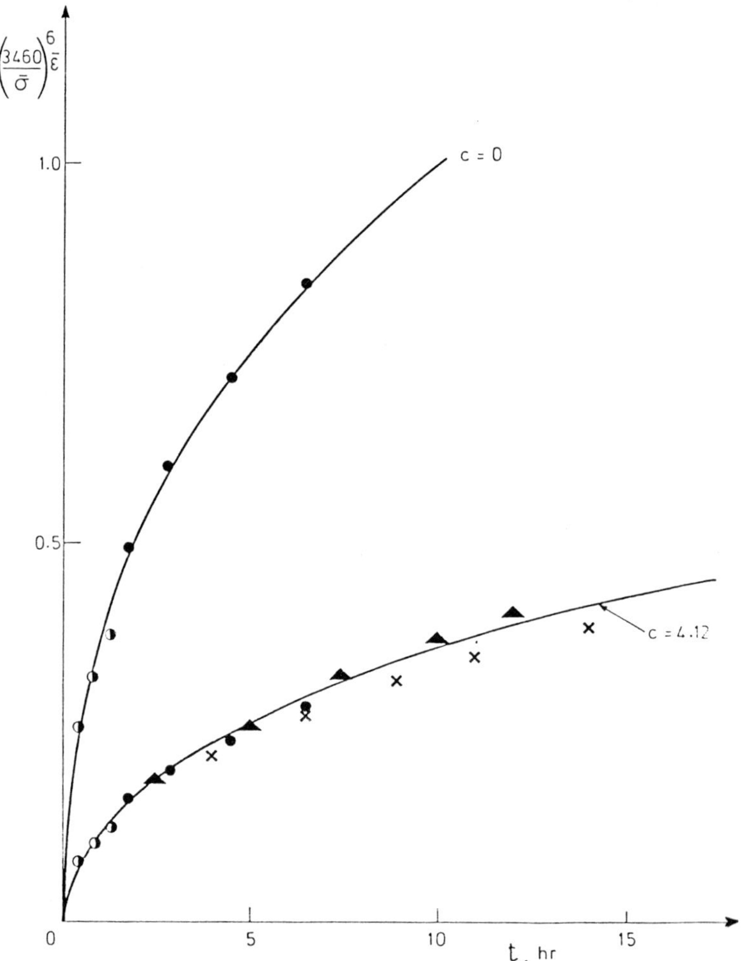

Fig.8 Correlations between short-term creep tests on thin-walled cylinders of aluminium at 250°C under, ✗ circumferential tension, ▲ axial compression and internal pressure at ● 197.2 and ◑ 218.6 bar (21). The normalised strain ordinate is established from uniaxial stress data in which $\bar{\sigma}$ and $\bar{\varepsilon}$ are defined according to (i) $f = J_2'$ and (ii) $f = J_2'^3 - cJ_3'^2$ for c = 4.12.

When the contribution to strain of thermally activated diffusion processes is significant it is likely that the creep
potential leads to unrealistic predictions. This is discussed in connection with void growth and the assumption of
incompressibility in the following section. There is, of
course, the added experimental difficulty that during internal pressure testing the stress and strain systems are modified by the onset of unstable bulging, which the present
author has observed at relatively low strains for thin tubes
with slight eccentricity in diameters or with insufficient
length to diameter ratio. It is possible, however, to
correlate the results in Fig. 8 from a homogenous potential
function in J_2' and J_3' because $J_3' \neq 0$ for tension. For
internal pressure $J_3' = 0$, $J_2' = \sigma_\theta^2/4$ and the second function
in Table I becomes $f = (\sigma_\theta^2/4)^3$. Combining this with
equation (11) the equivalent stress in creep under internal
pressure becomes

$$\bar{\sigma} = 3\sigma_\theta/2(27-4c)^{1/6} \tag{21}$$

and, from the work hypothesis for zero axial strain, the
equivalent strain is given by

$$\bar{\varepsilon} = \frac{\sigma_\theta}{\bar{\sigma}}\varepsilon_\theta = 2(27-4c)^{1/6}\varepsilon_\theta/3 . \tag{22}$$

Substituting equations (21) and (22) into equation (16) the
normalised strain now becomes

$$\left[\frac{2156}{\sigma_\theta}\right]^6 (27-4c)^{7/6}\varepsilon_\theta = F(t) . \tag{23}$$

Fig. 8 shows the much improved correlation between tests when
c = 4.12 in equation (23). The normalised strain method
is applicable to creep data for which the stress, time and
temperature dependenciesof the strain are separable functions
in the hardening law. A function of that form was recently
developed (22) to account for the presence of transient creep
strain in aluminium at room temperature during incremental
loading in tension. For a wide range of test conditions in
which stress, temperature and stress increment ($\Delta\sigma$) were
varied the total strain at time t corresponding to an applied
stress level was composed of the sum of an instantaneous
strain and creep strain. The latter was of the form

$$\varepsilon = s \, \exp(g\Delta\sigma) \, \exp(q/T)\sigma^k t^m$$

where s ,g,q,k and m are constants. A normalised equivalent
strain correlating parameter θ for all stress systems is
then defined as

$$\theta = \bar{\varepsilon}/s \, \exp(g\Delta\bar{\sigma}) \, \bar{\sigma}^k = \exp(q/T)t^m \tag{24}$$

in which $\bar{\varepsilon}$ and $\bar{\sigma}$, defined from the creep potential,are separated from the time and temperature dependence. When used with an associated constitutive relation, equation (24) allows for the prediction of primary creep rates resulting from short-time $0 < t < 15$ min. incremental loading over a wide range of stress and temperature.

6.   <u>KINEMATIC HARDENING</u>

When stress is not constant the resulting changes in shape of the dissipation potential are directly identified with those for the plastic potential. The latter is assumed to describe the progress of plastic deformation through the isotropic hardening rule in which the surface which is descriptive of initial yielding in the material expands to contain the stress vector while retaining its shape and orientation.   It has been shown (23) that the rule supplies realistic predictions of observed strain paths resulting from combined stresses in the plastic range when the loading path is radially outward.   However, when the direction of the loading path is changed, as, for example, when an account of the Bauschinger effect is required for reversed deformation, one of the available rules of kinematic hardening should be employed (24).   According to these rules the initial yield surface translates, and may expand or contract, with the stress vector.   When a translated surface is identified with a creep potential the anisotropy arising in creep due to previous plastic deformation is modelled (25).   Moreover, following a period of forward creep, the kinematic hardening model supplies a description of reversed creep deformation in which, allied to the Bauschinger effect, a change in the magnitude and direction of the strain rate vector occurs following the instantaneous recovery of elastic strain.   In this model the initial condition of the material, for zero time, is taken to be represented by an appropriate potential given in Table I.   The simplest kinematic representation of creep deformation and of the anisotropy that is induced by the strain response to any combination of stress, is provided when the initial potential surface is taken to translate rigidly with the stress vector whilst retaining its shape and orientation.   The motion ceases as the decreasing primary creep rates attain their steady secondary values. If the material is initially isotropic, for example, then rigid translations in subsequent dissipation potentials are expressed by

$$\dot{W} = f(\bar{J}_2', \bar{J}_3') = \text{constant},\qquad(25)$$

where
$$\bar{J}_2' = \tfrac{1}{2}(\sigma_{ij}' - \alpha_{ij}')(\sigma_{ij}' - \alpha_{ij}'), \quad \bar{J}_3' = \frac{1}{3}(\sigma_{ij}' - \alpha_{ij}')(\sigma_{jk}' - \alpha_{jk}')(\sigma_{ki}' - \alpha_{ki}'),$$

while, for an initially anisotropic material
$$\dot{W} = f(F_1, F_2, F_3) = \text{constant},\qquad(26)$$

where
$$F_1 = C_{ij}(\sigma_{ij}' - \alpha_{ij}'), \quad F_2 = \tfrac{1}{2}C_{ijkl}(\sigma_{ij}' - \alpha_{ij}')(\sigma_{kl}' - \alpha_{kl}')$$

$$F_3 = \frac{1}{3}C_{ijklmn}(\sigma_{ij}' - \alpha_{ij}')(\sigma_{kl}' - \alpha_{kl}')(\sigma_{mn}' - \alpha_{mn}').$$

By analogy with the plastic potential the translation tensor
α'_{ij} is taken to supply the current centre co-ordinators of
a potential with respect to the stress vector by one of three
available rules (23).   While further modifications may be
made to equations (25) and (26) in order to account for contraction, distortion and rotation the severe lack of experimental evidence at present, makes further theoretical advances uncertain.   The difficulty, as outlined by Drucker (26),
lies with the experimental determination of dissipation surfaces for creep.   More recently, however, Brown (27), in his
experimental confirmation of Rice's potential theory of continuum slip (5), has shown, for combined tension-torsion
deformation of aluminium at 250°C, that macroscopic dissipation potentials, for $\dot{W}$ = constant, do exhibit primarily a
kinematic translation following short-time creep deformation
history in the direction of a radial stress loading vector.
Fig. 9 shows translations for the locus $\dot{W} = 3 \times 10^4$ in-lb/in³ s
actually corresponding to separate radial loading paths OP in
(a) tension, (b) torsion and (c) combined tension-torsion,
from which it would appear that respective translations are,
in fact, accompanied by distortion, contraction and rotation.

In creep following a non radial loading path or creep under
non-steady stresses the accumulation of history effects, that
are apparent in successive translations, must be considered
in order to properly represent the current dissipation
potential in the flow rule.

7.   TERTIARY CREEP AND FRACTURE

The simplified Marin-Soderberg $f = J_2'$ isotropic theory
has been modified to allow for material deterioration and
the resulting increase in strain rate which occurs during the
tertiary stage of creep.   In Chapters 3 and 5 it is seen,
from a generalisation of the uniaxial single-state variable
Rabotnov-Kachanov equations to multiaxial states, that
equation (6) is written in the form

$$\dot{\varepsilon}_{ij} = 3A\bar{\sigma}^{n-1}\sigma'_{ij}/2(1 - \omega)^n \tag{27}$$

where the damage state variable boundary conditions are
clearly $\omega = 0$ in the undamaged steady state condition and
$\omega = 1$ at fracture.   The intermediate rate of damage accumulation depends upon the criterion of fracture for a particular
material (28).   It is seen from equation (27) that material
damage apparently does not affect the isotropic ratio between
the components of the strain rate tensor for material in the
undamaged state.   This is equivalent to a linear prediction
of component strain paths for creep life.   An examination of
Johnson's results has revealed approximately linear creep
strain paths (28).   Further confirmation of linearity is
provided by Figs. 1 and 5 for the longer time tests that
entered the tertiary region before the test specimen was
unloaded.   We might, therefore, further modify the potential

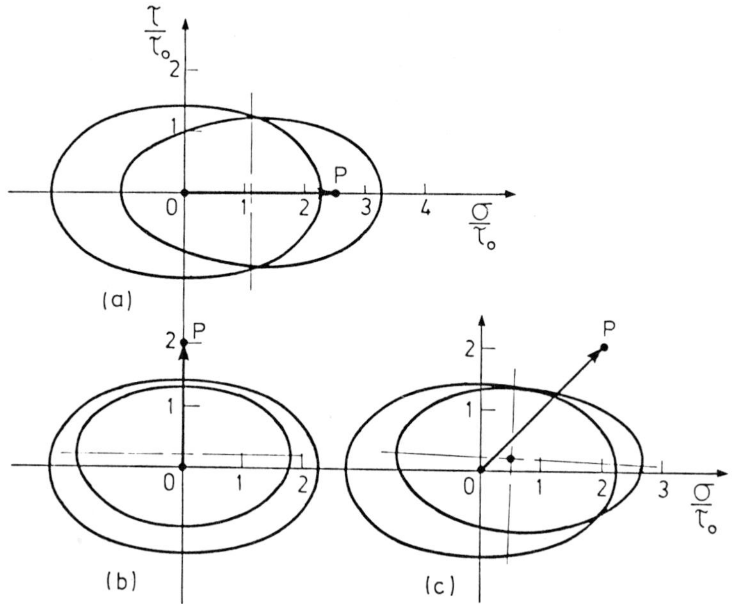

Fig. 9   Translations in the $\dot{W} = 3 \times 10^4$ in-lb/in³s
dissipation creep potential in tension-torsion
stress space for aluminium at 250°C following
loading vectors $\underset{\sim}{OP}$(a) in tension after 280 sec.
(b) in torsion after 250 sec. and (c) under
a radial path $\tau/\sigma = 1$ after 75 sec.   Stress
axes are normalised with respect to shear
stress $\tau_0 = 10$ k.s.i.   After Brown (27).

theory to account for increasing $\dot{\varepsilon}_{ij}$ due to tertiary creep
damage in materials which do not conform to J_2' control
either through the influence of J_3' or by the presence of
initial anisotropy.   For a dissipation potential $f(J_2',J_3')$
for example, in which ϕ in equation(15) depends upon f,the
Norton law is replaced by the Kachanov uniaxial damage growth
law

$$\dot{\varepsilon} = A \left\{ \frac{\bar{\sigma}}{1 - \omega} \right\}^n$$

to give

$$\dot{\phi} = \frac{A \, \bar{\sigma}^{\,n-5}}{\frac{2}{9}(1 - \frac{4c}{27})(1-\omega)^n}$$

Alternatively, we can more simply write from equation (3)

$$\dot{\varepsilon}_{ij} = \dot{\phi}'\{\frac{\partial f}{\partial J_2'}\sigma'_{ij} + \frac{\partial f}{\partial J_3'}t'_{ij}\}$$

where,

$$\dot{\phi}' = \dot{\phi}/(1-\omega)^n$$

and $\dot{\phi}$ remains the scalar defined by the Norton law for the undamaged state that is assumed to exist for the primary and secondary regions of creep deformation.

A similar concept can be applied to equation (7) when linear strain paths imply that initial anisotropy remains constant in subsequent creep deformation provided the applied stresses are transformed to reference directions, such as axes of orthotropy, in the material for which the uniaxial behaviour is known.

By such modifications the implication is that the potential theory, and the associated creep rates as supplied by the flow rule, may again be employed for the damaged state provided that the scalar multiplier $\dot{\phi}$ is modified for the softening effect during tertiary deformation. Moreover, the effect of increasing strain rate on the rupture life has been logically established when the isochronous rupture surface is taken to be given by a function whose form is governed by the effect of the stress state on $\dot{\phi}'$ (28). Isochronous rupture surfaces that embody kinematic translations have yet to be shown experimentally because test results are generally incomplete.

8. <u>HYDROSTATIC STRESS AND INCOMPRESSIBILITY</u>

The theory of creep in equation (3) implies that creep rates under multiaxial stresses are independent of the magnitude of the component of hydrostatic or mean stress and, furthermore, that they would remain unaffected by a superimposed hydrostatic pressure. In relation to this, the material is taken to be incompressible in creep. In support of incompressibility Boettner and Robertson (29) and Bowring, Davies and Wilshire (30) found from density measurements that the total increase in volume due to void growth under high temperature creep conditions was usually less than 1% at fracture. Volume changes were markedly less during secondary creep, which is the region of most interest to design practice. Moreover, Johnson showed (15) that the addition of the same order of hydrostatic stress to an applied combined tension-torsion stress system was ineffective in modifying the creep rates in Mg-alloy at 20°C. However, there have been noticeable

80

effects of high superimposed hydrostatic pressures on low
temperature plastic deformation (31) and in medium temperature
creep (32-34) when its effect is to suppress the nucleation
of cavities associated with the seemingly negligible volume
changes.   The experiments showed that a sufficiently high
hydrostatic pressure superimposed on a tensile creep test
markedly reduced the creep rate and increased the creep life
and strain at fracture for a given applied stress.   In the
potential or continuum theory the implication is that f cannot
be formulated from deviatoric invariants as the assumption of
volume constancy in deformation must then be abandoned by
writing the potential as a function of the absolute stress
invariants J_1, J_2 and J_3.   The associated volume change, which
is due entirely to the enlarging effect of the mean component
of stress acting around existing cavities in the structure, is
called continuum void growth.   A limit is set on the application of the macroscopic creep potential theory when an
additional or alternative mechanism of void growth arises at
higher temperatures where the grain boundary acts as both a
perfect source and a sink for vacancies which diffuse under
stress to boundaries that are normal to the stress axis (35).
Reviews of microstructural approaches to creep damage by cavitation and cracking have been given elsewhere (35,36).

9.   MICROSTRUCTURAL MODELLING

In this final section a justification for the inelastic
flow potential is given from an idealised model due to Rice
(5).   The decomposition of the overall rate of deformation
into the sum of elastic and inelastic components allows for
the construction of constitutive relations when elasticity is
taken to arise from lattice distortion, while plasticity and
creep deformation are attributable to slip on crystallographic
planes and dislocation motion.   Johnson and Gilman (37,38)
considered the latter to be inherently time dependent, a concept which was extended to general stress states and small
inelastic strains in the Rice model.   If the resolved shear
stress on a particular slip system is τ for which γ is the
permanent shear strain then the specific work rate by continuum slip on a physical volume V, for which V_{slip} point systems
operate, is

$$\sigma_{ij}\dot{\varepsilon}^{P}_{ij} = \frac{1}{V}\int_{V\ slip} \tau\dot{\gamma}dV \tag{28}$$

where σ_{ij} and $\dot{\varepsilon}^{P}_{ij}$ are the macroscopic stress and plastic strain
rate tensors respectively.   The positive work rate due to the
motion of discrete dislocation lines is found from the time
derivative of the increasing rate of change of slipped area
due to the normal velocity, v, of a dislocation line segment,
dL.   That is

$$\sigma_{ij}\dot{\varepsilon}^{P}_{ij} = \frac{1}{V}\int_{L\ disl} f\ v\ d\ L \tag{29}$$

where f is the force on a dislocation and L_{disl} refers to all dislocation lines within V. For the two idealisations of slip the macroscopic permanent strain ε^P_{ij} is given by the surplus $\varepsilon_{ij} - \varepsilon^e_{ij}$ where in a homogenous macroscopic strain field, ε_{ij}, the response to σ_{ij} which causes no slip is the elastic component ε^e_{ij}. Then ε^P_{ij} is the volume average of permanent internal strains. By the processes of continuum slip and dislocation motion that is, respectively

$$\varepsilon^P_{ij} = \frac{1}{V} \int_{V\ slip} \tfrac{1}{2}(n_i s_j + n_j s_i)\gamma dV, \quad \varepsilon^P_{ij} = \frac{1}{V}\int_{A\ slip} \tfrac{1}{2}(n_i b_j + n_j b_i)dA$$

where A_{slip} denotes the collection of active slip surfaces, n_i is the exterior unit normal to a particular slip plane, s_i is the permissible slip direction and b_i is the operative Burger's vector of magnitude $|b| = \gamma V/A_{slip}$, such that $\dot{\gamma} = |b|\rho v_m$ where $\rho = L_{disl}/V$ is the mobile dislocation density and v_m is the mean dislocation velocity. In the consideration of the function $\dot{\varepsilon}_{ij}(\sigma_{ij})$ which expresses the dependence of $\dot{\varepsilon}_{ij}$ on σ_{ij} for a current fixed slipped state corresponding to an infinitesimal macroscopic stress change, $d\sigma_{ij}$, in which only the elastic field is altered by $d\tau$ and df, equations (28) and (29) express the corresponding changes in specific work rates as

$$\dot{\varepsilon}_{ij}(\sigma_{ij})\ d\sigma_{ij} = \frac{1}{V}\int_{V\ Slip}\dot{\gamma}(\tau)d\tau dV, \quad \dot{\varepsilon}_{ij}(\sigma_{ij})d\sigma_{ij} = \frac{1}{V}\int_{L\ disl}v(f)df\ dL . \qquad (30\ a,b)$$

As both integrands in equations (30,a,b) have an exact differential there must exist a potential function of stress $f(\sigma_{ij})$ at each slipped state whose derivatives with respect to σ_{ij} supply components of the inelastic strain rate tensor. That is

$$\dot{\varepsilon}_{ij}(\sigma_{ij}) = \dot{\phi}\ \frac{\delta f(\sigma_{ij})}{\delta\sigma_{ij}} \qquad (31)$$

which, of course, is immediately recognised as the flow rule in which normality in $\dot{\varepsilon}_{ij}$ and convexity in f are implied. The form of the function f as we have seen is readily determined from macroscopic experiments which have clearly shown its further dependence upon the history of deformation. Microscopically f is given from equations (28), (29) and (31) in either one of the two double integrals

$$f(\sigma_{ij}) = \frac{1}{V}\int_{V\ slip}\int_o^\tau \dot{\gamma}(\tau)d\tau dV, \quad f(\sigma_{ij}) = \frac{1}{V}\int_{L\ disl}\int_o^f v(f)dfdL . \qquad (32,a,b)$$

As residual stresses R_{ij}, associated with the deformation history, increase preferentially in directions opposite to those of active slip planes the implication is that f translates with the stress vector so that Bauschinger and recovery effects are to be expected. Further confirmation of macroscopic translations in f was demonstrated by Brown (39) from experiments on prestrained aluminium under steady-state creep conditions.

82

An extension to the microscopic theory for modelling inelastic
finite strains caused by dislocation motion has been achieved
(40,41) through an irreversible thermodynamic formulation in
which internal state variables are made to represent specific
microstructural rearrangements.   When such arrangements move
at a rate governed by their associated forces in a so-called
normality structure it was shown that the consequence of this
inter-relationship was the existence of a macro flow potential
and normality of the strain rate vector.   Further discussion
has been made (42) on diffusion, phase – changes, micro-voids
and cracks in a general framework for transition from higher
temperature microscale processes to macroscopic strain.

10.   <u>CONCLUSIONS</u>

A creep potential, formulated from the second and third
stress deviator invariants, is valid in a flow rule for isotropic hardening creep in the low-medium temperature regime.
The function is capable of correlating creep strains under
different stress systems while creep rates, supplied by the
flow rule, are found to be in good agreement with the measured
rates from biaxial stress laboratory creep tests following
proportional loading.   The use of the second deviator stress
invariant as a creep potential (von Mises) can only be
expected to provide a first approximation to the creep rates
resulting from multiaxial stresses under service conditions.
Nevertheless, by its simplicity the J_2' function retains its
usefulness in design considerations.

When the initial condition of a material is known to be
anisotropic a corresponding dissipation potential, that is
composed of linear and quadratic terms in deviatoric stress,
provides a good description of the wide range of strain
behaviour that is observed in subsequent creep.   In common
with plastic anisotropy (43) the constants in a descriptive
quadratic potential for creep are conveniently found from
the measured strain ratios in uniaxial tests on specimens
whose axes are those of the principal axes of anisotropy
in the material.   Significant differences between the tensile
and compressive creep properties would be accountable through
the linear stress term (44).

The effects of creep strain history are modelled by the
rule of kinematic hardening irrespective of the initial condition of the material.   The rule is important when providing
a realistic description of the position of the dissipation
potential for the initial stages of creep following non-radial
loading, or, under changing stresses during creep.   Following
radial loading it is likely that the kinematic and isotropic
hardening potentials become tangential at the load point and,
from the normality rule, thereby supply identical strain-rate
directional response.   This fact would explain why the
simpler though less realistic isotropic hardening rule is

consistent with radially outward loading experiments.

In the tertiary region of creep the flow rule is modified to account for the accumulation of damage preceding
fracture.   The fact that the same potential would apply to
plastic deformation and throughout each stage of creep,
illustrates the great versatility the potential theory has in
providing the strain response from any combination of applied
stress, for the low-medium temperature deformation of polycrystalline materials.

11.   <u>REFERENCES</u>

1.   MARIN, J.   J. Appl. Mech., 1937, $\underline{4}$(2), 55.

2.   SODERBERG, C.R.   Trans ASME, 1936, $\underline{58}$(8), 733.

3.   ZIEGLER, H.   "A generalisation of Onsager's principal",
     Proc. I.U.T.A.M. "Irreversible Aspects of Continuum
     Mechanics", Eds. Parkus, H. and Sedov, L.I., Springer
     1968, 411.

4.   BESSELING, J.F. "Thermodynamic interpretation of the
     concept of ideal creep and plasticity", the Folke Odqvist
     Volume, "Recent Progress in Applied Mechanics", Eds.
     Broberg, B., Hult, J. and Niordson, F., Almqvist and
     Wiksell, Stockholm 1967, 45.

5.   RICE, J.R.   Trans ASME. J. Appl. Mech. 1970, $\underline{37}$,728.

6.   RABOTNOV, Y.N.   "Creep Problems in Structural Members",
     North-Holland, Amsterdam, 1969, Ch.5, 297.

7.   REES, D.W.A. "Initial and deformation induced anisotropy",
     in preparation for "Plasticity Today" Symp; Udine, Italy,
     June 1983.

8.   REES, D.W.A. "Biaxial creep and plastic flow of aniso-
     tropic aluminium", Ph.D. thesis C.N.A.A., U.K. 1976.

9.   JOHNSON, A.E.   Metallurgia, 1950, $\underline{42}$(252), 249.

10.   JOHNSON, A E.   Metallurgia, 1949, $\underline{40}$ (237), 125.

11.   TAPSEL, H.J. and JOHNSON, A.E.   Engineering 1940,$\underline{150}$, 24,
      61,104,134,164.

12.   JOHNSON, A.E.   The Engineer, 1949, $\underline{188}$, 126, 138, 165,189.

13.   MARIN, J., FAUPEL, J.H. and HU, L.W. Proc. ASTM, 1950, $\underline{50}$,
      1054.

14.   DHARMARAJAN, S. and SIDEBOTTOM, O.M. Expl. Mech. 1963,
      July, 153.

15.   JOHNSON, A.E.   Proc. I. Mech. E., 1951, $\underline{164}$(4), 432.

16.   STOWELL, E.Z. and GREGORY, R.K.   J. Appl. Mech. 1964.
      Paper No. 64-WA1 APM-19.

17.   KENNEDY, C.R., HARMS, W.O. and DOUGLAS, D.A. Trans ASME,
      J. Basic Eng., 1959, $\underline{81}$, 599.

18.   TILLY, G.P. and HARRISON, G.F.   J. Strain Analysis, 1972,
      7(3), 163.

19.   REES, D.W.A.   Proc. Roy.Soc. Lond., 1982, $\underline{A383}$, 333.

20.   CROSSLAND, B., PATTON, R.G. and SKELTON, W.J.   "Advances
      in Creep Design", Eds. Smith, A.I. and Nicolson, A.M.
      Applied Science, London, 1970, Ch. 8,129.

84

21. FINNIE, I. "An experimental study of multiaxial creep in tubes", Proc. Jnt. Int. Conf. on Creep, I.Mech.E., London 1963, 2, 21.

22. REES, D.W.A. Proc. R.Ir. Acad. 1981, 81A(2), 167.

23. REES, D.W.A. J. Strain Analysis, 1981, 16(4), 235.

24. REES, D.W.A. J. Strain Analysis, 1981, 16(2), 85.

25. REES, D.W.A. "Effects of plastic prestrain on the creep of aluminium under biaxial stress", Proc. "Creep and Fracture of Engineering Materials and Structures", Eds. Wilshire, B., and Owen, D.R.J. Pineridge, Swansea, 1981, 331.

26. DRUCKER, D.C. "On time-independent plasticity and metals under combined stress at elevated temperature" The Folke Odqvist Volume, "Recent Progress in Applied Mechanics" Eds. Brogerg, B., Hult,J. and Niordson, F., Almqvist and Wiksell, Stockholm, 1967, 209.

27. BROWN, G.M. J. Mech. Phys. Solids, 1970, 18 367.

28. LECKIE, F.A. and HAYHURST, D.R. Acta Metallurgica, 1977, 25, 1059.

29. BOETTNER, R.C., and ROBERTSON, W.D. Trans Met.-Soc., AIME, 1961, 221, 613.

30. BOWRING, P., DAVIES, P.W. and WILSHIRE, B., Met. Sci.Jl. 1968, 2, 168.

31. HU, L.W. Proc. 2nd Symp. Naval Struct. Mech."Plasticity", Pergamon, New York, 1960, 194.

32. HULL, D., and RIMMER, D.E. Phil. Mag. 1959, 4, 673.

33. RATCLIFFE, R.T. and GREENWOOD, G.W. Phil Mag. 1965,12, 59.

34. WADDINGTON, J.S. and WILLIAMS, J.A. Acta Met. 1967, 15 1563.

35. DYSON, B.F. "A unifying view of the kinetics of creep crack growth", Proc. "Creep and Fracture of Engineering Materials and Structures", Eds. Wilshire, B., and Owen, D.R.J., Pineridge, Swansea, 1981, 235.

36. BOLTON, C.J., DYSON, B.F., and WILLIAMS, .R. "Review of metallographic methods of determining residual life", Proc. "Determination of Life Expectancy of Plant which has operated at High Temperatures", I. Mech.E., Nov.1978.

37. JOHNSON, W.G., and GILMAN, J.J. Jl. of Appl. Phys.1959, 30, 129.

38. GILMAN,J.J. "Progress in the microdynamical theory of plasticity", Proc. 5th U.S. Nat. Cong. of Appl.Mech., ASME 1966, 385.

39. BROWN, G.M. J. Mech. Phys. Solids, 1970, 18, 383.

40. KESTIN, J. and RICE, J.R. "Paradoxes in the application of thermodynamics to strained solids", Proc. 'A Critical Review of Thermodynamics', Eds. Stuart, E.B. et al., Mono Book Corp.,Baltimore, 1970, 275.

41. RICE, J.R. J. Mech. Phys. Solids, 1971, 19, 433.

42. RICE, J.R. "Mechanics and thermodynamics of plasticity", in 'Constitutive Equations in Plasticity', Ed. A.S. Argon, MIT Press, Cambridge, Mass., 1975, Ch.2, 23.

43. REES, D.W.A. Expl. Mech. 1981, 21(7), 245.

44. REES, D.W.A. Acta Mechanica, 1982, 43, 223.

Chapter 3

ON THE RÔLE OF
CREEP CONTINUUM DAMAGE
IN STRUCTURAL MECHANICS

D.R. Hayhurst

Department of Engineering, The University,
Leicester, LE1 7RH, U.K.

SUMMARY

Experimental techniques are discussed for the reliable
measurement of deformation and rupture lifetimes of metallic
alloys which undergo uni-axial and multi-axial stress creep.

Results of high-temperature creep tests are presented for
multi-axial states of stress and it is shown that creep damage,
in the form of grain boundary defects, is uniformly distributed throughout regions of homogeneous stress and that these
observations have led to the usage of the term "Continuum
Damage". It is shown how constitutive equations may be used
to describe the growth of continuum damage and the accompanying deformations, and how the apparently different approaches
of the engineering scientist and of the materials scientist
are equivalent for proportional loading conditions.

The numerical solution of creep deformation and rupture
problems in structures is discussed and it is shown how numerical methods have been developed for this class of problem.
Such methods of solutions have been used together with constitutive equations to study the behaviour of plane stress
plates, containing stress concentrators, axi-symmetrically
notched tension bars and plane strain cracked tension bars.
In all cases it is shown that tertiary creep results in stress
redistribution which can effectively change the character of
the original structure. The effect of the multi-axial stress
rupture criterion is shown to cause an interaction between
material and structure which can strongly influence the
character of the growth of continuum damage.

The theory of continuum damage mechanics is shown to be
capable of predicting the deformation and rupture behaviour
of simple plates, notched bars and cracked structural members.

1.0 INTRODUCTION

When metals operate for long periods, typically 10^5h, at temperatures in excess of $0.4\ T_m$, where T_m is the melting temperature of the base metal in degrees Kelvin, and at stresses below one half of the uni-axial yield stress consideration must be given to the possibility that the lifetimes of structural components may be limited either by excessive creep deformation or by failure due to creep rupture. In pressure vessels, or in bolted assemblies, where creep resistant materials are employed, lifetimes may be governed by creep rupture. Under these circumstances it is necessary to perform calculations which take account of the effects upon structural performance of two aspects of material behaviour. These are: the increased strain rates which occur during tertiary creep and the form of the multi-axial stress rupture criterion of the material.

Since the 1930s design standards have formally neglected the effects of tertiary creep on structural behaviour and estimates of lifetimes have been made using the maximum stress obtained from a stationary-state creep analysis, in conjunction with uni-axial tensile rupture data. In complex structures the stationary-state stress distribution is usually found from a full transient creep analysis using numerical methods. The transient analysis determines the change in the stress field from the initial elastic response to the final stationary-state condition without consideration of the tertiary creep process. Because of the non-linear form of the creep law the maximum stationary-state stress is significantly less than the maximum elastic value.

This approach does not take account of the additional stress redistribution which takes place due to the increased strain rates during tertiary creep and of the influence of the multi-axial stress rupture criterion of the material. Before the significance of these phenomena can be assessed in terms of the overall structural performance it is first necessary to develop constitutive equations which accurately describe the creep deformation and rupture of the material and to develop techniques for structural analysis which allow one to model the progressive material degeneration which occurs during creep rupture.

In the development of constitutive equations two approaches have principally been used. The first is the so-called phenomenological method [1,2] which has been shown to be equivalent to a single state damage variable theory [3]. The second is the approach of the materials scientist in which the physics of the microstructural processes are described either by experimental measurement [4,5] or by the usage of suitable material models [6,7]. The effectiveness of both approaches has to be judged by their ability to describe the strain rates and lifetimes measured in long-term tests under both uni-axial and multi-axial stress.

Given an accurate description of material behaviour one requires the use of numerical methods for structural analysis in order to study how structural components behave when large regions of the components undergo tertiary creep. In a discussion of the development of such techniques it is convenient to consider two types of analysis. The first is the so-called stationary state analysis which neglects the effects of tertiary creep; and the second takes account of the increase in strain rate during tertiary creep and its stress-state dependence. In both cases the strain rates are non-linear functions of stress and, for structures with high stress concentrations, the differential equations to be solved tend to be stiff in character. As a consequence novel techniques have been developed for the solution of this class of problem.

Given an accurate description of the material behaviour and validated methods for the numerical analysis of structures, they can then be used to identify and to study the physical features which control the creep rupture of structures. The types of structures in which creep rupture is important, and is likely to provide a limitation on service lifetimes, are those where stress concentrations occur due to changes in geometrical form: e.g. at holes, notches, fillet radii and at sharp defects or cracks present within the material.

The purpose of the paper is three-fold: firstly to describe how the processes of creep deformation and rupture can be described for homogeneously stressed materials under isothermal conditions; secondly to study the time-variation of stress, strain and creep damage in structural components which contain stress concentrations of increasing severity; and thirdly to identify those microstructural phenomena which govern the failure of structures due to creep.

The first part of the paper deals with the creep rupture of homogeneously stressed material and with the development of constitutive equations for both uni-axial and multi-axial stress conditions. The concept of continuum damage is introduced at this stage. The second part of the paper deals with the development of techniques for the numerical analysis of structural components. It is shown how stationary-state solutions are obtained and how the creep rupture behaviour of structural components may be studied using the concept of continuum damage. A series of structures is considered in which the severity of the stress concentration is progressively increased and finally consideration is given to the behaviour of components in which initial defects are present in the form of sharp cracks. The common theme in both the description of the material behaviour and the study of the rupture of structural components is the concept of continuum damage.

2.0   BEHAVIOUR OF METALS UNDER STEADY STATES OF HOMOGENEOUS STRESS

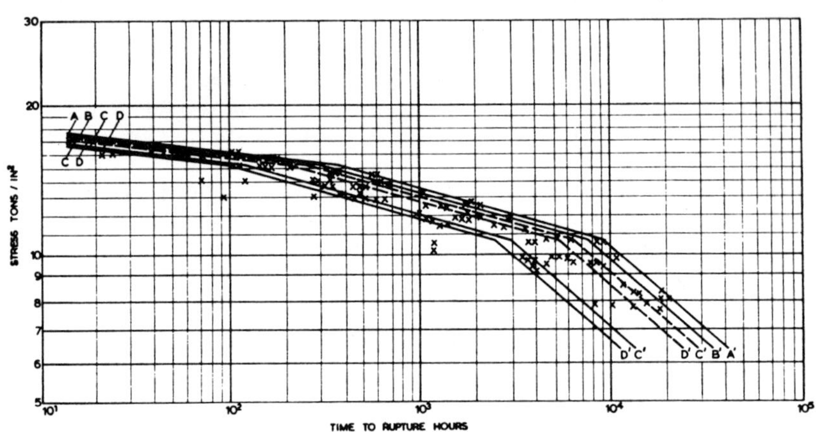

Fig.1.   Stress-rupture data for silicon-killed steels
collected using solid circular tension specimens
by Glen et al (8)

2.1   High-temperature testing techniques

The uni-axial creep tension test is probably the most
common test carried out at high-temperatures and it is probably
regarded as the most simple test to perform.   However,

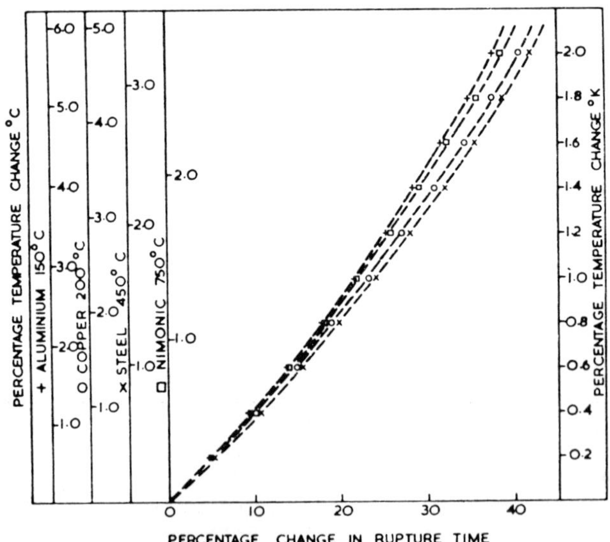

Fig.2.   Percentage change of rupture life due to a
percentage change of test temperature

examination of many test results carried out by Glen et al (8)
on a single melt of metal have shown large amounts of scatter
to be present in the rupture data, despite the procedures recommended by British Standards (9) being carefully followed.
The data obtained by Glen et al is reproduced in Fig.1, where
it may be seen that for a given stress the values of lifetimes
can differ by up to a factor of ten; similar degrees of scatter
are frequently obtained in measurements of creep strain rates.
Such variation or scatter is often attributed to material variation.   While this explanation is plausible to a degree it is
important to point out that variations in stress state and in
temperature level can have a dramatic effect on creep rate and
on lifetime.   It is well known, for example, that the temperature dependence of rupture life t_f is given by
$t_f = K \exp(Q/RT)$, where K is a constant, R is the gas constant, T is the absolute temperature and Q is the activation energy for grain boundary diffusion.   The effect of

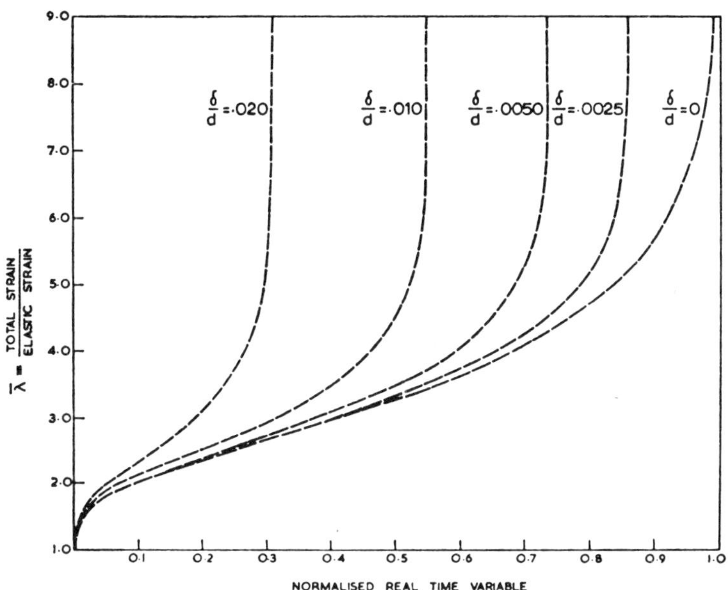

Fig.3.   Effects of eccentric loading
on creep rupture test (n=5.0, X=3.5)

errors in the set level of test temperature, often permitted in
test standards (9), can have a pronounced effect on specimen
lifetimes.   This is shown in Fig.2; for example, if Nimonic is
inadvertently tested at 760°C instead of at 750°C then a reduction in lifetime of 22% results.   Similar effects due to
stress state may be observed.   For example, if a uni-axial
specimen is subjected to tension plus superimposed bending instead of pure tension then the effects on strain rate and lifetime can be pronounced.   A complete analysis of this situation
has been carried out (10) using the following material laws:

90

$$dv/dt = B(\sigma/(1-\omega))^n K(t), \quad \text{and}$$

$$d\omega/dt = G(\sigma X/(1-\omega)^\phi)K(t) ,$$

where v is the creep strain, ω is a parameter which represents the extent of damage or tertiary creep, $K(t)$ is a
time function which models primary-secondary creep and B,G,n,
ϕ and X are material constants. Some typical results are
presented in Fig.3 for the extreme case where the eccentricity
of loading δ , measured from the geometrical centre line of the
specimen, is maintained at a constant fraction of the test bar
diameter d . The time variable has been normalised in such a
way that the lifetime for zero eccentricity is unity. The
effect, for example, of a typical combined error of $\delta/d=0.01$,
for $d=5$ mm, due to setting up the specimen in the machine
will result in a reduction of life to 55% of the zero eccentricity lifetime. The combined effects of errors in temperatures and in axiality of loading can be dramatic. The results presented in Figs.2 and 3 can be used to assess the importance of temperature and stress-state control in the experimental determination of lifetimes. For example in Fig.1 the
line of maximum strength AA' can be attributed to tests
carried out with zero eccentricity of loading at the lowest temperature within the measured range. The line BB'
corresponds to tests carried out at the true temperature
$(450°)$ with zero eccentricity. The unbroken line CC' corresponds to tests carried out at the true temperature with an
eccentricity of $\delta/d=0.015$, typical of the type of apparatus
used. The unbroken line DD' corresponds to tests with an
eccentricity value of $\delta/d=0.015$, carried out at the maximum
temperature within the measured range. The significance of
the broken lines CC' and DD' has been described and discussed by Hayhurst (10).

Probably the most important requirement for repeatable
creep testing is to establish procedures for control (10) of
the test variables. This can be achieved economically by
using modern electronic temperature controllers and by using
special purpose specimen grips to reduce eccentricity of
loading.

In all the experiments reported here temperatures have
been controlled to within $\pm 1°C$ and better, for the lower test
temperatures; and specimen bending, defined by

$$\% \text{ Bending } = (\varepsilon_1 - \varepsilon_2)100/(\varepsilon_1 + \varepsilon_2) ,$$

where ε_1 and ε_2 are surface strains measured on opposing
surfaces of the specimen, has been maintained below seven per
cent.

2.2 <u>Uni-axial stress</u>

In this section consideration is given to some aspects of the physical processes which govern tertiary creep under uni-axial stress. At stresses near to the yield stress Hoff (11) has shown that creep rupture can be due to deformation processes alone. Under these conditions rupture takes place due

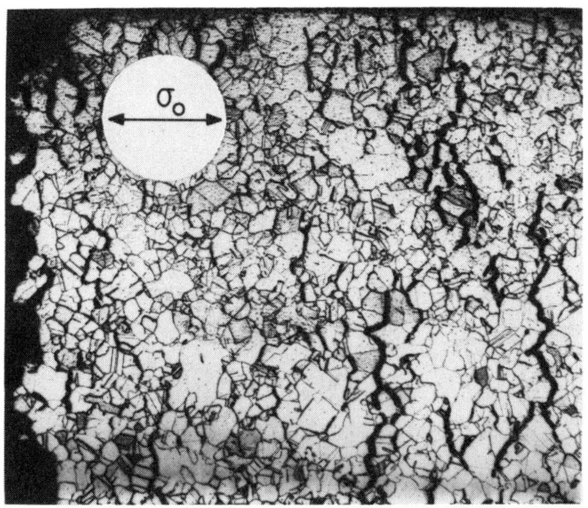

Fig.4.   Micrograph taken from; uni-axial tension specimen
in copper tested at 250°C; plane of failure is at the
left-hand edge.   (Mag. x 6)

to the formation of a neck with little or no evidence of microstructural change having taken place (12). At the lower stresses found in design applications, typically less than $0.3\,\sigma_y$ where the lifetimes are in excess of 10^5h , microstructural changes take place. Measurements of the changes in volume which have been made during creep deformation (13) indicate that defects begin to grow in primary creep. The growth of defects ultimately causes a monotonic increase in strain rate from a constant value, during secondary creep, to an infinite value at failure. This part of the process, referred to as tertiary creep, can last for up to fifty per cent of the lifetime and is therefore an important consideration in the design of structural components.

An example of a uni-axial copper[+] specimen which has undergone long term creep at 250°C under low stress is given in Fig.4. The micrograph is of a diametral section taken after failure has occurred. Rupture has taken place due to the initiation and time dependent growth of grain boundary

[+]Tough pitch high conductivity copper of 99.9% purity manufactured to British Standard Specification B.S.2873-CIDI.

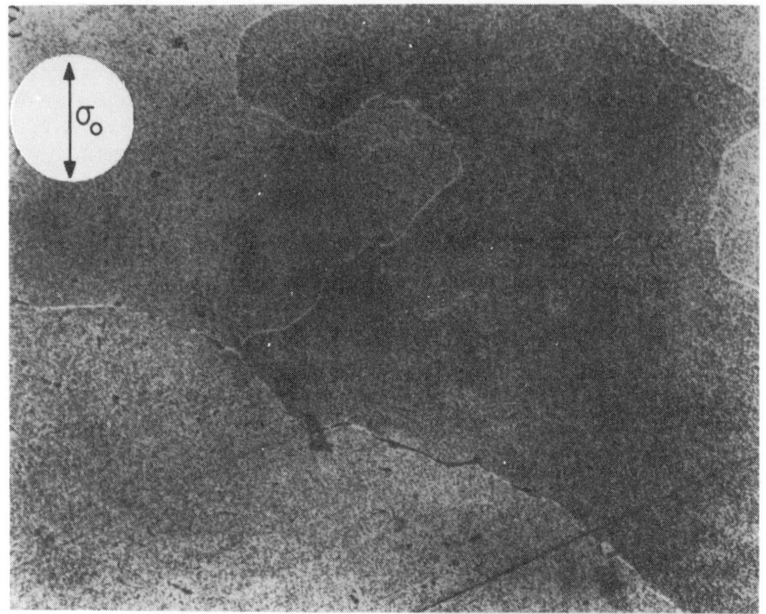

Fig.5.   Micrograph taken from a uni-axial tension specimen
in an aluminium alloy tested at 210°C after
95% of the lifetime.   (Mag. x 105)

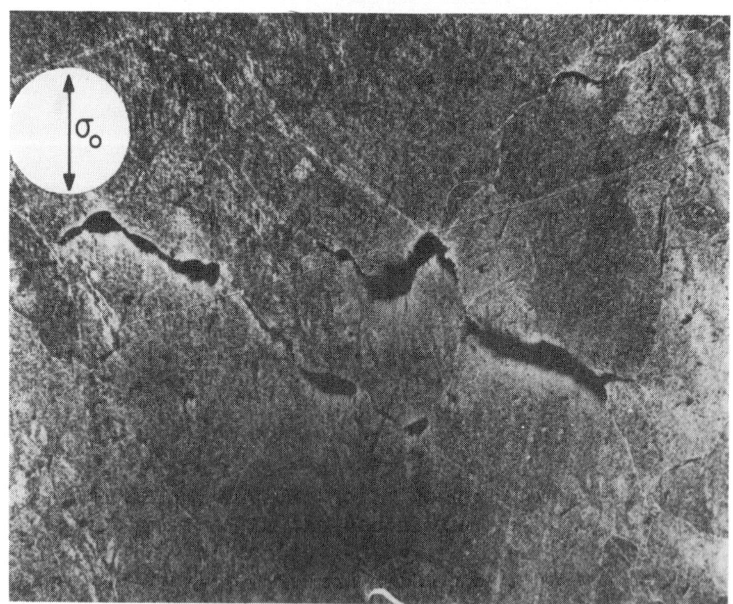

Fig.6.   Micrograph taken from an aluminium alloy
uni-axial specimen tested at 210°C.   The test was terminated
immediately before rupture.   (Mag. x 105)

cavities; the cavities grow and eventually link to form short
grain boundary cracks, which in turn grow and link up to form
macrocracks. The striking feature about the micrograph in
Fig.4 is that the density of cracking or level of creep rupture is uniformly distributed throughout the homogeneously
stressed region. Final separation of the specimen has taken
place on a plane which is perpendicular to the direction of
the applied stress.

An example of failure in an aluminium alloy[+] subjected to
steady-load creep at 210°C is shown in the micrograph of Fig.5.
The character of the grain boundary damage is noticeably different than that for copper shown in Fig.4. Long narrow microcracks have formed on the grain boundaries with no sign of the
formation of discrete bubbles, as in copper. The grain boundary fissures form on planes which are perpendicular to the
direction of the maximum principal tension stress; they are
uniformly distributed throughout the homogeneously stressed
region. The final mode of failure differs from that of copper
in that it occurs on a shear plane, inclined at 45° to the
stress axis. Fig.6 shows how the grain boundary fissures,
which have grown on planes perpendicular to the stress axis,
link up by a shear instability mechanism during the final
moments of the lifetime.

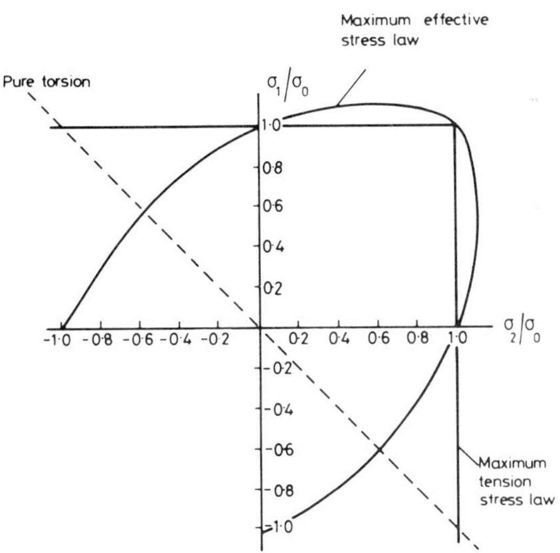

Fig.7.   Plane stress isochronous loci

[+] Aluminium, Cu, Fe, Ni, Mg and Si alloy manufactured to
British Standard Specification B.S. 1472

Although the mechanisms which cause ultimate failure are different for the two materials, the processes by which defects grow exhibit a common feature (14): namely that regions of homogeneous stress experience uniform states of material degradation.

2.3 Multi-axial stress

Probably the most notable experimental work on bi-axial stress creep has been carried out by A.E. Johnson et al (15). The majority of his work was carried out on thin tubular specimens subjected to combined tension and torsion, hence his results were confined to the tension-compression quadrants of the plane principal stress space ($\sigma_3=0$) shown in Fig.7. In this figure are plotted loci of constant rupture time (isochronous loci) (14); the rupture lifetime is that of a uni-axial test carried out at the normalising stress σ_0. Johnson tested several metals in the tension-compression quadrants and concluded that the extreme types of behaviour could be categorised by two criteria: the maximum principal tension stress criteria σ_1 and the maximum effective stress criteria $\sigma_e(=\{((\sigma_1-\sigma_2)^2+(\sigma_2-\sigma_3)^2+(\sigma_3-\sigma_1)^2)/2\}^{\frac{1}{2}})$. The behaviour of copper is characterised principally by the former criterion and that of aluminium alloys by the latter criterion. Since the behaviour of components made from materials which satisfy the extreme criteria is of interest, the results of tests conducted on copper and on aluminium alloys will be discussed here. Three types of test will be described: firstly, tests under equal tension-tension loading, to rectify the gap in Johnson's data; secondly, tests under pure torsion, to reinforce Johnson's conclusions; and thirdly, tests under tri-axial stress, to study material behaviour under crack-tip conditions. The aluminium alloy and the copper materials tested are those described in the previous section.

2.3(a) Bi-axial Tension

Equal bi-axial tension tests were carried out on plate specimens of the type shown in Fig.8. The specimen design is based on that due to Mönch and Galster (16). The principal difference is that the central region of the specimen has been uniformly thinned (17,18) to give a higher stress and to ensure that failure takes place in the central region. The specimens were tested in special purpose machines which have been described in detail elsewhere (18).   Rupture of the specimen invariably occurred at its centre. The lifetimes of the aluminium specimens (14) were predicted predominantly by the σ_e criterion and those of the copper specimens (14) showed a slight weakening when judged against the σ_1 criterion; that is the isochronous locus fell within the σ_1 locus by approximately six per cent of the σ_1 stress. However the behaviour of copper may be described predominantly by the σ_1 criterion;

Fig.8.   Bi-axial tension specimen

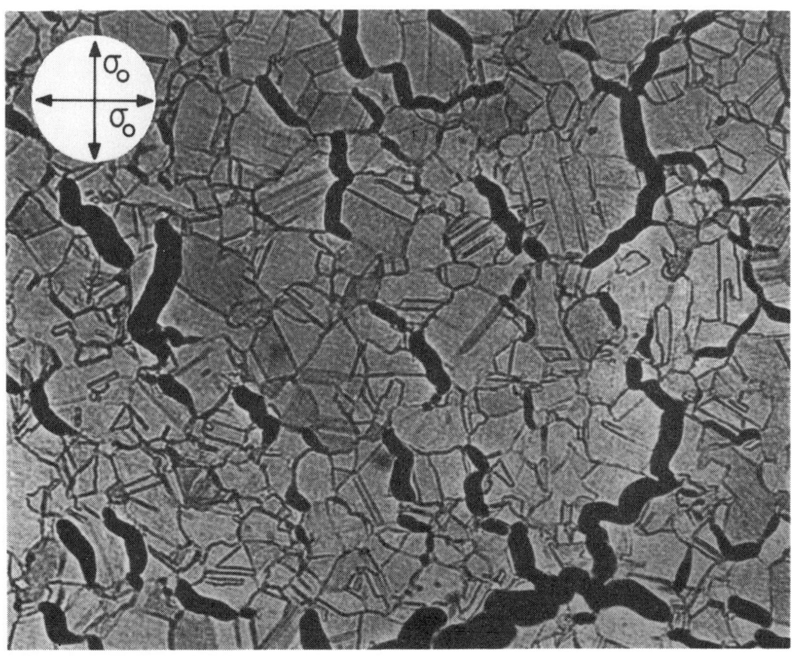

Fig.9.   Micrograph taken from an equal bi-axial tension
copper specimen tested at 250°C close to failure.(Mag. x 26)

96

the precise form of the description will be discussed at a
later stage in the paper.

   Metallographic studies were made on the plate specimens
and the results are now briefly discussed. A mid-thickness
micrograph of a copper plate specimen, close to failure, is
shown in Fig.9. The mode of failure is similar to that shown
in Fig.4 except that it is not possible to assign a 'preferred'
direction for the growth of damage. A similar type of micrograph for the aluminium alloy is shown in Fig.10.   The
grain boundary cracks are similar to those shown in Fig.5, for
uni-axial tension. As in the case of the bi-axial tension
test on copper it is not possible to identify a direction in

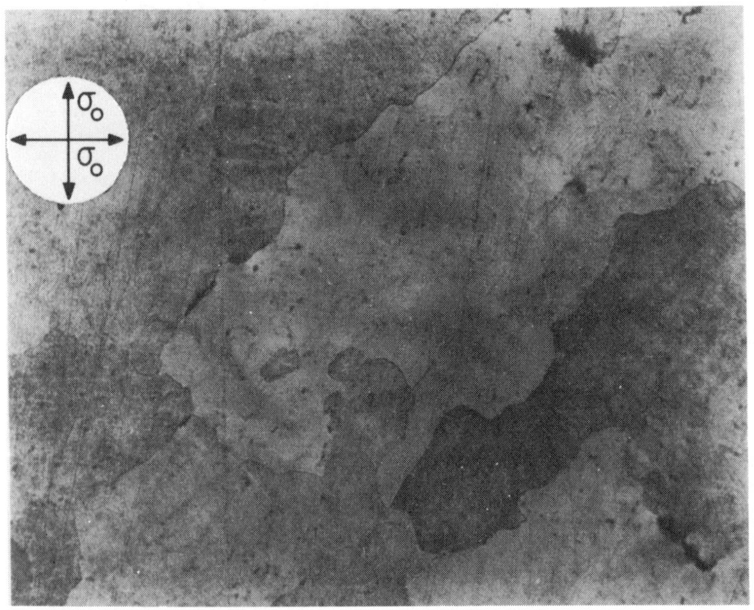

Fig.10.   Micrograph taken from an equal bi-axial tension
aluminium alloy specimen tested at 210°C
close to failure. (Mag. x 105)

which preferred growth has taken place. The mechanisms by
which final rupture occurs are similar to those for the uni-
axial specimens; the copper plates separate on planes normal
to the plane of the plate and the aluminium alloy plates fail
by a shear instability mechanism on planes inclined at 45° to
the plane of the plate.

2.3(b) <u>Pure torsion or shear</u>

   The tubes tested by Johnson et al (15) were designed to
be thin in order to achieve a uniform shear stress through the
thickness of the specimen. Such tubes may be prone to

Fig.11.   Andrade shear discs
(Top: shear failure in an aluminium alloy disc)
(Bottom: copper disc after failure)

98

buckling instabilities.   It has been questioned (19) whether
failure in this type of tube is due to the growth of creep
alone or to creep buckling accelerated by material degeneration.   In an attempt to overcome the buckling problem Hayhurst and Storåkers (19) have tested thicker material sections
using the Andrade shear disc, shown in Fig.11.   The disc is
subjected to a steady torque, about an axis perpendicular to
the plane of the disc and which passes through its centre;
the thickness of the disc decreases with increasing radius so
as to maintain a constant shear stress.

A segment has been cut from the aluminium disc shown in
Fig.11 for microstructural examination but the circumferential
failure line can be seen; it falls within the region of uniform stress.   The lifetimes of the aluminium discs tested
were closely predicted by the σ_e criterion.

Fig.12.   Micrograph taken from a mid-thickness plane
of a copper disc.   The base line of the figure
corresponds to a radial direction at failure.   (Mag. x 8)

The copper disc underwent large deformations prior to
failure; this may be observed from the white logarithmic
spiral markings on the surface of the disc, these markings
were initially radial.   In the interpretation of the behaviour of the discs it was necessary to complete a finite deformation analysis of the disc, which showed that, to a reasonably
good approximation, the lifetime could be described by the σ_1
criterion.

A mid-thickness micrograph taken from a copper disc is
shown in Fig.12.   It can be seen that damage has formed on
those grain boundaries which are inclined at 45° to the radial

direction.   Since the stress state is one of pure shear, or
the equivalence of equal tension and compression in mutually
perpendicular directions, the damage planes correspond to
those on which the maximum principal tension stress acts.
The grain boundary damage is uniformly distributed over the
region of the micrograph.   Final rupture of the specimen
occurs on planes which are perpendicular to the plane of the
disc, cf. Fig.11.

2.3(c) Tri-axial tension

The multi-axial tests discussed so far have been carried
out under plane stress conditions.   Data derived from such

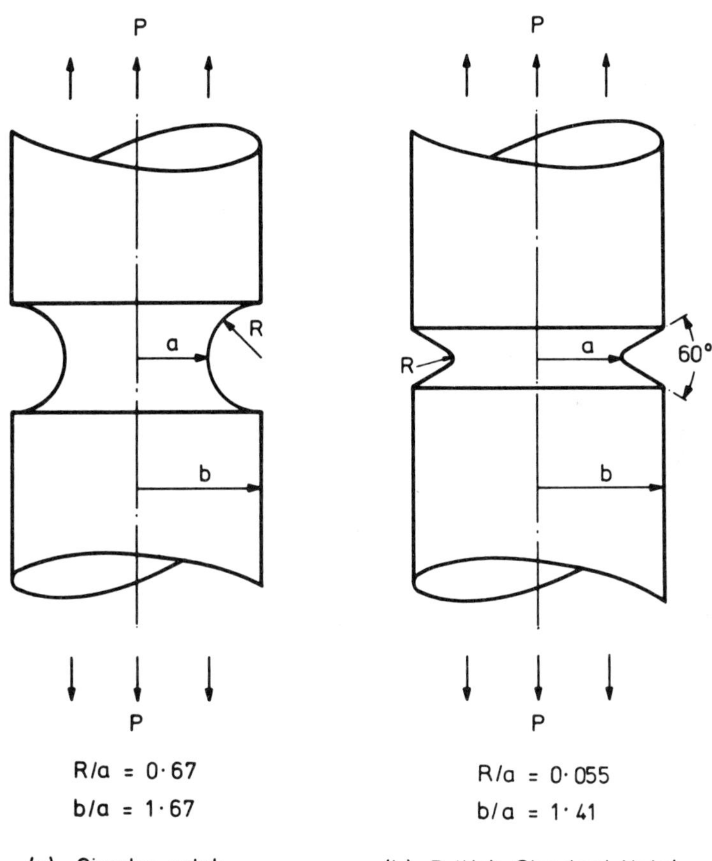

Fig.13.   Circumferentially notched circular
bars under uniform tension

100

tests is relevant to a wide range of pressure vessel and pipe
structures, but may not be appropriate for those situations in
which high constraint (J_1/σ_e, where $J_1=\sigma_1+\sigma_2+\sigma_3$) occurs, for
example in structures where severe changes of section are
found and where sharp defects or cracks are present.   In
order to be able to design reliably under these conditions it
is necessary to have data available for tri-axial tensile
conditions.   Test specimens have not been designed which
allow material elements to be subjected to homogeneous states
of tri-axial tension stress.   This difficulty has been overcome by making use of axi-symmetrically notched tension bars,
examples of which are given in Fig.13.   The circular notched
specimen is due to Bridgman (20) who used it in studies of
large plastic deformations in necking bars.   His perfectly
plastic analysis showed that the centre of a notched bar is
subjected to a high degree of triaxial tension.   The same geometry, and that of the British Standard (9) notched creep
specimen, have been studied extensively under creep conditions.
Some results of elastic and stationary-state analyses (21),
carried out on both specimens, are given in Fig.14 for a stress
index, in Nortons creep law, of n=5.   For both notch geometries the elastic distributions are non-uniform, but the
stationary-state distributions differ; in the case of the
circular notch both the axial and the effective stress components change by relatively small amounts with the normalised
radial distance $\bar{r}$ (=r/a), in the case of the B.S. notch the
axial and effective stress concentrations are non-uniform.

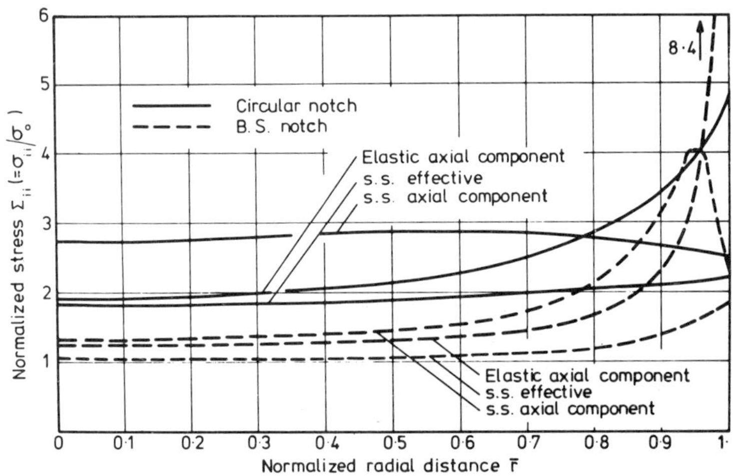

Fig.14. Comparison of the spatial variations of the normalised components of elastic and stationary-state (s.s.) axial
stresses with effective stresses at stationary-state for the
circular and B.S. notches at the minimum section (n=5).

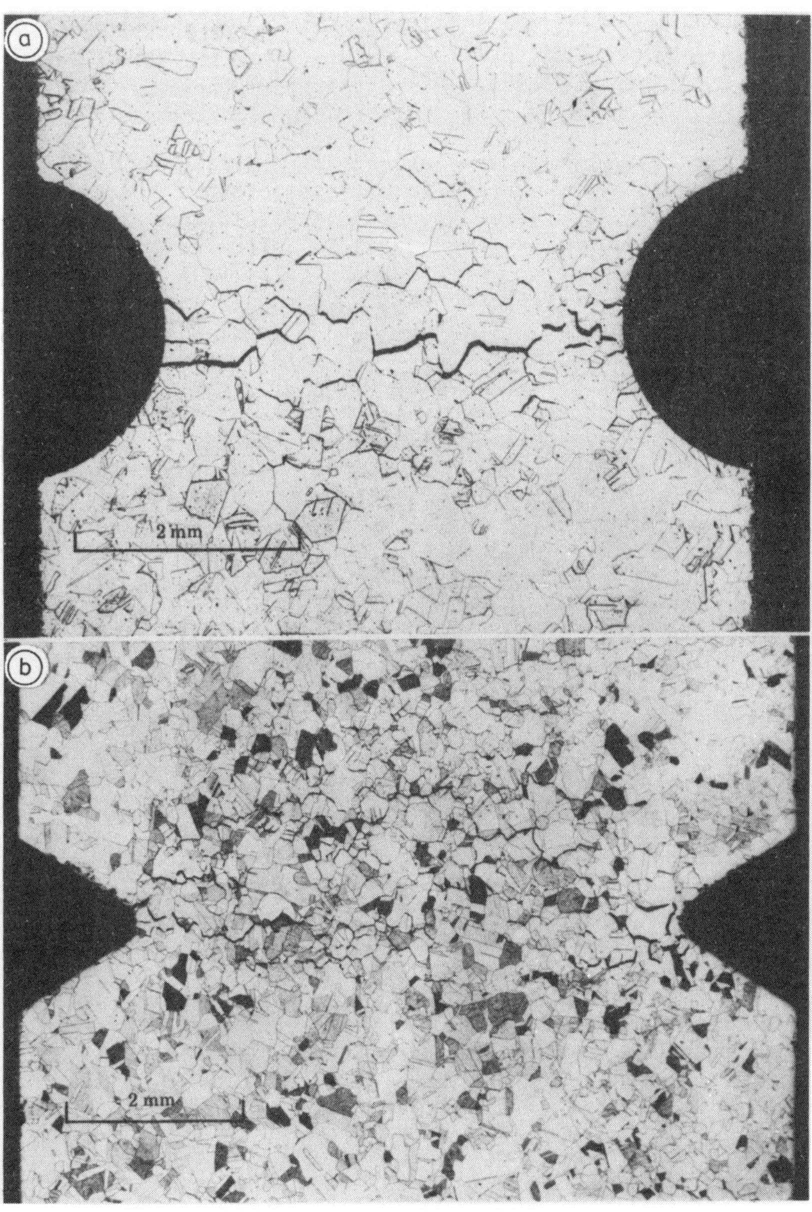

Fig.15.   Micrographs taken from diametral planes of copper circumferentially notched circular bars immediately before failure.   (a) circular notch,   (b) B.S. notch.

102

Under creep rupture conditions these stress distributions are
responsible for the circular notch behaving essentially as a
homogeneously stressed specimen, with all points across the
section failing at the same time, and for the B.S. notched
specimen behaving in a manner which resembles the slow propagation of a crack-like zone of ruptured material (22,23).

The circular notched specimen has been used extensively
(22,24,25) to subject material to high-temperature creep rupture.   In most cases the stationary-state solutions due to
Hayhurst et al (21,26) have been used to determine the stress-
states and stress-levels.   Micrographs taken from diametral
planes of circular and B.S. notched specimens, close to failure in copper tested at 250°C, are presented in Fig.15.
The interruption of specimens close to failure has been
achieved by testing specimens with two, well-separated, identical notches in the same specimen; when one of the notches
fails the second is automatically interrupted close to failure.
The micrograph of Fig.15(a), for the circular notch, shows that
the creep damage is uniformly distributed across the minimum
section of the specimen.   The creep damage shown in Fig.15(b),
for the B.S. notch, is more heavily concentrated at the tip of
the notch, but despite the non-uniform stresses shown in
Fig.14 the centre of the specimen is relatively uniformly damaged.   In both cases creep damage has formed on the grain
boundary planes which are perpendicular to the direction of
the applied stress.   Final rupture of both specimens
occurs essentially on the plane of the minimum section of the
notch.

2.3(d) <u>Continuum damage</u>

In this section tests have been described which have been
carried out under different homogeneous states of multi-axial
stress.   In all of the micrographs presented no evidence has
been found of a single crack which has grown preferentially;
instead, it has been shown that if the stresses are spatially
uniform then creep damage forms during tertiary creep which is
also spatially uniform.   When the stress fields are non-
uniform, as in the notch-tip region of the B.S. notch shown in
Fig.15(b), then the distribution of damage tends to be non-
uniform.

It is evident therefore that the creep damage is distributed in a manner which has similar characteristics to the
stress-field.   Alternatively it may be said that creep damage
possesses a field or continuum property similar to that of
stress, hence the term CONTINUUM DAMAGE.   The development of
constitutive equations is now described which model the growth
of time dependent strains and the formation of continuum damage.

2.4. <u>Constitutive equations for steady load and temperature</u>

Constitutive equations have been developed for primary and secondary creep deformation under both uni-axial and multi-axial stress conditions. The equations are well proven particularly for steady-load conditions. These equations will merely be referred to here and the majority of the section will be devoted to showing how uni-axial primary-secondary-tertiary creep behaviour can be described using a single state damage variable theory and how this theory can be generalised for steady multi-axial stress conditions.

2.4(a) <u>Primary-secondary creep</u>

The uni-axial creep strain rate is usually described by Norton's law,

$$\dot{v}/\dot{v}_o = (\sigma/\sigma_o)^n K(t) , \tag{1}$$

where n is a material constant, $\dot{v}_o$ is the uni-axial strain rate due to the stress σ_o and $K(t)$ is a function of time selected to describe the decrease in strain rate with time.

The generalisation of equation (1) to multi-axial stresses has been carried out by Odqvist (27) and results in the following equation

$$\dot{v}_{ij}/\dot{v}_o = (3/2)(\sigma_e/\sigma_o)^{n-1}(S_{ij}/\sigma_o)K(t) \tag{2}$$

where $S_{ij} = \sigma_{ij} - \delta_{ij}\sigma_{kk}/3$. Both equations (1) and (2) have been verified by the test results of Johnson et al (15).

2.4(b) <u>State variable description for uni-axial continuum</u> <u>damage</u>

If, for convenience of discussion, the primary portion of the creep curve is neglected then the strain time curve in a constant stress test will have the form shown in Fig.16. The creep rate increases from the initial steady state value as damage causes an advance into the tertiary portion of the curve. The steady-state creep rate can be expressed as a function of the applied stress σ alone. In order to account for the increase in strain rate during a constant stress test it is necessary to introduce a new variable into the strain equation. Since the increase in strain rate is the result of a damage process the variable ω introduced is referred to as the damage state variable. The strain rate equation then takes the form

$$dv/dt = f(\sigma,\omega) , \tag{3}$$

where f is a function which has yet to be defined. However,

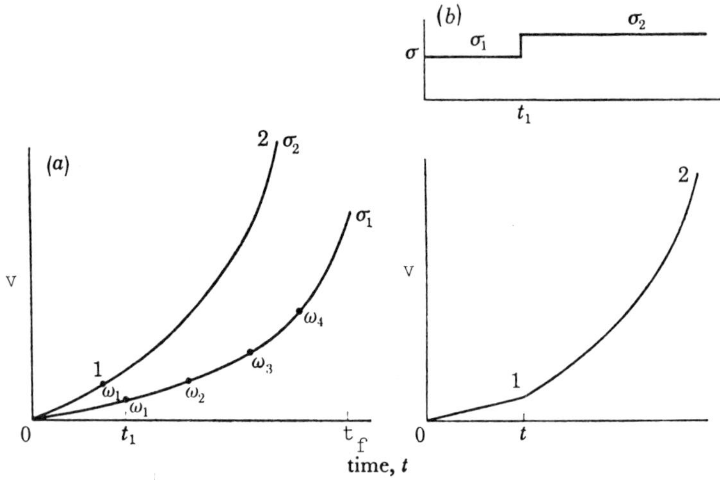

Fig.16.   Uni-axial creep test for state variable description

in the undamaged state when ω = 0 the equation must reduce to
the form observed for steady state conditions.   To define the
strain rate it is necessary that the value of ω as well as
the stress σ be known and consequently an equation must be
introduced which defines the growth of the damage state variable.   The assumption that the damage rate depends both on the
current state of stress and damage gives the equation

$$d\omega/dt = g(\sigma,\omega) \; . \tag{4}$$

The problem is how to find the functions f and g in a
systematic manner.   This may be achieved by following a testing procedure suggested by Leckie and Hayhurst (3) which makes
use of the results of tests with step changes in stress.
Neglecting the primary portion of the creep curve, the
strain-time curves for two constant stress tests carried out
at stresses σ_1 and σ_2 would have the form shown in Fig.16
(a).   Suppose in the constant σ_1 test that ω varies
between zero at time t = 0 and unity at the rupture time t_f.
Select an arbitrary variation of ω along the σ_1 curve
(Fig.16(a)).   Now perform a test in which the stress σ_1 is
applied for time t_1 when the stress is increased to σ_2 and
maintained constant at this value to give the creep curve
shown in Fig.16(b).   Since a single state variable is sufficient to define the creep curves it follows that section 12 of
the curve 012 must be identical in shape with a portion of the
constant stress creep curve σ_2 .   In this way the state ω_1
may be defined on the σ_2 curve.   By performing similar
tests, constant ω contours may be constructed as shown in
Fig.17.   This information can then be used to predict the

variation of strains and damage in a result of variable stress
histories (3).

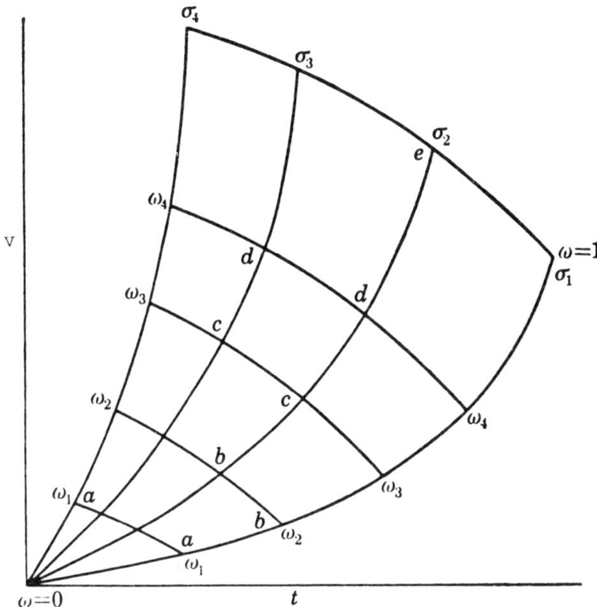

Fig.17.   Contours of constant damage

The procedures described above are applicable provided
that the state of damage can be described by a single damage
parameter.   If a single state parameter suffices it might
still be necessary to modify the equations (3)and (4) for
loading and unloading structures.   This is certainly the case
for creep deformations in the primary region when it is found
that the growth laws of dislocation density differ for loading
and unloading.   This point has been discussed by Leckie and
Ponter (28).

2.4(c) Rabotnov-Kachanov equations for uni-axial creep -
        A phenomenological approach

Unfortunately it is difficult and time consuming to con-
duct experiments described in the previous section and the
variable stress experiments which have been performed are limited to the prediction of rupture life under cyclic conditions
of loading. Faced with this difficulty Rabotnov (29) proposed
modifications of constitutive equations first suggested by
Kachanov (30) which described existing experimental results.
For uni-axial stress tests the growth laws are assumed to have
the simple form:

$$\dot{v}/\dot{v}_o = (\sigma/\sigma_o)^n \, K(t)/(1-\omega)^n , \quad \text{and} \qquad (5)$$

$$\dot{\omega}/\dot{\omega}_o = (\sigma/\sigma_o)^X/(1+\phi)(1-\omega)^\phi \, , \qquad (6)$$

where X, ϕ and $\dot{\omega}_o$ are constants. In a constant stress test the value of ω monotonically increases from zero to a value of unity at failure. When $\omega = 0$ equation (5) reduces to Norton's law, given by equation (1), and when $\omega = 1$ the strain rate is infinite. Integration of equation (6) between the limits $\omega = 0$, at $t = 0$, and $\omega = 1$, at $t = t_f$, yields the experimentally observed expression for the rupture lifetime

$$t_f = 1/\dot{\omega}_o \, (\sigma/\sigma_o)^X \, . \qquad (7)$$

Equations (5), (6) and (7) have been shown (18) to accurately describe uni-axial creep behaviour.

2.4(d) <u>Generalisation of state variable description</u>

The generalisation of equations (5) and (6) for multi-axial stresses has been achieved by making the assumption that the influence of continuum damage on the deformation rate processes is scalar in character. Equation (2) can then be written as:

$$\dot{v}_{ij}/\dot{v}_o = (3/2)(\sigma_e/\sigma_o)^{n-1}(S_{ij}/\sigma_o)K(t)/(1-\omega)^n. \qquad (8)$$

In the generalisation of equation (6) the stress-state effects observed from the isochronous loci must be reflected; this is achieved by the introduction of the homogeneous stress function $\Delta(\sigma_{ij}/\sigma_o)$ as follows:

$$\dot{\omega}/\dot{\omega}_o = \Delta^X(\sigma_{ij}/\sigma_o)/(1+\phi)(1-\omega)^\phi; \qquad (9)$$

for copper $\Delta(\sigma_{ij}/\sigma_o) = \sigma_1/\sigma_o$ and for aluminium alloys $\Delta(\sigma_{ij}/\sigma_1) = \sigma_e/\sigma_o$. Integration of equation (9) for the conditions $\omega = 0$, $t = 0$ and $\omega = 1$, $t = t_f$ yields the experimentally observed result $t_f = 1/\dot{\omega}_o \, \Delta^X(\sigma_{ij}/\sigma_o)$, which may be normalised to give

$$t_f/t_o = 1/\Delta^X(\sigma_{ij}/\sigma_o) \, ; \qquad (10)$$

substitution of $t_f = t_o$ gives the equation of the isochronous surface $\Delta(\sigma_{ij}/\sigma_o) = 1$.

Equation (8) has been verified using the results of Johnson et al (15) for tension-torsion tests. If in such a test, γ is the shear strain, v is the axial strain and γ_f and v_f denote their values at failure, then it may be shown from equation (8) that the ratio $(\gamma/\gamma_f)/(v/v_f)$ must remain constant throughout the test. The experimental verification of this may be seen in Fig.18, where the results of the tests

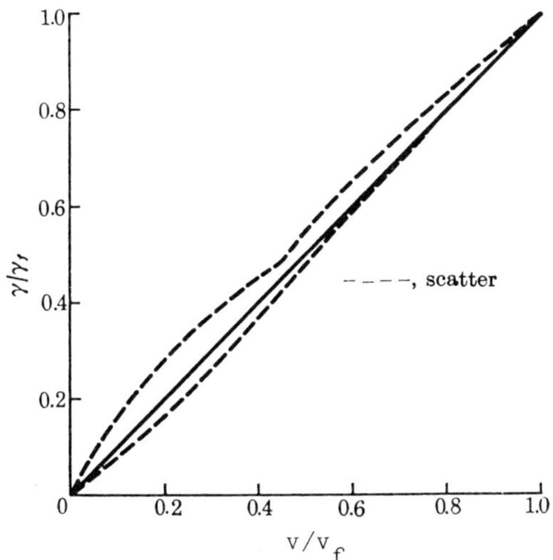

Fig.18.   Variation of axial and shear creep strains
for copper and aluminium alloys

by Johnson et al on copper and aluminium have been plotted.
The implication of this result is that the state variable ω
has a scalar interpretation in the strain rate equation.   In
addition equation (9) is capable of describing the multi-axial
rupture behaviour.

2.4(e) Isochronous rupture surface

The concept of an isochronous rupture surface has been
introduced in section 2.3, where an isochronous loci in plane
stress space is shown in Fig.7, and also in section 2.4(d)
where the dependence of the damage growth rate upon stress
state is expressed by equation (9).   At this stage it is
necessary to discuss the different types of multi-axial behaviour and the techniques used for their description in
structural analysis.

A review of bi-axial creep rupture behaviour, carried out
by Hayhurst (14), indicates that a general representation must
exhibit two qualities.   First, it must represent the sensitivity of the micro-structural changes to the maximum principal tensile stress; and secondly, express the stress sensitivity of different materials to the final collapse or failure
mechanism.   A general relationship which satisfies these
conditions is $\qquad t = h(J_1, J_2, J_3)$, $\qquad\qquad$ (11)
where J_1, J_2, J_3 are the invariants of the stress tensor

and h is a homogeneous algebraic function of degree $-\chi$ in stress. The form of the function h must be such that on parametric change it is capable of describing different types of material behaviour.   Before discussing the appropriate form of h some particular forms of the function will be examined which have been used to describe the behaviour of individual materials.

It has been shown by Hayhurst (2) that for bi-axial stress conditions the behaviour of an aluminium alloy can be closely approximated by an octahedral shear stress criterion and it is not surprising that providing the function h (see equation (11)) is not strongly dependent upon J_1 a close approximation to the experimental data can be obtained.   The rupture behaviour of copper when subjected to bi-axial stresses approximately satisfies a maximum principal tensile stress criterion; in the tension-compression quadrant good agreement with experiment has been obtained, but in the tension-tension quadrant the rupture times are less than the predictions given by this criterion.   In many respects the rupture behaviour of copper and aluminium alloys can be regarded as representing the two limiting types of stress sensitive rupture.   The behaviour of other alloys and in particular commercial creep-resistant alloys falls between the two extremes.   Sdobyrev (31) in attempting to describe the multi-axial behaviour of this type of material, has proposed the relationship

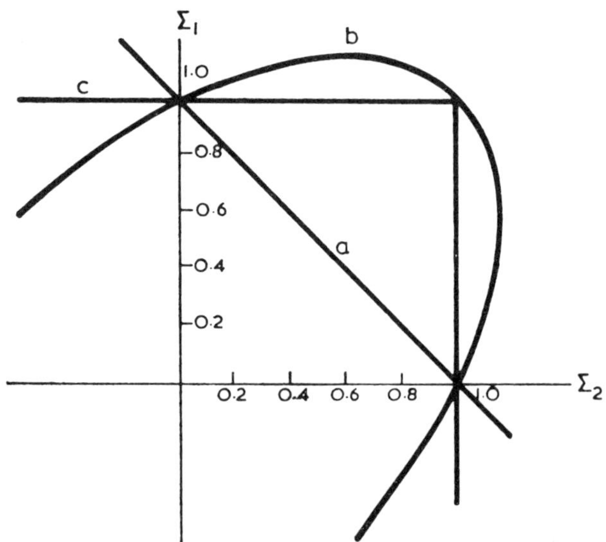

Fig.19.   Loci of constant rupture time for values of
(α,β,γ): maximum, (a) hydrostatic stress (0,1,0),
(b) octahedral stress (0,0,1), and (c) tensile stress (1,0,0).

$$t = M\{\lambda\sigma_1 + (1-\lambda)J_2'^{\frac{1}{2}}\}^{-X} \quad (\lambda = 0.5), \tag{12}$$

where J_2' represents the second deviatoric stress invariant and σ_1 is the maximum principal tension stress. Both of the relationships given by equations (11) and (12) can be used to represent accurately the behaviour of copper alloys in the tension-compression quadrant but the predictions of rupture time are in error for bi-axial tension stresses. The bi-axial tension stress levels predicted by equations (11) and (12) are too high and too low respectively. It is clear that a general relationship which is capable of expressing the

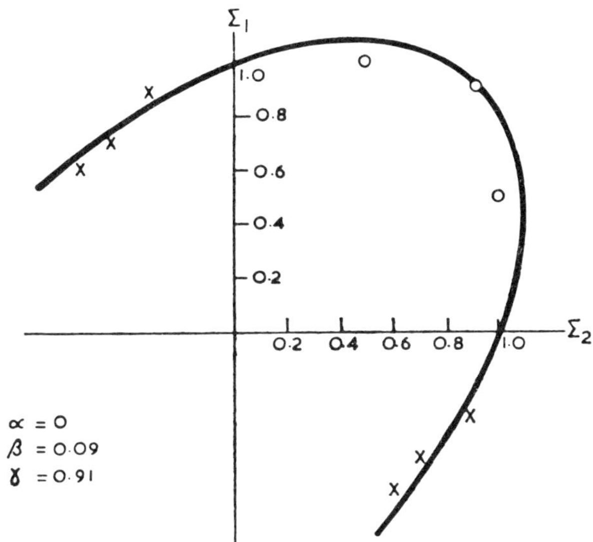

Fig.20.   Comparison of creep rupture data for aluminium alloys with the locus of constant rupture time computed from equation (17); X Johnson et al. (16), 0 Hayhurst (15) .

different types of behaviour must exhibit the main features of equations (11) and (12). The following linear combination of the maximum principal tensile stress and the first and second stress invariants is proposed:

$$t = M\{(\alpha\sigma_1 + \beta J_1 + \gamma J_2'^{\frac{1}{2}}\}^{-X}, \tag{13}$$

where M is a constant independent of stress, α, β and γ are constants, and J_1 and J_2' in terms of principal stresses are given by

$$J_1 = \sigma_1 + \sigma_2 + \sigma_3, \quad J_2' = \frac{1}{6}[(\sigma_1 - \sigma_2)^2 + (\sigma_2 - \sigma_3)^2 + (\sigma_3 - \sigma_1)^2] .$$

By writing equation (13) for principal bi-axial stresses, the time to rupture is given (for $\sigma_1 > \sigma_2 > \sigma_3$) by

$$t = M\{\alpha\sigma_1 + \beta(\sigma_1^2 + \sigma_2^2 + 2\sigma_1\sigma_2)^{\frac{1}{2}} + \gamma(\sigma_1^2 + \sigma_2^2 - \sigma_1\sigma_2)^{\frac{1}{2}}\}^{-\chi}. \tag{14}$$

By writing $\Sigma_i = \sigma_i/\sigma_0$ and $T = t/t_0$, where t_0 is the time to rupture in a uni-axial test conducted at a stress σ_0 , equation (14) can be written in the normalized form

$$T = \{\alpha\Sigma_1 + \beta(\Sigma_1^2 + \Sigma_2^2 + 2\Sigma_1\Sigma_2)^{\frac{1}{2}} + \gamma(\Sigma_1^2 + \Sigma_2^2 - \Sigma_1\Sigma_2)^{\frac{1}{2}}\}^{-\chi}. \tag{15}$$

By expressing the stress Σ_2 as a ratio ζ of the stress Σ_1 , equation (15) can be rewritten to give an expression for the time to rupture T in terms of the maximum principal tensile stress, viz.

$$T = \{\alpha + \beta(\zeta^2 + 2\zeta + 1)^{\frac{1}{2}} + \gamma(\zeta^2 - \zeta + 1)^{\frac{1}{2}}\}^{-\chi}\Sigma_1^{-\chi} , \tag{16}$$

where $\alpha + \beta + \gamma = 1$. Equation (16) can now be seen to be a special form of equation (11). For constant values of α , β and γ a family of constant rupture times can be plotted in stress space; the maximum hydrostatic stress criterion is represented by the straight line ($\alpha = 0, \beta = 1, \gamma = 0$), the maximum octahedral shear stress criterion ($\alpha = 0, \beta = 0,$ $\gamma = 1$) by the ellipse, and the maximum principal stress criterion ($\alpha = 1, \beta = 0, \gamma = 0$) by the two intersecting straight lines as shown in Fig.19. Equation (16) can now be used to investigate the magnitudes of the bi-axial stresses required to give a rupture time equal to the uni-axial rupture time. By setting the normalized rupture time T equal to unity, the magnitude of stress Σ_1 can be determined for different values of the bi-axiality ratio ζ . Equation (16) can now be written

$$\Sigma_1 = \{\alpha + \beta(\zeta^2 + 2\zeta + 1)^{\frac{1}{2}} + \gamma(\zeta^2 - \zeta + 1)^{\frac{1}{2}}\}^{-1} , \tag{17}$$

which implies that the shape of the locus of points in the (Σ_1, Σ_2)-plane having the same rupture time is independent of χ . In order to determine the shape of the rupture curves at least three sets of rupture tests must be conducted, one set of uni-axial tests and two sets of bi-axial tests. The most severe stress systems in the tension-compression and tension-tension quadrants are given by the conditions of pure shear and equal bi-axial tension respectively; pure shear being the most severe condition and representing a decrease in the maximum principal tensile stress of 50 per cent for the same rupture time computed by the maximum principal tensile stress criterion and the maximum octahedral shear stress criterion. In Fig.20 the results of bi-axial creep tests conducted on an aluminium alloy by Johnson et al (15), together with results by Hayhurst (14) are compared with the rupture locus corresponding to $\alpha = 0, \beta = 0.09, \gamma = 0.91$. Although most of the points were obtained from very few test results, a reasonable degree of correlation has been achieved. The low values of α and β reinforce the conclusion of Johnson et al. (15) that the time to rupture for the aluminium alloy tested can be closely correlated to the applied stress system by a J_2' or σ_e theory.

In practice the β-term is usually small and the behaviour of most materials can be represented using $\beta = 0$. Under these conditions $\alpha + \gamma = 1$ and equation (13) can be written as

$$ t = M \{ \alpha\sigma_1 + (1-\alpha) J_2'^{(\frac{1}{2})} \}^{-\chi} , $$

or, to a close approximation by

$$ t = M \{ \alpha\sigma_1 + (1-\alpha) \sigma_e \}^{-\chi} . \qquad (18) $$

The term in brackets corresponds to the function $\Delta(\sigma_{ij})$ introduced in section 2.4(d). The corresponding damage rate equation is

$$ \dot{\omega} = \{ \alpha\sigma_1 + (1-\alpha) \sigma_e \}^{\chi} / (M(1+\phi)(1-\omega)^{\phi}) . \qquad (19) $$

The material constants M, ϕ and χ may be determined from the results of uni-axial tests but the determination of the constant α requires the results of rupture tests to be obtained under two different states of stress. The tests usually carried out to determine α are uni-axial and pure shear tests ($\sigma_1 = -\sigma_2$, $\sigma_3 = 0$) .

2.5. The directional character of creep damage

As shown earlier creep damage, characterised by the damage state variable ω , grows on grain boundaries perpendicular to the direction of maximum principal tension stress, even for aluminium alloys. The question then arises as to whether equations (8) and (9) are capable of describing the deformation and rupture behaviour of copper and aluminium alloys in situations where the stress field rotates relative to the material element during deformation. This type of loading is characteristic of a class of loading conditions found in practice. To test the validity of equations (8) and (9) under these conditions tests have been carried out on thin tubes where the magnitudes of σ_e and σ_1 are maintained constant but where the direction of σ_1 rotates through approximately $34°$. The loading conditions of the test are shown schematically in Fig.21; loading proceeds from 0 to A , the load is maintained at A for a period, unloading to 0 then occurs and is followed by reloading to B , where the load is maintained constant until rupture. The magnitude of the stresses have been arranged to give equal strain rates in order to avoid large shear strains and associated buckling effects. The tests were carried out in a specially developed testing machine (32).

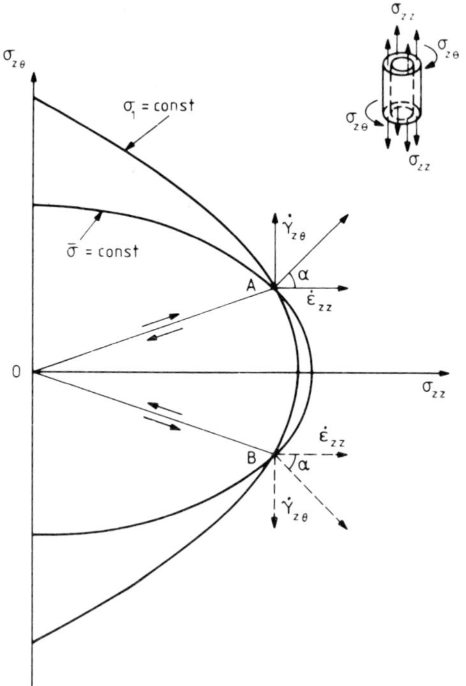

Fig.21.   Schematic representation of loading
conditions for copper tube.

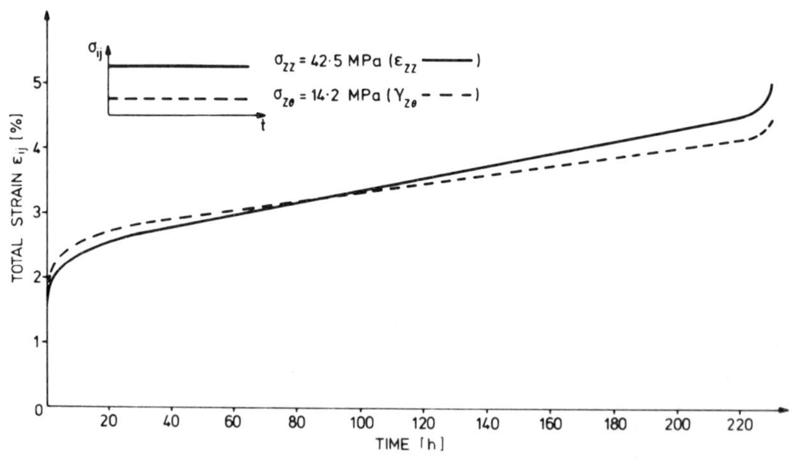

Fig.22.   Steady load tension-torsion
test on copper tube.

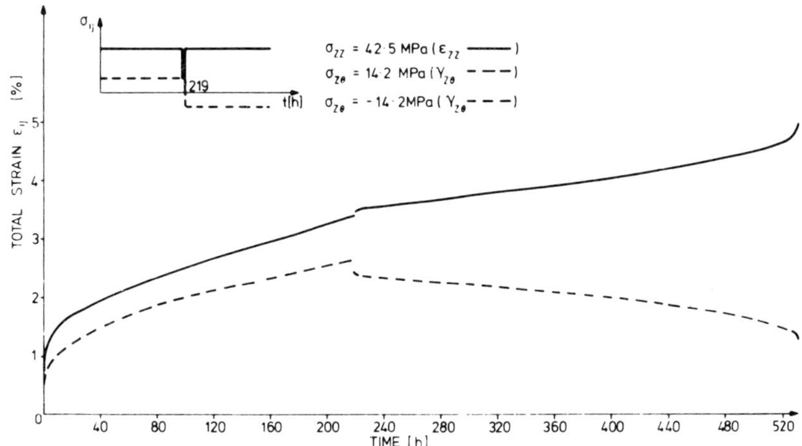

Fig.23.   Single reverse torsion, steady tension
load test on a copper tube.

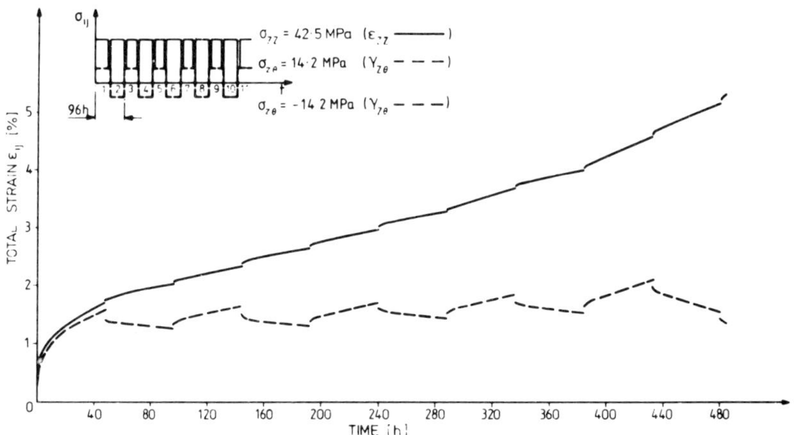

Fig.24.   Multiple reverse torsion,
steady tension load test on a copper tube.

A steady load test (loading to OA only) was carried out
first in copper; the observed behaviour is shown in Fig.22,
the lifetime was 230 h.   A single reverse torsion test was
then performed, Fig.23, at the same stress level, the stress
reversal took place just prior to the steady rupture.   The
measured lifetime was 530 h, which is slightly more than
twice the steady load lifetime.   A multiple reverse torsion,
constant tension, test was carried out in which the direction
of the torsion was reversed every 48 h.   The results of such
a test are presented in Fig.24.   The measured lifetime was
490 h, which compares favourably with twice the lifetime of
the steady load torsion-tension test, 2 x 230 = 460 h.

Fig.25.   Mid-thickness micrograph of a copper tube
tested to failure under multiple reverse torsion
steady tension loading. (Mag. x 50)

These results have been discussed in detail elsewhere
(33) and metallographic studies have revealed the formation of
two distinct sets of damage planes which do not interact, as
shown in Fig.25, despite their normals being separated by
only $34°$.   The measured axial and shear strain rates are almost equal, as arranged using equation (8).   In the single
and multiple reverse torsion tests a reduction in strain rate
occurred after the first reversal and on subsequent reversals
to that particular direction.   This had the effect of producing a lower shear strain rate in one direction than in the
other; the difference in the strain rates was small compared
with the scatter on uni-axial data.

Similar tests were conducted on aluminium tubes.   For
these tests the lifetimes of the steady load tests and of the
single reverse load tests were almost equal.   This result is
the one predicted by equations (8) and (9).

Equations (8) and (9) can then be used to predict lifetimes of copper and aluminium alloys in which the principal
stress directions rotate.   For aluminium alloys the damage is
included in both equations as a scalar and no special treatment is required.   For copper the damage level on the most
highly damaged plane determines the strain rate behaviour in
equation (8) and the damage histories on all possible planes

must be integrated, failure takes place on the plane where the failure criterion is first satisfied.

2.6 The approach of the materials scientist

In this section an attempt is made to present the approach of the materials scientist and to identify those features which are common with the single state damage variable approach described earlier. A general formulation for the growth of damage at grain boundaries has been given by Ashby and Raj (34). The growth of damage is expressed in terms of two mechanisms. One mechanism, referred to as nucleation, gives a measure of the rate at which grain boundary voids are formed. The other mechanism, referred to as growth, gives a measure of the rate of growth of void size.

The nucleation rate is represented by dn/dt and is measured in the number of voids formed per unit time on unit area of grain boundary. Holes nucleate at points of strain concentration which can occur at inclusions contained in the grain boundary or at the extremities of slip bands within the crystal. The growth of holes can be the result of the diffusion of vacancies along the grain boundary or of concentration of creep strains at inclusions.

Since voids are forming and growing continuously it is necessary to use the integral relation derived by Ashby and Raj to obtain a measure of the total damage. Suppose at time τ the nucleation rate is $\dot{n}(\tau)$ so that in time interval $d\tau$ the number of new voids formed is $\dot{n}(\tau)d\tau$. At a later time, t , these holes will be growing and the rate of increase of cross sectional area of the voids formed at time τ is $\dot{a}(t,\tau)$. During the time interval $(t - \tau)$ the holes formed at time τ will then have a cross sectional area

$$\dot{n}d\tau \int_{\tau}^{t} \dot{a}(t,\tau)dt \ .$$

The total area of the voids at time t_f is then

$$A(t_f) = \int_{\tau=0}^{\tau=t_f} \dot{n}(\tau)d\tau \int_{t=\tau}^{t=t_f} \dot{a}(t,\tau)dt \ . \tag{20}$$

Instead of using A as a measure of damage, some authors prefer to use the volume of the voids. Using the same method, illustrated above, the total volume V of the voids is

$$\bar{V}(t_f) = \int_{\tau=0}^{\tau=t_f} \dot{n}(\tau)d\tau \int_{t=\tau}^{t=t_f} \dot{\bar{v}}(t,\tau)dt, \tag{21}$$

where $\dot{\bar{v}}$ is the volume growth rate of the voids.

The formulation is quite general but it implies that to calculate damage it is necessary to keep track of the size of each void as it is formed. This is equivalent to a multi-state variable theory in which the number of state variables corresponds to the number of voids and is increasing with time. It is unlikely that the integral can be calculated in closed form except for simple loading histories or for special forms of the rate equations.

Various studies have been made with the objective of the formulation of equations which define the nucleation and void growth rates. Two particular examples are now discussed against the background of the phenomenological procedures already discussed.

2.6(a) The Greenwood equations

Greenwood (5) has studied the growth of creep damage in copper at a temperature of $500°C$. The tests were performed under conditions of constant uni-axial stress and measurements made of the number and size of voids. The visual observations indicate that the voids are all approximately the same size which implies that when a void is first formed it grows rapidly in size until it catches up with the voids formed earlier. It is possible to express this observation in a suitable mathematical form (35) which can be substituted in equation (20), but it is easy to see that the total cross-sectional area is simply $A = na$. No attempt was made to study the effect of damage on strain rate although it appears that the tertiary strains were little greater than those predicted by steady state theory.

Suppose that a uni-axial test is conducted at constant stress σ_0 . Let the rupture time be t_0 when the strain is v_0 , the hole density n_0 and the average volume of the holes $\bar{v}_0$. With reference to these physical values the growth equations for a uni-axial stress σ are

$$d(n/n_0)/dt = (\sigma/\sigma_0)^2 d(v/\dot{v}_0)/dt \; ;$$
$$d(\bar{v}/\bar{v}_0)/dt = (\sigma/\sigma_0)/t_0 \; ;$$
$$d(v/v_0)/dt = (\sigma/\sigma_0)^5/t_0 \; . \qquad\qquad (22a\text{-}c)$$

In the first of these equations the nucleation rate is proportional to the strain rate but is, in addition, proportional to the square of the applied stress. In the second the volume rate of the voids is assumed to be controlled by a diffusion process which is proportional to the stress. The third equation illustrates that the strain rate is proportional to the fifth power of applied stress and the effect of damage on strain rate is neglected.

The damage is defined as

$$\omega = n\bar{v}^{2/3}/n_o \bar{v}_o^{2/3} = A/A_o \ , \qquad (23)$$

and the rupture condition as $\omega = 1$. This is the same condition as proposed by Greenwood (5) who used the concept of a critical area fraction at failure. For multi-axial states of stress Leckie and Hayhurst (35) suggested that equations (22) and (23) take the form

$$d(n/n_o)/dt = (\sigma_1/\sigma_o)^2 \, d(v_e/v_o)/dt \ ; $$
$$d(\bar{v}/\bar{v}_o)/dt = (\sigma_1/\sigma_o)/t_o \ ; $$
$$d(v_e/v_o)/dt = (\sigma_e/\sigma_o)^5/t_o \ ; \qquad (24a\text{-}c)$$

and the damage is $\qquad \omega = n\bar{v}^{2/3}/n_o \bar{v}_o^{2/3} .$

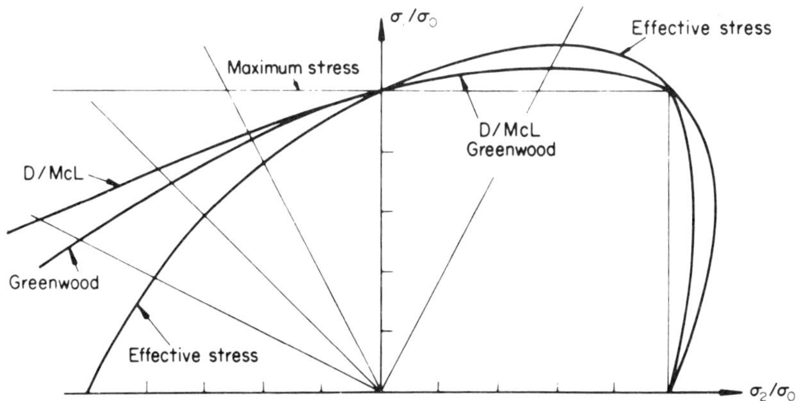

Fig.26.   Predictions of isochronous rupture loci

Integration of the above equations, for the rupture condition $\omega = 1$, gives the isochronous surface for rupture time t_o:

$$\Delta(\sigma_{ij}/\sigma_o) = (\sigma_1/\sigma_o)^{8/23}(\sigma_e/\sigma_o)^{15/23} = 1 \ . \qquad (25)$$

The corresponding isochronous rupture surface is shown in Fig.26. For constant stress, σ , the equations are easily re-expressed to give the following expression for damage rate:

$$d\omega/d(t/t_o) = (5 \, \Delta^{23/5} \, (\sigma_{ij}/\sigma_o)13)\omega^{2/5} \ , \qquad (26)$$

For the uni-axial case, $\sigma = \sigma_o$, the variation of ω with time is given in Fig.27. If now the damage rate equation is written in the following particular Rabotnov-Kachanov form

$$d\omega/d(t/t_o) = 2 \, \Delta^{23/5} \, (\sigma_{ij}/\sigma_o)/3(1 - \omega)^{1/2} \ , \qquad (27)$$

with the rupture condition $\omega = 1$,
it is found that this gives the damage-time graph shown in

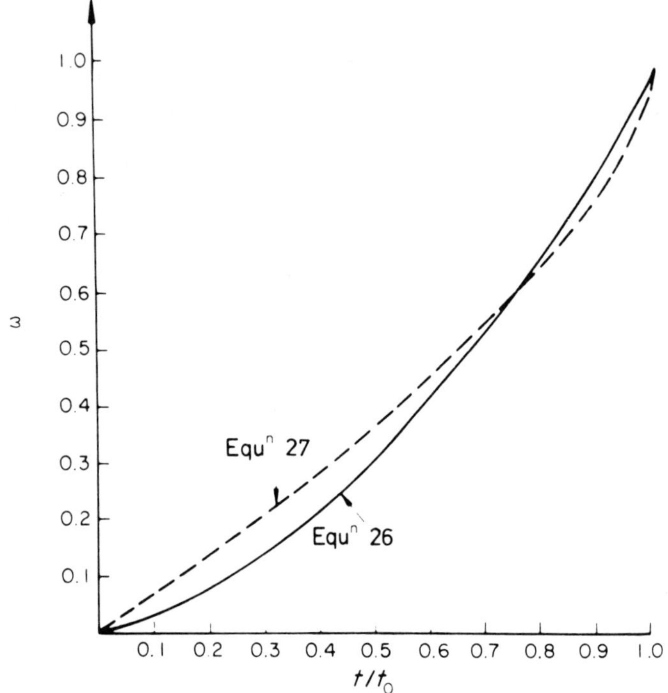

Fig.27.　Comparison of predictions from
damage growth laws for copper

Fig.27.　It can be seen that the two equations (26) and (27)
give similar results.　The corresponding strain rate expression to equation (27) is given by

$$d(v_e/v_o)/d(t/t_o) = (\sigma_e/\sigma_o)^5 . \tag{28}$$

These equations are a particular form of the Rabotnov-Kachanov
equations in which the damage state variable ω is the normalised form of the damage expressed by equation (20).

2.6(b) The Dyson-McLean equations

Dyson and McLean (4) have performed constant uni-axial
tension and torsion tests on Nimonic 80 A at 700°C.　In addition to measuring tertiary strains and times to rupture,
measurements were also made of the number of voids per unit
area of grain boundary and of the total volume of voids.

For constant uni-axial stress tests carried out at stress
σ_o , the rupture time is t_o at which time the void density
is n_o and the total volume of the voids is $\bar{V}_o$.　The constitutive equations proposed by Dyson and McLean (4) give
expressions for the rate of void nucleation and of void growth.
Damage is expressed as the total volume of voids and the

effect of damage on the creep rate is also included. For multi-axial states of stress the proposed rate equations may be written in the form

$$d(v_e/v_o)/dt = 3(\sigma_e/\sigma_o)^4 (\bar{V}/\bar{V}_o)^{4/9}/t_o \; ;$$

$$d(n/n_o)/dt = \frac{1}{2}(\sigma_1/\sigma_e)^2 (v_o/v_e)^{\frac{1}{2}} d(v_e/v_o)/dt \; ;$$

$$d(\bar{v}/\bar{v}_o)/dt = (\sigma_1/\sigma_e)^{0.7} d(v_e/v_o)/dt \; ; \qquad (29a\text{-}c)$$

where
$$\bar{v}_o = 3\bar{V}_o/2n_o \; .$$

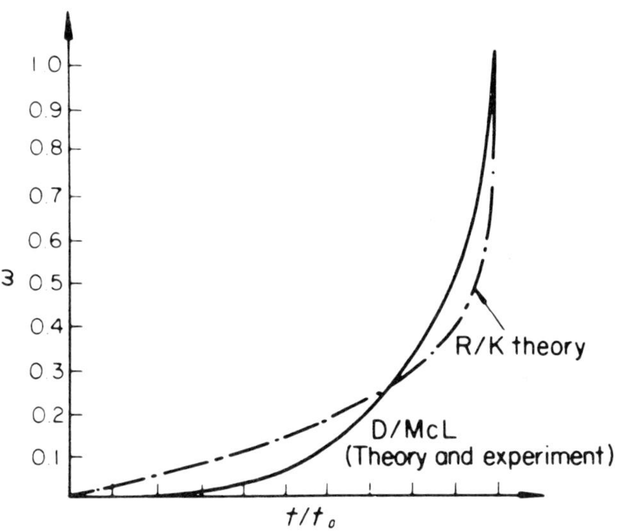

Fig.28.   Comparison of predictions from
damage growth laws for Nimonic 80 A.

The damage is defined in terms of the total volume of voids which are defined by the integral relation of equation (21).   Then

$$\omega = \frac{\bar{V}}{\bar{V}_o} = \frac{3}{4} \int_{v_o=o}^{v_1=v} (\sigma_1/\sigma_e)^2 (v_o/v_1) d(v_1/v_o) \int_{v_1}^{v} (\sigma_1/\sigma_e)^{0.7} d(v/v_o) \; .$$

$$(30)$$

For proportional loading, the ratio (σ_1/σ_e) is constant and the general damage equation (30) can be integrated and the constitutive equations (29) reduced to the single state variable form:

$$d(v_e/v_o)/d(t/t_o) = 3(\sigma_e/\sigma_o)^4 \omega^{4/9} \qquad \text{(31a-b)}$$

$$d\omega/d(t/t_o) = 9 \Delta^4(\sigma_{ij}/\sigma_o) \omega^{7/9}/2$$

where, $\qquad \Delta(\sigma_{ij}/\sigma_o) = [\sigma_1^{1.8} \sigma_e^{2.2}/\sigma_o^4]^{1/4},$

and the rupture condition is

$$\omega = \bar{V}/\bar{V}_o = \sigma_o/\sigma_1 ; \qquad (32)$$

in addition, it may be shown, for constant stress, that:

$$\frac{\bar{V}}{\bar{V}_o} = \left(\frac{\sigma_1}{\sigma_e}\right)^{2.7} \left(\frac{v_e}{v_o}\right)^{3/2} . \qquad (33)$$

For the uni-axial case, $\sigma = \sigma_o$, the variation of ω with time is given in Fig.28. If now the damage rate equation is written in the Rabotnov-Kachanov form:

$$d\omega/d(t/t_o) = \Delta^4(\sigma_{ij}/\sigma_o)/5(1-\omega)^4 , \qquad (34)$$

and used with the rupture condition $\omega = 1$, it is found that this gives the damage-time graph shown in Fig.28. The two equations (31b) and (34) give similar results. The corresponding strain rate expression to equation (31a) is given by

$$d(v_e/v_o)/d(t/t_o) = (\sigma_e/\sigma_o)^4/(1-\omega)^4 . \qquad (35)$$

When $\Delta(\sigma_{ij}/\sigma_o) = 1$ the isochronous surface is obtained for the rupture time t_o , Fig.26.

It would appear that for proportional stress or loading histories the phenomenological constitutive equations developed in Section 2.4b using a single damage state damage variable satisfy the macroscopic observations and are also capable of physical interpretation. However, for non-proportional loading this approach indicates that it is necessary to provide rate equations for both nucleation and void growth and in these circumstances at least two state variables will be required. Fortunately, in many practical components, while stresses vary they do remain sensibly proportional. Nevertheless problems will arise in which the stress fields have large rotations when the sequence of loading is applied. Growth and nucleation will occur at different rates in particular directions and it is to be expected therefore that the description of damage will be tensorial in form. It will also be necessary to understand how and if damage introduces important anisotropic effects.

2.7 Creep under non-proportional loading in Nimonic 80 A.

Equations (29a) and (33) may not be applied to non-proportional loading situations unless the nucleation growth and fracture mechanisms all have scalar properties. Should any one of the mechanisms have directional properties then it is necessary to integrate equation (21) in a piecesise manner for each of the critical directions corresponding to each phase of the non-proportional loading. The integration is tedious and is best carried out numerically. Before such calculations may be performed it is necessary to assign a specific physical interpretation to nucleation, growth and fracture. To gain some insight into the physical significance of equations (29) consider a single reverse torsion test carried out on a thin tube. The role of V/V_O in the deformation process has been established by Leckie and Hayhurst (36) to be scalar in nature and one would expect this to be the case for non-proportional loading. A number of possible interpretations of equations (29b) and (29c) can be made; these are:

1. $\bar{v}/\bar{v}_O$ vector and n/n_O scalar with $\bar{V}/\bar{V}_O$ being interpreted as either a scalar or a vector in the failure criterion.

2. $\bar{v}/\bar{v}_O$ vector and n/n_O vector with $\bar{V}/\bar{V}_O$ being interpreted as either a scalar or a vector in the failure criterion.

The possibility of $\bar{v}/\bar{v}_O$ being a scalar has been excluded because there is overwhelming microstructural evidence (14) to confirm that in most materials damage grows on planes orthogonal to the maximum principal stress direction. To the author's knowledge no conclusive microstructural evidence exists from which it is possible to determine the directional characteristics of the processes of nucleation. Hence, the possibility of n/n_O being either a scalar or a vector must be considered. The total volume fraction at failure $\bar{V}/\bar{V}_O$ may be interpreted in two ways. Firstly if the damage accumulated on two independent planes is V_1 and V_2 then for a scalar interpretation of damage at failure the rupture condition expressed by equation (32) becomes $|\bar{V}_1/\bar{V}_O| + |\bar{V}_2/\bar{V}_O| = \sigma_O/\sigma_1$. Secondly, if a vectorial interpretation of the damage at failure is taken then the failure condition is $|\bar{V}_1/\bar{V}_2, \bar{V}_2/\bar{V}_O| = \sigma_O/\sigma_1$. The form of the shear strain-time curve which ensues for single reverse torsion tests may be anticipated for the interpretations given above. The theoretical creep curves for a steady load and a single reverse load torsion tests are shown in Fig.29 for $\sigma_e = 297$ MPa. The theoretical creep curve for steady torsion is represented by ABC. This curve has been determined from equations (31) using the failure condition expressed by equation (32). Consider the torsion load being reversed at point B. If n/n_O is scalar $\bar{v}/\bar{v}_O$ is vectorial and $\bar{V}/\bar{V}_O$ is scalar then the curve BC will result; the curve BC is a mirror image of the steady load

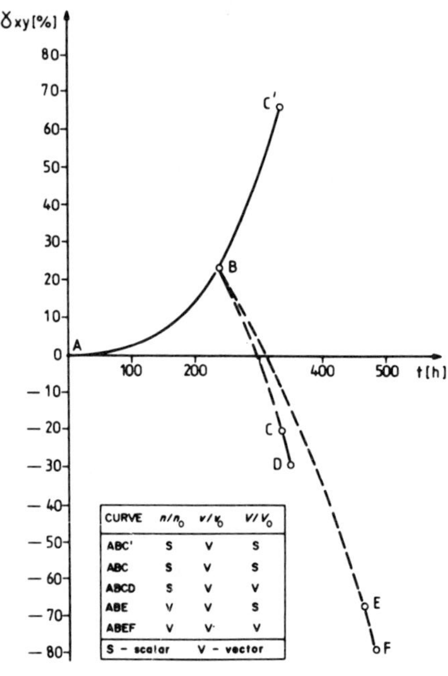

CURVE	n/n_0	v/v_0	V/V_0
ABC'	S	V	S
ABC	S	V	S
ABCD	S	V	V
ABE	V	V	S
ABEF	V	V·	V
S - scalar		V - vector	

Fig.29. Theoretical shear strain-time curves for a single reverse torsion test on virgin material (σ_e=297 MPa)

curve BC' about the horizontal axis through B. If n/n_0 is scalar, $\bar{v}/\bar{v}_0$ is vectorial and $\bar{V}/\bar{V}_0$ is vectorial then the curve BCD will result. The additional curve CD is due to the vectorial fracture condition. The curve ABE will result if n/n_0 is vectorial, $\bar{v}/\bar{v}_0$ is vectorial and $\bar{V}/\bar{V}_0$ is scalar. The curve BE must be to the right of the tangent to the curve BC at B since $\bar{V}/\bar{V}_0$ influences the strain rates in a scalar manner. The curve ABEF will result if n/n_0 is vectorial, $\bar{v}/\bar{v}_0$ is vectorial and the fracture condition $\bar{V}/\bar{V}_0$ is vectorial. By piecewise integration of equation (21) it may be shown that the time function which determines the shape of BEF if different from the $(t/t_0)^3$ function which determines the shape of ABC' and BCD. Similar arguments may be applied to multi-reverse torsion tests. It is clear that the results of single and multi-reverse torsion tests may be used to highlight the differences between different physical interpretations of the theory. In the following section of the paper results are presented for steady torsion and reverse torsion tests on thin tubes and are compared with theoretical predictions.

The Nimonic 80 A was prepared in the same manner as in an earlier investigation (37). The as-received 16 mm bar was cold-rolled to 13.5 mm, annealed for 2 h at 1050°C to give a 20 μm grain size, air cooled, and followed by the standard commercial strengthening treatment of 16 h at 700°C.

Thin-walled tubular torsion test pieces were machined with a gauge length of 16 mm , 0.8 mm wall thickness and outer diameter of 8.00 mm. The torsion creep tests were carried out in creep machines manufactured by W.H. Mayes Ltd., and modified by the author to improve the degree of homogeneity of the stress field. Specimen bending defined by $\{(\varepsilon_1-\varepsilon_2)/(\varepsilon_1+\varepsilon_2)\}100$ was measured (32), where ε_1 and ε_2 are principal tension surface strains at diametrically opposite locations. In the

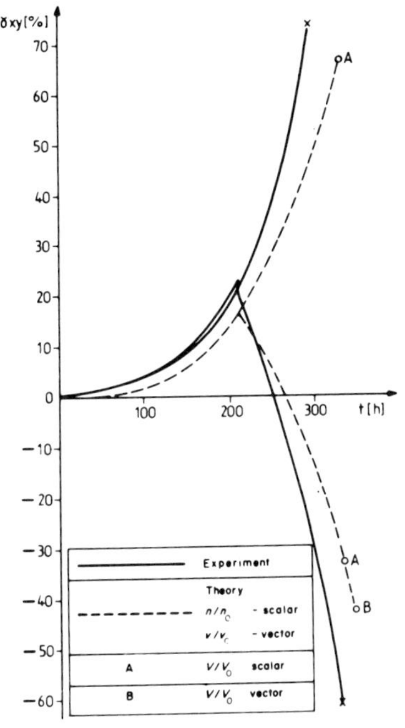

Fig.30.   Comparison of shear
strain-time curves for a
single reverse torsion test
and constant torsion test
with theoretical results for
$\sigma_e = 297$ MPa.

tests carried out the magnitude of specimen bending was maintained below 7%. The specimen temperature was maintained at 750 $\pm 1°$C.

The results of a single reverse torsion test conducted at $\sigma_e = 297$ MPa and of a steady torsion test, carried out at the same effective stress, are presented in Fig.30 and compared with theoretical results. The material constants used in the theoretical predictions are $\sigma_0 = 309$ MPa, $t_0 = 93$ h, $v_0 = 9.6\%$.

In the test result given in Fig.30, the torsion load has been reversed at $\gamma_{xy} = 22.5\%$. The experimental curve for the reverse torsion test is to the left of the theoretical curve. The theoretical curve has been computed assuming that n/n_0 is a scalar, $\bar{v}/\bar{v}_0$ is a vector and $\bar{V}/\bar{V}_0$ is scalar (point A) and n/n_0 is a scalar, $\bar{v}/\bar{v}_0$ is a vector and $\bar{V}/\bar{V}_0$ is a vector (point B). Comparison of the results presented in Fig.29 with those of Fig.30 clearly show that n/n_0 and $\bar{v}/\bar{v}_0$ should be interpreted in equations (29b) and (29c) as a scalar and as a vector respectively and that the fracture condition, given by equation (32) is a vector.

More extensive studies on the creep of Nimonic 80 A have been carried out under non-proportional loading (38). From the experimental results it is clear the cavity nucleation and growth laws, developed for steady load conditions, may be used to predict creep deformation and rupture under non-proportional loading, provided that the directional characteristics are correctly modelled.

2.8. Discussion and conclusions

In Section 2 a review is presented of the experimental techniques which have been developed to achieve repeatable

124

values of creep strain, creep strain-rate and creep rupture
lifetime. The results of metallographic examinations have
been presented for specimens subjected to homogeneous multi-
axial states of stress; from these results it has been shown
that creep damage or tertiary creep possesses a field or continuum property. It has been shown how the growth of the
field quantity and the accompanying strains can be described by
constitutive equations; in particular, how a range of multi-
axial behaviour can be described using an isochronous rupture
surface in stress space. The work of the materials scientist
on creep rupture is reviewed and an attempt is made to show
the commonality between their approach and that of the engineer.
When engineering structures are subjected to non-proportional
loading it is then necessary to understand the directional
characteristics of the processes of cavity nucleation, growth
and rupture.

The constitutive equations which are necessary to carry
out the analysis of structural components have been presented
in this section. In the following section the development of
numerical methods for the analysis of creeping structures will
be outlined and it will be shown how the field property of
continuum damage may be used to study the creep rupture behaviour of a class of structures which contain stress concentrators of increasing severity.

3.0 BEHAVIOUR OF STRUCTURAL COMPONENTS

In this section a review is presented of the methods of
structural analysis used to study the time dependent behaviour
of structural components. Consideration is given first to the
techniques used to obtain stationary-state solutions and then
to the methods used to study the growth of continuum damage in
structures containing stress concentrators. Emphasis will be
given to the identification of those physical characteristics
which determine the overall structural performance.

3.1 <u>Analysis of the time-dependent behaviour of structural</u>
<u>components</u>

Consider the convex function

$$\Phi(\dot{e}_{ij}) = \int_{o}^{\dot{e}_{ij}} \dot{\sigma}_{ij} d\dot{e}_{ij} \ ,$$

where $\dot{\sigma}_{ij}$ and $\dot{e}_{ij}$ are the rates-of-change of the stresses
and elastic strains respectively. Since $\dot{\sigma}_{ij}$ can be
expressed in terms of $\dot{e}_{ij}$ by

$$\dot{\sigma}_{ij} = C_{ijkl}\dot{e}_{kl} \ ,$$

where C_{ijkl} is the generalized elasticity tensor, the function $\Phi(\dot{e}_{ij})$ can be written $\Phi(\dot{e}_{ij}) = \frac{1}{2}C_{ijkl}\dot{e}_{kl}\dot{e}_{ij}$. For any pair of elastic strain-rates $\dot{e}_{ij},\dot{e}^{*}_{ij}$, the convexity of the function $\Phi(\dot{e}_{ij})$, which follows from its positive definiteness, can be expressed by

$$\Phi(\dot{e}^{*}_{ij}) - \Phi(\dot{e}_{ij}) \geq \dot{\sigma}_{ij}(\dot{e}^{*}_{ij} - \dot{e}_{ij}) . \tag{36}$$

Consider now the boundary-value problem for a body of volume V and on whose surface A the traction rates T_i are given on a part A_T and the velocities $\dot{u}_i$ are given on a part A_u . Although A_T and A_u are not exclusive in a mixed-mixed boundary-value problem, together they make up the whole surface A . Let $\dot{\sigma}_{ij}$ denote the rate-of-change of stress, $\dot{e}_{ij},\dot{v}_{ij},\dot{\varepsilon}_{ij}$ the elastic, creep and total strain-rates respectively, and $\dot{u}_i$ the actual velocity field. The artificial total strain-rate field is denoted by $\dot{\varepsilon}^{*}_{ij}$ and is compatible with the velocity field $\dot{u}^{*}_i$, which satisfies the conditions $\dot{u}^{*}_i = \dot{u}_i = 0$ on A_u . For a body in a state of creep it is assumed that the total strain-rates are given by the sum of the elastic and creep strain-rates:

$$\dot{\varepsilon}_{ij} = \dot{e}_{ij} + \dot{v}_{ij} . \tag{37}$$

Equation (36) can be integrated over the volume of the body to give

$$\int \Phi(\dot{\varepsilon}^{*}_{ij}-\dot{v}_{ij})dV - \int \dot{\sigma}_{ij}\dot{\varepsilon}^{*}_{ij}dV \geq \int \Phi(\dot{\varepsilon}_{ij}-\dot{v}_{ij})dV - \int \dot{\sigma}_{ij}\dot{\varepsilon}_{ij}dV. \tag{38}$$

On application of the principle of virtual velocities to the stress- and traction-rates for the body and to the strain- and displacement-rates, equation (38) becomes

$$\int \Phi(\dot{\varepsilon}^{*}_{ij}-\dot{v}_{ij})dV - \int T_i\dot{u}^{*}_i \, dA_T \geq \int \Phi(\dot{\varepsilon}_{ij}-\dot{v}_{ij})dV - \int \dot{T}_i\dot{u}_i \, dA_T. \tag{39}$$

The left-hand side of (39) has a minimum value equal to the right-hand side when $\dot{u}^{*}_i = \dot{u}_u$. The problem of determining the total strain-rate field for a given creep strain-rate field becomes one of minimizing the functional

$$\Omega(\dot{e}^{*}_{ij}) = \int \Phi(\dot{e}^{*}_{ij}-\dot{v}_{ij})dV - \int \dot{T}_i\dot{u}^{*}_i \, dA_T \tag{40}$$

with respect of the values of the assumed velocity field $\dot{u}^{*}$. Procedures are now discussed for obtaining numerical solutions.

Close form numerical solutions for the stress and strain rate fields in the boundary value problem cannot be obtained because of the high non-linearity of the material-structural

problem.   Two approaches have been used.   In the first, the
problem is formulated in terms of the constitutive equations
and the equations of equilibrium and compatibility.   For some
structural problems these equations can be solved, at a given
time, within the period of interest, by the numerical evalua-
tions of simple integrals.   The solution over the entire time
period can then be obtained by numerical integration of the
field variables in time.   In the second, the spatial variations of the field variables within the body are expressed
in terms of a set of discrete variables; the solution of the
boundary value problem, at a particular instant, can then be
achieved numerically by minimisation of the functional, defined in equation (40), with respect to the set of discrete
variables.   The solution over the entire period of interest
can then be obtained by numerical time integration.

Although solutions obtained using the former method will
be described later, it is the latter method, of which the
finite element displacement method is a particular case, which
will be described and discussed in more detail here.

The finite element procedure assumes that the entire volume is split up into small regularly-shaped discrete volumes
or elements.   The approximate velocity field will be assumed
to vary as a linear function of the spatial co-ordinates and
be described in terms of the nodal velocities $\dot{\hat{u}}_r^*$, where r
is an appropriate discrete variable.   Since the velocity
field is linearly dependent upon the spatial co-ordinates the
strain-rate field is constant within the elements.   The problem now becomes one of determining the velocity field $\dot{\hat{u}}^*$
which, within the imposed constraints, minimizes the function-
al $\Omega(\dot{e}_{ij}^*)$, i.e. $\partial\Omega(\dot{e}_{ij}^*)/\partial\dot{\hat{u}}_r^* = 0$ for $r = 1$, ...,2n and
for a positive second variation of Φ_1 where n is the total
number of nodes.   The minimum condition is then expressed by

$$\partial\Omega(\dot{e}_{ij}^*)/\partial\dot{\hat{u}}_r^* = (\partial/\partial\dot{\hat{u}}_r^*)\left[\int \frac{1}{2} C_{ijkl}\dot{e}_{kl}^*\dot{e}_{ij}^* dV - \int \dot{T} \dot{u}_i^* d A_T\right],$$

or, in terms of total creep strain-rates, by

$$\partial\Omega(\dot{e}_{ij}^*)/\partial\dot{\hat{u}}_r^* = \int C_{ijkl}(\dot{\varepsilon}_{kl}^*-\dot{v}_{kl})(\partial\dot{\varepsilon}_{ij}^*/\partial\dot{\hat{u}}_r^*)dV - \int \dot{T}_i(\partial\dot{u}_i^*/\partial\dot{\hat{u}}_r^*)d A_T.$$

$$(41)$$

Since the stress and strain fields are constant within each
element, the integration in equation (41) can be made by sum-
ming individual terms for each element, and the equation can
then be expressed for the minimum condition as

$$\sum_s \{(\partial\dot{\varepsilon}_{ij}^*/\partial\dot{\hat{u}}_r^*)C_{ijkl}(\dot{\varepsilon}_{kl}^*-\dot{v}_{kl})\tilde{V}\}_s = \dot{T}_r A_T , \qquad (42)$$

where s refers to the element number and $\tilde{V}$ is the element
volume.   For m degrees-of-freedom at each node, equation

(42) gives m x r linear simultaneous equations. The elastic problem can be formulated in the same manner by setting the creep-rate term to zero and omitting the time-derivatives:

$$\sum_s \{(\partial \varepsilon^*_{ij}/\partial \hat{u}^*_r)C_{ijkl}\varepsilon^*_{kl}\tilde{V}\}_s = T_r A_T \ . \qquad (43)$$

The finite element formulation for a creeping body, subjected to external loads, is given in terms of the rates of change of the field quantities; for this reason the formulation is known as the RATE METHOD. An alternative formulation has been used which can be described using (42) and omitting the time-derivatives:

$$\sum_s \{(\partial \varepsilon^*_{ij}/\partial u^*_r)C_{ijkl}(\varepsilon^*_{kl}-v_{kl})\tilde{V}\}_s = T_r A_T \ . \qquad (44)$$

The principal difference between this formulation and that for the rate method is that the boundary value problem is solved in terms of absolute field quantities and not their rate equivalent. For this reason the formulation given by (44) is known as the ABSOLUTE METHOD. A comparison of the usage of the two methods is now presented.

3.2   Comparison of the rate and absolute methods

A comparitive study of the rate and absolute methods of solution for time dependent problems is best carried out in the context of the solution of a particular problem. Consideration is given here to the use of both methods to obtain the stationary-state creep solution for the three bar problem shown in Fig.31. The finite element method of solution will not be used in this example; instead, the problem is formulated in terms of the constitutive equations and the equations of equilibrium and compatibility.

The structure is composed of three tension members of lengths l_1, l_2 and l_3 which have cross-sectional areas A_1, A_2 and A_3, respectively. One end of each member is connected to a rigid abutment and the other end to a rigid block whose motion is constrained to be vertical as shown in

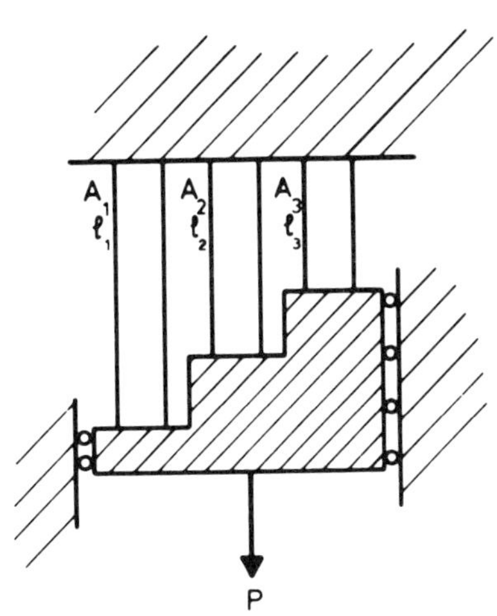

Fig.31. Three bar structure

128

Fig.31.   A vertical load P is applied to the rigid block.
The bars are assumed to creep according to equation (1), where
the time function $K(t) = t^m$ and m is a material constant.
It will be assumed that the total strain in bar i may be
expressed in terms of the elastic strain $e_i (= \sigma_i/E$, where E
is Young's modulus) and the creep strain v_i by

$$\varepsilon_i = e_i + v_i \ . \tag{45}$$

3.2(a)   <u>Problem formulation</u>

For equilibrium of forces it is required that

$$\sum_{i=1}^{s} \sigma_i A_i = P \ . \tag{46}$$

Compatibility of displacements requires that

$$\varepsilon_i l_i = \Delta \ , \tag{47}$$

where Δ is the vertical displacement of the rigid moveable
block.   The stress in the ith bar may be expressed using
equation (45) in terms of the total strain and length of the
 jth bar by

$$\sigma_i = E \left\{ \frac{l_j}{l_i} \left(\frac{\sigma_j}{E} + v_j \right) - v_i \right\} \ . \tag{48}$$

Substitution of equation (48) into equation (46) enables the
following expression to be derived for the stress in the   jth
bar:

$$\sigma_j = \frac{\left[P + E \sum_{i=1}^{3} v_i A_i \right]}{l_j \sum_{i=1}^{3} A_i / l_i} - E v_j \ . \tag{49}$$

Differentiation of equation (49) with respect to time gives
the following expression for the stress rates:

$$\dot{\sigma}_j = \frac{E \sum_{i=1}^{3} \dot{v}_i A_i}{l_j \sum_{i=1}^{3} A_i / l_i} - E \dot{v}_j \ . \tag{50}$$

Substitution of equation (1) into equation (50) gives the
following relation for the stress rates in terms of the
current stresses:

$$\dot{\sigma}_j = K t^m \left\{ \frac{E \sum_{i=1}^{3} \sigma_i^n A_i}{l_j \sum_{i=1}^{3} A_i / l_i} - E \sigma_j^n \right\} \ . \tag{51}$$

At time $t = 0$ the creep strains v_i are zero, and from equation (49) the initial elastic stresses are given by

$$\sigma_j = p \left/ \left\{ 1_j \sum_{i=1}^{3} A_i / 1_i \right\} \right. .\tag{52a}$$

The theoretical values of the stationary-state stresses, determined from the condition that $\dot{\sigma}_j = 0$, are given by

$$\sigma_i = P \left/ \left\{ \sum_{i=1}^{3} A_i (1_j / 1_i)^{1/n} \right\} \right. .\tag{52b}$$

3.2(b) <u>Absolute method</u>

The stresses at time $t = 0$ are given by equation (52). At any subsequent time t the stresses can be obtained by solving the first-order system of equations

$$\dot{v}_j(t) = Kt^m \left\{ \frac{\left(P + E \sum_{i=1}^{3} v_i A_i \right)}{1_j \sum_{i=1}^{3} A_i / 1_i} - Ev_j \right\}^n , \tag{53}$$

with initial conditions $v_j(0) = 0$, and using equation (49). The term absolute method has been used because the absolute form of the equilibrium equation (46) is invoked at all times in the derivation of equation (53).

3.2(c) <u>Rated method</u>

The stresses at time $t = 0$ are given by equation (52). The stresses at any subsequent time are given by the solution of

$$\dot{\sigma}_j(t) = Kt^m \left\{ \frac{E \sum_{i=1}^{3} \sigma_i^n A_i}{1_j \sum_{i=1}^{3} A_i / 1_i} - E\sigma_j^n \right\} , \tag{54}$$

with the initial conditions given by equation (52). The term rate method has been used because the derivation of equation (54) makes use of the rate form of the equilibrium equation (46):

$$\sum_{i=1}^{3} \dot{\sigma}_i A = 0 .$$

3.2(d) Time integration errors

Absolute Method

The numerical integration of creep strains to be carried out in the absolute method (equation (53)) is usually based on the Taylor series

$$v_j(t+\delta t) = v_j(t) + \dot{v}_j(t)\delta t + \sum_{n=2}^{\infty} \frac{1}{n!} \frac{\partial^n v_j(t)}{\partial t^n} \delta t^n \qquad (55)$$

for v_j at t . In practice the first two terms of equation (55) are frequently taken to calculate v . By virtue of equation (49), whatever numerical technique is adopted for the integration of equation (53), the computed stresses will satisfy the equilibrium condition. Truncated error which may occur in the numerical evaluation of v , and consequently in σ, do not lead to a violation of the equilibrium equation (46).

Summation of the Taylor series

$$\sigma_j(t+\delta t) = \sigma_j(t) + \dot{\sigma}_j(t)\delta t + \sum_{n=2}^{\infty} \frac{1}{n!} \frac{\partial^n \sigma_j(t)}{\partial t^n} \delta t^n \qquad (56)$$

for σ_j at t over all three bars in the structure gives the following equation:

$$\sum_{j=1}^{3} \sigma_j(t+\delta t)A_j = \sum_{j=1}^{3} \sigma_j(t)A_j + \sum_{n=1}^{\infty} \sum_{j=1}^{3} \frac{1}{n!} \frac{\partial^n \sigma_j(t)}{\partial t^n} A_j \delta t^n .$$

$$(57)$$

It is known from the equilibrium equation (46) that

$$\sum_{j=1}^{3} \sigma_j(t)A_j = P , \qquad (58)$$

$$\sum_{j=1}^{3} \frac{\partial^n \sigma_j}{\partial t^n} A_j = 0 ,$$

and from equation (57) the equilibrium condition

$$\sum_{j=1}^{3} \sigma_j(t+\delta t)A_j = P \qquad (59)$$

is satisfied. Hence, it may be seen that truncation of the series given in equation (56) leads to an error in $\sigma_j(t+\delta t)$, but the stresses still satisfy the equilibrium equation (46).

Rate method

In the rate method the first two terms of equation (56) are frequently used to calculate the stresses. Although truncation of the series does not lead to a violation of

equation (46) at t+δt, any round-off error in $\sigma_j(t)$ will;
such errors will propagate throughout the solution for times
> t since $\sigma_j(t+\delta t)$ are calculated from $\sigma_j(t)$.

3.2(e) Usage of the methods

It can now be seen that truncation errors will arise in
σ and v when using both the rate and absolute methods.
The rate method suffers from the disadvantage that cumulative
errors lead to a violation of the equilibrium equations.
The absolute method has the advantage that although the stresses need not be exact, they always satisfy the equilibrium
equations.

It is clear from the foregoing discussion that the two
methods will yield identical results provided that perturbations of the stresses do not occur and that the calculations
are carried out to a sufficiently high degree of accuracy.
Although the two methods are identical, irrespective of the
magnitude of the time step selected in the integration procedures, the accuracy and stability of the numerical methods
will be strongly dependent upon the method used to select the
time step. In the past, several methods have been used to
determine the time steps. These techniques have been studied
in detail elsewhere (21, 23, 39) and are not discussed here.
However, an example will now be given of the consequences of
failure to maintain appropriate time-step control.

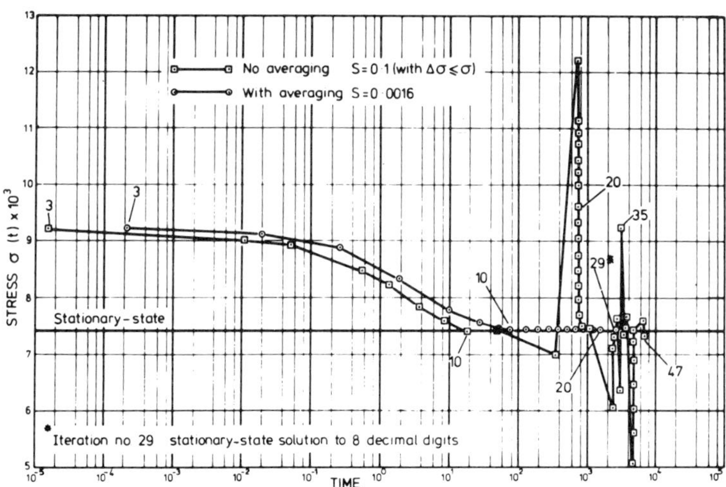

Fig.32.   Variations of stress with time in bar number 3
for the absolute and rate methods.

In the consistent units described by Hayhurst and Krzeczkowski
(40) the dimensions of the structure are

$$A_1 = A_2 = A_3 = 5 \quad \text{and} \quad l_1 = 20, \; l_2 = 15, \; l_3 = 10 \; .$$

The time step has been determined from the, frequently used, equation,

$$\delta t = \frac{1}{S} \left\{ \min \left| \frac{\sigma_i(t)}{\dot{\sigma}_i(t)} \right| \right\}, \tag{60}$$

where S is a constant. Two types of integration procedure have been used: the first in which the first two terms of the Taylor expansion are used in conjunction with equation (60), and the second in which an iterative or averaging procedure is used over each time step, to achieve convergence of the procedure. The variation of stress with time in bar number 3, computed using the absolute and rate methods with and without averaging respectively, are presented in Fig.32. The restriction of a one per cent change in stress per iteration was relaxed, and the stress allowed to change by one hundred per cent per iteration, in the rate method. In Fig.32, the square symbols denote the results obtained using the rate method without averaging and the circular symbols denote the results obtained using the absolute method. The instability in the stationary-state solution shown in Fig.32 may be attributed to the method of time step determination. Hayhurst and Krzeczkowski (40) have shown how the use of equation (60) is inappropriate for this class of problem and how its deficiencies may be overcome. The averaging technique referred to in Fig.32 is one such possibility; it produces the rapid and monotonic convergence of the stress to the stationary-state value shown in Fig.32.

For simple problems, such as the one described here, satisfactory results may be obtained from both the rate and the absolute methods provided that suitable time-step control is maintained. However, when the two techniques are to be used in the solution of large structural problems by the finite element method it can be expected (21,41) that the deficiency of the rate formulation will manifest itself by the failure of the method to satisfy equilibrium to a sufficiently high degree of accuracy.

The results of numerical studies, obtained using the finite element displacement method, are presented later.
To provide the necessary background the formulation of the techniques used is presented in terms of both the rate and displacement methods.

3.3 Solution of creep problems using the finite element method

3.3(a) Stationary-state solutions-rate method

Consider a body to be idealised using triangular finite elements whose displacements are linear functions of the spatial coordinates. Thus for the ith element the displacements $\underline{u}^i(\underline{x})$ are related to a global displacement vector $\underline{U}$ by $\underline{u}^i = \underline{N}^i\underline{U}$, where $\underline{N}^i$ is the displacement matrix which is dependent upon element geometry. In the same way the elastic strains are given by $\underline{e}^i = \underline{B}^i\underline{U}$, and the isotropic stress-strain relationship can be written as

$$\hat{\underline{\sigma}}^i = \underline{D}\,\underline{e}^i = \underline{D}\underline{B}^i\underline{U} \ , \tag{61}$$

where $\hat{\underline{\sigma}}$ denotes the elastic stresses and D is a matrix of elastic constants.

The creep strain rates $\dot{v}_{ij}$ are assumed to depend only on the current stress and are expressed by equation (2). The total strain rate $\dot{\underline{\varepsilon}}^i$ is composed of an elastic component $\dot{\underline{e}}^i$ and a creep component $\dot{\underline{v}}^i$:

$$\dot{\underline{\varepsilon}}^i = \underline{B}^i\underline{U} = \dot{\underline{e}}^i + \dot{\underline{v}}^i \ .$$

Since $\underline{\sigma}^i$ is the stress due to the elastic strain $\underline{e}^i$, then

$$\underline{D}\underline{B}^i\dot{\underline{U}} = \dot{\underline{\sigma}}^i + \underline{D}\dot{\underline{v}}^i \ . \tag{62}$$

To solve the stationary-state creep problem it is convenient to introduce a residual stress field $\underline{\rho}^i$, for element i , which varies with time t , giving the stress at t to be

$$\underline{\sigma}^i = \hat{\underline{\sigma}}^i + \underline{\rho}^i \ . \tag{63}$$

Differentiation of equation (63) with respect to time and substitution into equation (62) yields

$$\dot{\underline{\rho}}^i = \dot{\underline{\sigma}}^i = \underline{D}\underline{B}^i\dot{\underline{U}} - \underline{D}\dot{\underline{v}}^i \ . \tag{64}$$

Now consider the boundary value problem for which the nodal load vector $\underline{f}$ is given in terms of the applied load vector $\underline{w}$ by

$$\underline{f} = \sum_i \int_{S_i} (\underline{N}^i)^T \, \underline{w} \, dS \ .$$

S_i denotes the element of surface corresponding to the ith finite element. The principle of virtual work may now be applied to the boundary value problem:

$$\underline{U}^T \frac{d\underline{f}}{dt} = \sum_i \int_{V_i} \tilde{\underline{\varepsilon}}^{iT} \, \dot{\underline{\sigma}}^i \, d\tilde{V} \ ,$$

$$= \sum_i \int_{V_i} (\underline{B}^i\underline{U})^T (\underline{D}\underline{B}^i\dot{\underline{U}} - \underline{D}\dot{\underline{v}}^i) d\tilde{V} \ ,$$

$$= \sum_i \underline{U}^T \left[\left(\int_{V_i} \underline{B}^{iT}\underline{D}\underline{B}^i \, d\tilde{V} \right) \dot{\underline{U}} - \left(\int_{V_i} \underline{B}^{iT}\underline{D} d\tilde{V} \right) \dot{\underline{v}}^i \right] , \tag{65}$$

134

where $\tilde{V}_i$ denotes the volume of element i . For steady applied loads $d\underline{f}/dt$ is zero and on substitution of

$$\underline{K} = \sum \int \underline{B}^{iT} \underline{DB}^i d\tilde{V} \quad \text{for the stiffness matrix, and using}$$

$$A^i = \int \underline{B}^{iT} \underline{D} d\tilde{V} \text{ , equation (65) becomes}$$

$$\underline{K}\dot{\underline{U}} = \sum_i A^i \dot{\underline{v}}^i \ . \tag{66}$$

The stress rates are then given by equation (64). The elastic stresses at $t = 0$ may be derived from the non-rate equivalent of equation (65):

$$\underline{U}^T \underline{f} = \sum_i \underline{U}^T \left[\left(\int_{V_i} \underline{B}^{iT} \underline{DB}^i d\tilde{V} \right) \underline{U} \right]$$

or $$\underline{f} = K\underline{U} \ . \tag{67}$$

Equations (66) and (67) are the same as equations (42) and (43) respectively but are expressed in matrix notation. The stationary-state creep solution is obtained by first calculating the elastic stresses from equation (67) and equation (61), and then by time integration of equation (66). The time integration can be carried out satisfactorily using a number of methods, e.g. the Euler method, (39), i.e. the first two terms of a Taylor series, and the Runge-Kutta higher order method with Merson's error estimate (21). Time step selection in such methods has been carried out using equation (60), but in more recent work with the Runge-Kutta method (21) acceptance or rejection of the time increment has been carried out using an error criterion. The parameter C , defined by

$$C = \frac{\max_i \left\{ \| \underline{DB}^i \dot{\underline{U}} - \underline{D}\dot{\underline{v}} \|_\infty \right\}}{\max_i \left\{ \| \underline{DB}^i \dot{\underline{U}} \|_\infty , \| \underline{D}\dot{\underline{v}} \|_\infty \right\}} \ ,$$

where $\| \underline{a} \|_\infty = \max_j \{ |a_j| \}$ denotes the max-norm of vector $\underline{a}$, was introduced to determine when the stationary-state had been reached. C varies between 0 and 2 : the closer to zero it is the nearer one is to the stationary-state.

3.3(b) <u>Creep damage solutions - absolute method</u>

The solutions to problems in which continuum damage grows and propagates have been obtained using both the rate (41) and the absolute methods (42). In this class of problem it is convenient to remove from the numerical model material elements which have failed; when this is carried out it inevitably leads to perturbations in the field quantities. It has been found (23,40) that under these circumstances the absolute method is more well-conditioned than the rate method. For this reason the absolute method is described here.

The idealisation of the boundary value problem follows the same procedure as that outlined in the previous section for the rate method.

The creep strain rates $\dot{v}_{ij}$ are dependent on the current stress and damage and are expressed by equations (8) and (9). The total strain $\underline{\varepsilon}$ is composed of an elastic component $\underline{e}$ and a creep component $\underline{v}$, and is given for the ith finite element by

$$\underline{\varepsilon}^i = \underline{B}^i \underline{U} = \underline{e}^i + \underline{v}^i \; .$$

Since $\underline{\sigma}$ is the stress due to the elastic strain $\underline{e}$ then

$$\underline{D} \, \underline{B}^i \underline{U} = \underline{\sigma}^i + \underline{D} \, \underline{v}^i \; . \tag{68}$$

Now consider the boundary value problem for which the nodal load vector $\underline{f}$ is given in terms of the applied load vector $\underline{w}$ by

$$\underline{f} = \sum_i \int_{S_i} (\underline{N}^i)^T \, \underline{w} \; dS \; ,$$

where S_i denotes the element of surface corresponding to the ith finite element.

The principle of virtual work may now be applied to the boundary value problem:

$$\underline{U}^T \underline{f} = \sum_i \int_{V_i} \underline{\varepsilon}^{iT} \, \underline{\sigma}^i \; d\tilde{V} \; ,$$

$$= \sum_i \int_{V_i} (\underline{B}^i \underline{U})^T (\underline{D} \, \underline{B}^i \underline{U} - \underline{D} \, \underline{v}^i) d\tilde{V} \; ,$$

$$= \sum_i \underline{U}^T \left[\left\{ \int_{V_i} \underline{B}^{iT} \, \underline{D} \, \underline{B}^i \; d\tilde{V} \right\} \underline{U} - \left\{ \int_{V_i} \underline{B}^{iT} \, \underline{D} \; d\tilde{V} \right\} \underline{v}^i \right] \; . \tag{69}$$

On substitution of $\underline{K} = \sum \int \underline{B}^{iT} \, \underline{D} \, \underline{B}^i \; d\tilde{V}$ for the overall

stiffness matrix of the structure, and using $A^i = \int \underline{B}^{iT} \, \underline{D} \; d\tilde{V}$.

and $\underline{f}^c = \sum_i A^i \, \underline{v}^i$ (69) becomes

$$\underline{f} = \underline{K} \, \underline{U} - \underline{f}^c \quad \text{or} \quad \underline{U} = \underline{K}^{-1} [\underline{f} + \underline{f}^c] \; . \tag{70}$$

The elastic stresses at $\tau = 0$ may be derived from equation (70) with $\underline{f}^c = 0$ or

$$\underline{f} = \underline{K} \, \underline{U} \quad \text{and} \quad \underline{U} = \underline{K}^{-1} \underline{f} \; . \tag{71}$$

The creep strains and associated damage values at any time may

136

be obtained from the constitutive equations (8) and (9) using
an appropriate integration procedure and the solution to the
boundary value problem given by equations (70) and (71).

The integration of creep strain and damage has been
carried out accurately and efficiently using the Runge-Kutta
method with Merson's error estimate (23).

In the rupture calculation, failure of an element is
assumed to have occurred when $\omega \geq 0.999$. When the damage in
an element has exceeded this value, the overall stiffness matrix and the boundary conditions are reformulated. Procedures
for effecting these changes are now discussed.

When the rupture condition for an element is satisfied,
the material is assumed to be no longer capable of trans-
mitting or withstanding force. The overall stiffness matrix
is reassembled with the stiffness matrix for the failed element omitted and the new inverse $\underline{K}^{-1}$ is formed and stored.
The force and displacement conditions for nodes which are
associated with several elements that have failed may require
inclusion in the boundary conditions. This was achieved by
forming the appropriate inverse of the stiffness matrix $\underline{K}^{-1}$
and by setting to zero the displacement components of nodes no
longer associated with unfailed elements.

It is conceivable that regions of failed elements can
form around smaller regions which have not failed. Checks
were carried out to test for the presence of these regions,
and if such areas existed the elements were assumed to satisfy
the rupture conditions and were treated as failed elements.

The results of studies carried out on a range of structure
using these and other methods of numerical solution are now
reported.

3.4 <u>Stationary-state creep in a uniformly stretched plate containing a circular hole</u>

In the earlier development of numerical techniques for
the solution of creep problems it was necessary to select
problems which were tractable in a numerical and in an
experimental sense, since the question to be answered was: can
constitutive equations developed for homogeneous states of
stress be used in conjunction with numerical techniques to
accurately predict the results of experiments carried out on
structural components made of the same material? With the
aim of providing the answer to this question in mind a uniformly stretched plate containing a circular hole was selected.
The analysis for this structure, shown in Fig.33, will now be
presented.

3.4(a) <u>Analysis</u>

In common with the elastic problem the radial and circumferential stress (Fig.33) must satisfy the equilibrium equation

$$\frac{\partial}{\partial r}(r\sigma_r) - \sigma_\theta = 0, \quad (72)$$

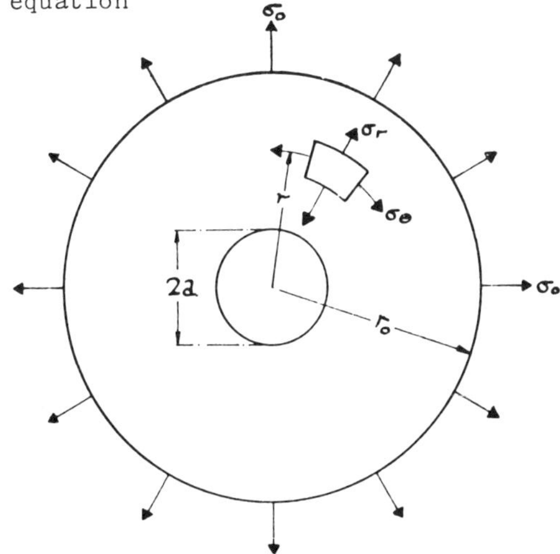

Fig.33.   Circular hole in a plate subjected to uniform tension

The strains are taken to be composed of elastic and creep components and are related to the radial displacement u as follows

$$\varepsilon_r = \frac{(\sigma_r - \nu\sigma_\theta)}{E} + v_r = \frac{\partial u}{\partial r}$$

$$\quad (73)$$

$$\varepsilon_\theta = \frac{(\sigma_\theta - \nu\sigma_r)}{E} + v_\theta = \frac{u}{r}$$

where E and ν are Young's modulus and Poisson's ratio respectively.

Inversion of equation (73) gives for the stresses

$$\sigma_r = \frac{E}{1-\nu^2}\left[\left(\frac{\partial u}{\partial r} + \nu\,\frac{u}{r}\right) - (v_r + \nu v_\theta)\right]$$

$$\sigma_\theta = \frac{E}{1-\nu^2}\left[\left(\frac{u}{r} + \nu\,\frac{\partial u}{\partial r}\right) - (v_\theta + \nu v_r)\right] \quad (74)$$

whereupon equation (72) becomes

$$\frac{\partial^2 u}{\partial r^2} + \frac{1}{r}\,\frac{\partial u}{\partial r} - \frac{u}{r^2} = \frac{f'}{r} + \frac{\partial g'}{\partial r} \quad (75)$$

where

$$f' = (1-\nu)(v_r - v_\theta) \quad (76)$$
$$g' = (v_r + v_\theta)$$

Introduction of the non-dimensional variables U = u/a, X = a/r and substitution into the equilibrium equation (75) yields

$$\frac{\partial}{\partial X}\left[X^3\,\frac{\partial}{\partial X}\left(\frac{U}{X}\right)\right] = \frac{f'}{X} - \frac{\partial g'}{\partial X}. \quad (77)$$

Thus if the plate is initially free from plastic strains (f' = g' = 0) the equations governing the behaviour of the plate during creep are

138

$$\frac{\partial}{\partial X}\left[X^3\frac{\partial}{\partial X}\left(\frac{U}{X}\right)\right] = 0 \quad (\tau = 0) \tag{78a}$$

$$\frac{\partial}{\partial X}\left[X^3\frac{\partial}{\partial X}\left(\frac{\dot{U}}{X}\right)\right] = \frac{\dot{f}}{X} - \frac{\partial\dot{g}}{\partial X} \quad (\tau \geq 0) \tag{78b}$$

where

$$\dot{f} = \frac{\partial f}{\partial \tau} = (1 - \nu)(\dot{\lambda}_\tau' - \dot{\lambda}_\theta') \quad,$$

$$\dot{g} = \frac{\partial g}{\partial \tau} = (\dot{\lambda}_\tau' + \nu\dot{\lambda}_\theta') \quad,$$

$$\tau = E\sigma_o^{m-1}\int_{\substack{o \\ t=o}}^{t} F(t)dt \text{ is a non-dimensional time}$$

measure (39), $\lambda \, (= \varepsilon/e_o)$ and $\lambda' (= v/e_o)$ denote the normalized total and creep strains respectively and $e_o = \sigma_o/E$.

Equation (78a) is appropriate to the elastic problem, the solution for which is

$$U = -\frac{A}{2X} + BX \quad (\tau = 0) \tag{79a}$$

while values for the rate of change of U with τ are found from the following solution of equation (78b)

$$\dot{U} = -\frac{A'}{2X} + B'X + I_1X \quad (\tau \geq 0) \quad, \tag{79b}$$

where

$$I_1 = -\frac{1}{2X^2}\int_1^X \frac{\dot{f}}{X}\,dX - \frac{1}{2}\int_1^X \frac{(2\dot{g}-\dot{f})}{X^3}\,dX \quad. \tag{80}$$

The constants of integration A,B,A',B' are found by satisfaction of the boundary conditions, as discussed later.

The stress and strain rates obtained from equations (73), (74) and (79b) are given by

$$\dot{\Sigma}_r = \frac{\dot{\sigma}_r}{\sigma_o} = \frac{1}{1-\nu^2}\left[-(1+\nu)\frac{A'}{2} - (1-\nu)B'X^2 + I_2 + I_3\right]$$

$$\dot{\Sigma}_\theta = \frac{\dot{\sigma}_o}{\sigma_o} = \frac{1}{1-\nu^2}\left[-(1+\nu)\frac{A'}{2} + (1-\nu)B'X^2 + I_2 - I_3 - (1-\nu^2)\dot{\lambda}_\theta'\right]$$

$$\dot{\lambda}_r = -X^2\frac{\partial\dot{U}}{\partial X} = -\frac{A'}{2} - B'X^2 + \frac{I_2}{1+\nu} + \frac{I_3}{1-\nu} + \dot{g}$$

$$\dot{\lambda}_\theta = \dot{U}X = -\frac{A'}{2} + B'X^2 + \frac{I_2}{1+\nu} - \frac{I_3}{1-\nu} \tag{81}$$

where

$$I_2 = -\frac{(1+\nu)}{2} \int_1^X \frac{\dot{f}}{X}\, dX = -\frac{(1-\nu^2)}{2} \int_1^X \frac{(\dot{\lambda}'_r - \dot{\lambda}'_\theta)}{X}\, dX$$

$$I_3 = \frac{(1-\nu)}{2}\, X^2 \int_1^X \frac{(2\dot{g}-\dot{f})}{X^3}\, dX = \frac{(1-\nu^2)}{2}\, X^2 \int_1^X \frac{(\dot{\lambda}'_r + \dot{\lambda}'_\theta)}{X^3}\, dX \quad (82)$$

For a plate with a stress free hole at its centre and a constant homogeneous state of stress at its outer edge the corresponding boundary conditions are:

$$\Sigma_r = \dot{\Sigma}_r = 0 \quad \text{at} \quad X = 1$$

$$\Sigma_r = \Sigma_o, \ \dot{\Sigma}_r = 0 \quad \text{at} \quad X = X_o \quad (\text{where } r = r_o) . \quad (83)$$

Substitution of equation (83) into the first of equation (81) yields for the integration constants:

$$(1+\nu)\,\frac{A'}{2} = -(1-\nu)B' = \frac{I_o}{(1-X_o^2)} , \quad (84)$$

where

$$I_o = (I_2 + I_3)_{X = X_o} .$$

The results for stress- and strain-rates can now be written in terms of the current stresses

$$\dot{\Sigma}_r = \frac{1}{1-\nu^2}\left[I_2 + I_3 - \frac{I_o}{(1-X_o^2)}\,(1-X^2) \right]$$

$$\dot{\Sigma}_\theta = \frac{1}{1-\nu^2}\left[I_2 + I_3 - \frac{I_o}{(1-X_o^2)}\,(1+X^2) - (1-\nu^2)\dot{\lambda}'_\theta \right]$$

$$\dot{\lambda}_r = \frac{I_2}{1+\nu} + \frac{I_3}{1-\nu} - \frac{I_o}{(1-\nu)(1-X_o^2)}\left[1 - \frac{(1+\nu)}{(1-\nu)}\,X^2 \right] + \dot{g}$$

$$\dot{\lambda}_\theta = \frac{I_2}{1+\nu} - \frac{I_3}{1-\nu} - \frac{I_o}{(1+\nu)(1-X_o^2)}\left[1 + \frac{(1+\nu)}{(1-\nu)}\,X^2 \right] . \quad (85)$$

For completeness, the integrals I_2 and I_3 are re-expressed in terms of the current stress:

$$I_2 = -\frac{3(1-\nu^2)}{4} \int_1^X \frac{\Sigma_e^{n-1}(\Sigma_r - \Sigma_\theta)}{X}\, dX$$

$$I_3 = \frac{(1-\nu^2)}{4}\, X^2 \int_1^X \frac{\Sigma_e^{n-1}(\Sigma_r + \Sigma_\theta)}{X^3}\, dX . \quad (86)$$

These were derived from the constitutive equation (2).

At zero time, when elastic loading is complete, the values of Σ_r and Σ_θ required in equation (86) are:

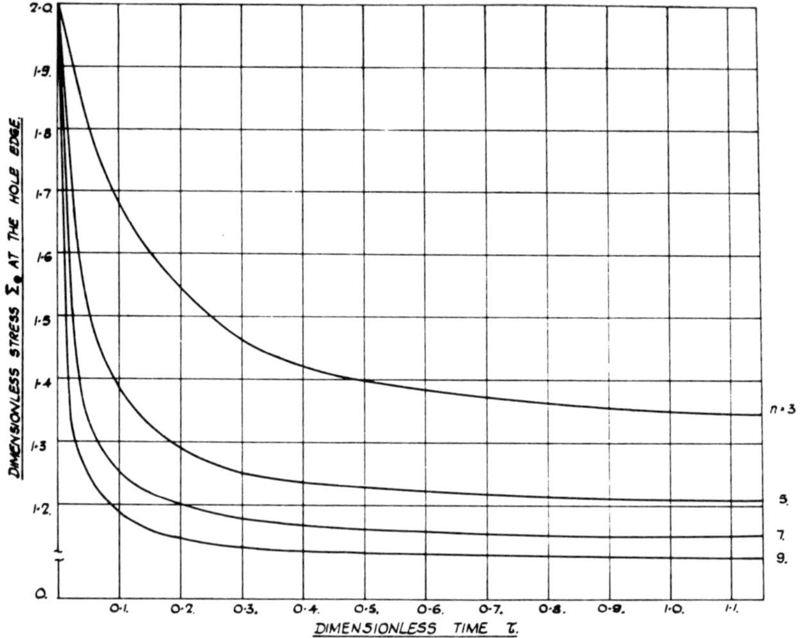

Fig.34.   Variation of normalised tangential stress
at edge of hole with normalised time

$$\Sigma_r = \frac{\Sigma_o}{1-x_o^2} (1-x^2) \quad , \quad \Sigma_\theta = \frac{\Sigma_o}{1-x_o^2} (1+x^2) \quad . \qquad (87)$$

The rate of change of stress and total strain at a particular
time may be determined using the current stress and equations
(85).   These rate equations may be integrated using the techniques discussed in Section 3.3 to give the variations of
normalised stress and strain with time.   Such variations are
shown in Fig.34 for the edge of the hole.   These results show
that the higher the stress exponent   n   the higher are the
initial stress rates and the lower are the stationary state
stresses.

The present stationary-state stress fields can be expected
to be very similar to plasticity solutions for   n-power
hardening materials having the same flow rule as that implicit
in equation (2); however it must be recognised that the
creep solutions are of the deformation type.   This need not
pose a severe limitation since Budiansky (43) has shown that
when the loading is not radial good results may be expected
from the deformation theory.   In Fig.35 the spatial variations
of stationary-state stress, for different values of   n , are

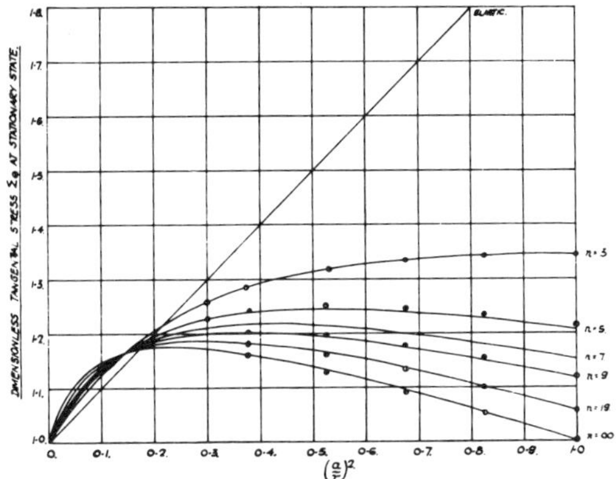

Fig.35.   Comparison of stationary-state stress
distributions with plasticity solutions of Reference (44).
Results from present calculations ——————
results from Reference 3. ⊙

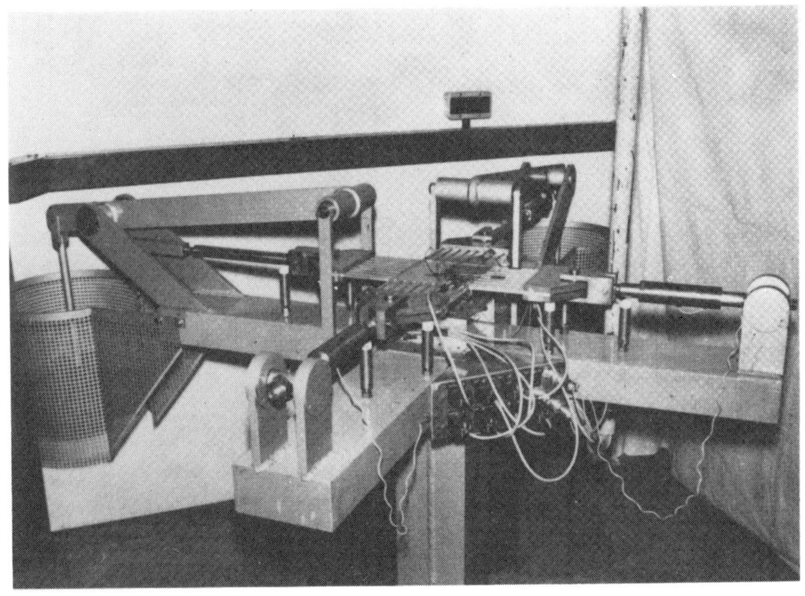

Fig.36.   View of plate testing rig.

compared with the stress distributions in a monotonically
loaded plate whose material deforms plastically according to
an n-powered strain hardening law.   The solutions, due to
Budiansky and Mangasarian [44] showed that the loading, at all
points in the plate, was almost radial.   Good agreement may
be observed in Fig.35 between the two sets of results.

3.4(b) Experimental verification

A dead load testing machine, shown in Fig.36, was designed
and built for testing plates in uniform tension.   The plates

Fig.37.   Cruciform test specimen

tested were cruciform in shape (Fig.37) with slotted arms to
provide uniform stressing in a central square of at least
115 x 115 mm.   The plates tested contained central circular
holes of 12.7 mm dia. and the strain at the edge of the hole
was measured using a special purpose electrical resistance
strain gauge [45].   In addition conventional strain gauges
were used to measure strains at several radial locations.   The
plate material was commercially pure aluminium in the half hard
condition.   The central thickness of the plate (1.575 mm) was
machined from plate of thickness 6.35 mm.   The theoretically
determined time variations of strains are compared with experimental values in Fig.38.   Good agreement may be observed until
250 h when the temperature of the room exceeded the limits of
$20 \pm \frac{1}{2}°C$ due to a failure of the air conditioning plant.

Hence it was possible to use uni-axial creep data to

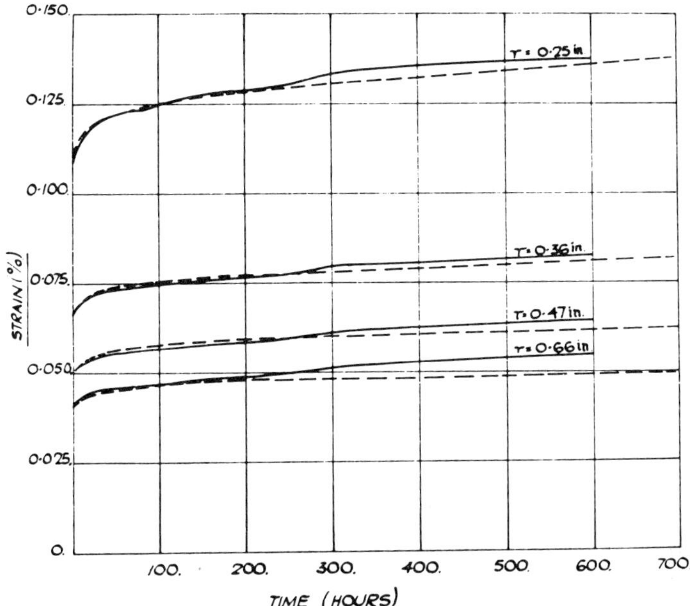

Fig.38.   Comparison of theoretical
and experimental strain histories.

Theoretical - - - - , Experimental ———

develop constitutive equations, given by equation (2), and
to use these in conjunction with numerical procedures to
predict the results of tests carried out on a model structure.

The usage of similar procedures to study the time-dependent degradation of structures is now discussed.

3.5   Growth of continuum damage in plane stress structures

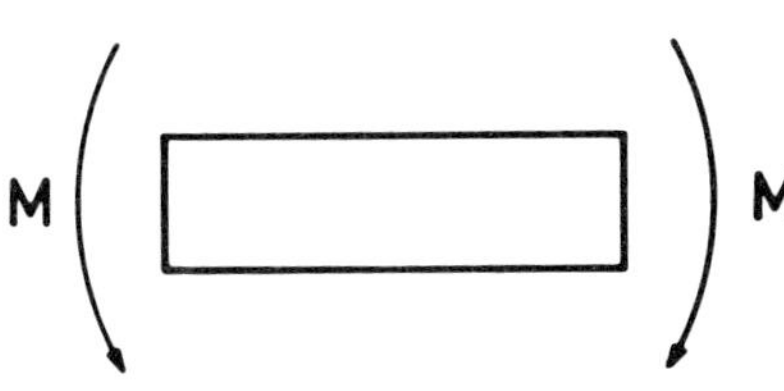

Fig.39.   Beam element
subjected to constant
bending moment.

The principal difference
between the growth of continuum damage in laboratory specimens and in structural
components is that in the
former the stress fields are
designed to be uniform and in
the latter they are highly in-
homogeneous.   The influence
of the inhomogeneity of the
stress field on the growth of
continuum damage is well
illustrated in studies (46)

Fig.40.   Mid-thickness micrograph of
a copper beam close to failure after
creep due to a steady moment at 250°C.
Black top edge denotes tensile surface
(Mag. x 9).

of beams, of rectangular cross-section, which have been subjected to uniform bending moments. A schematic representation of this situation is given in Fig. 39. On elastic loading the stress varies linearly through the depth of the beam from a tensile value at the top to a compressive value at the bottom. As creep and the growth of damage occur the stress distribution becomes progressively more non-linear, but the stresses still preserve their original signs.

Aluminium alloy and copper beams subjected to steady load conditions showed similar damage distributions. In Fig.40 it may be seen that grain boundary fissures and cracks have grown most rapidly nearer to the surface of the copper beam. The effect of the lower stresses can be observed some distance away from the surface of the beam; there is little evidence of the formation of damage close to the centre of the beam where the stresses are small.

Micrographs of aluminium beams which had been subjected to reverse loading were similar to that shown in Fig.40. Evidence of material deterioration was only clearly visible in regions of the structure which had been subjected to tension immediately prior to examination. This observation conflicts with the notion that material damage occurs equally in tension and in compression in aluminium alloys. However the most likely explanation is that the defects have the same characteristic dimension as the grain boundaries and the effect of a compressive stress is to close up the defects as they grow and to render them unobservable in the optical microscope.
The effect of load reversals on the growth of grain boundary cavities in copper was investigated in a single

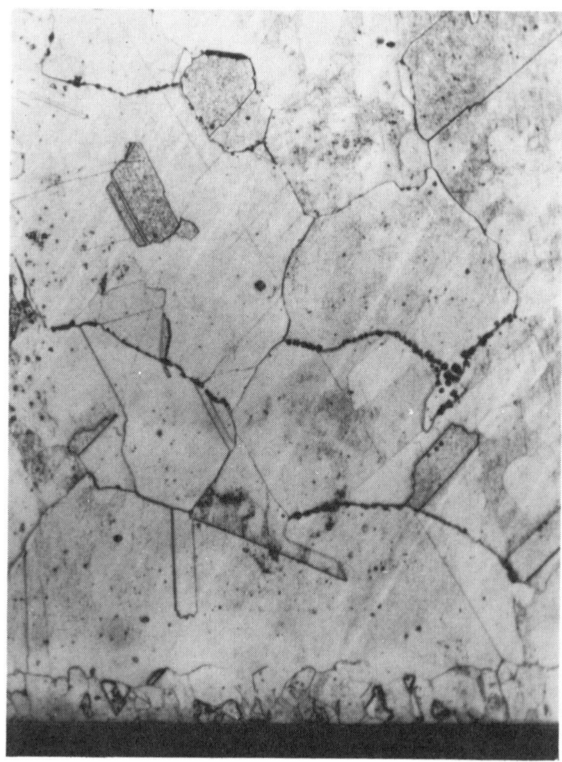

Fig.41.   Mid-thickness micrograph
showing a portion of the depth of a
copper beam which had been subjected
to tension for 150 h and the remainder
of its life (910 h) under compression
at 250°C.   Black bottom edge
denotes surface finally under compression
(Mag. x 85)

micrograph shown in Fig.41. The section of the beam was initially subjected to tension for 150 h and for the remainder of its lifetime (910 h) was subjected to compression. The cavities, which had originally grown on grain boundaries perpendicular to the stress direction, can be observed on grain boundaries parallel to the stress direction.

From the micrographs it may be seen that the field property of stress is reflected in the distributions of continuum damage in the beam. The quantification of this property is now examined together with the prediction of the structural lifetime for the uniformly stretched plate structure, with a central circular hole, discussed in the Section 3.4.

3.5(a) Continuum damage solution for a uniformly stretched plate containing a circular hole

The reason for the selection of this structure is that the analysis described in Section 3.4 can be used to study the growth of continuum damage with little modification. The principal difference is that the growth law for continuum damage, equation (9), has to be used and the creep rate equation (2) is replaced by equation (8). The uni-axial data for the material may be well represented by equations (8) and (9), as shown for example in Fig.42, for copper tested at 250°C.

The solution to the boundary value problem can then be obtained using the procedures already described except that it

146

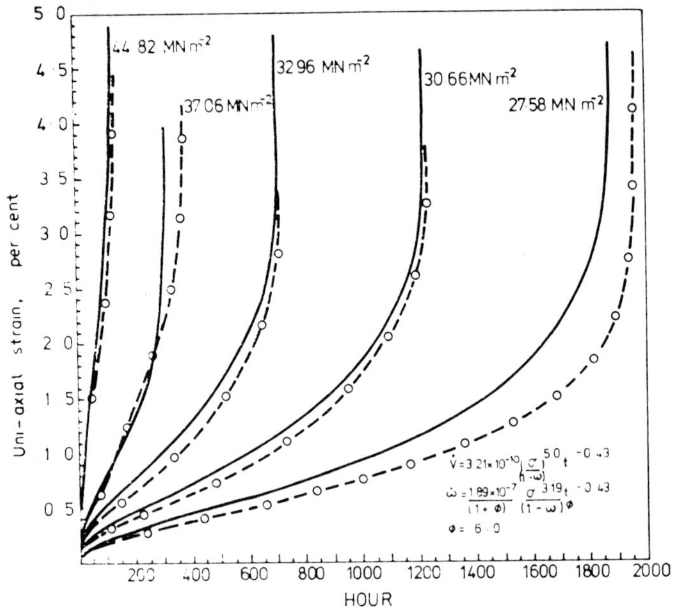

Fig.42.   Comparison of experimental values of creep strain
with those computed from equations (8) and (9)
for copper at 250°C (— o — experiment)

is necessary to allow for the propagation of a zone of ruptured material.   Rupture was said to have occurred when the
value of ω exceeded 0.6.   This value was determined by experimentation and values of ω exceeding 0.7 were set equal
to 0.7.   This was found to be necessary, at the time (18),
since if values of the damage parameter were allowed to exceed
0.7 difficulties were encountered with numerical instabilities.
The time integration of the field variables (i.e. σ_{ij}, ε_{ij}
and ω) was carried out using the Euler method.

The time variation of the spatial distributions of the
tangential stress Σ_θ are shown in Fig.43.   Comparison of
this figure with the stationary-state stress distributions
shown in Fig.35 shows that effect of the growth of continuum
damage is to cause considerable further stress redistribution.
It may be observed that the normalised stress level in the
region of failed material is typically 0.5, this is due to ω
being set at the level of 0.7.   This is not a severe limitation since the stress field adjacent to the edge of the hole
is predominantly uni-axial and the additional time for a

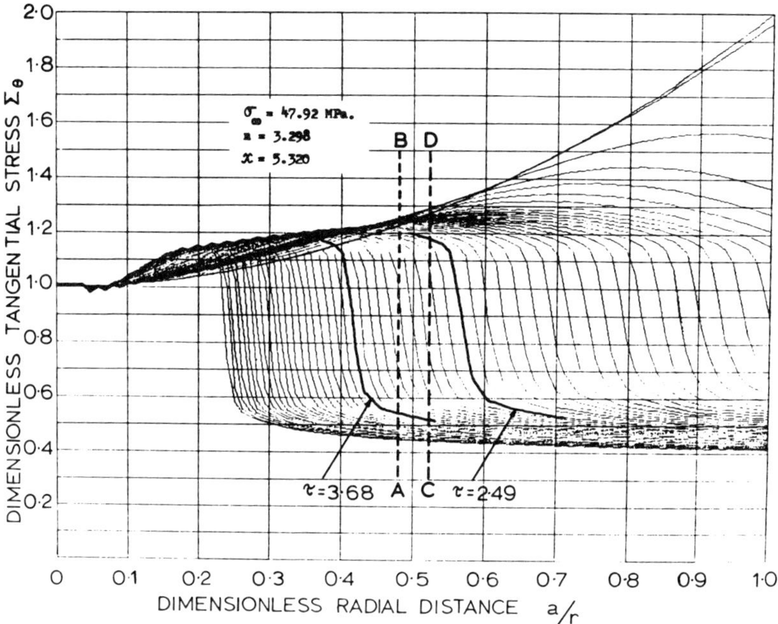

Fig.43.   Time variation of the spatial distribution of tangential stress in an aluminium plate tested at 210°C.

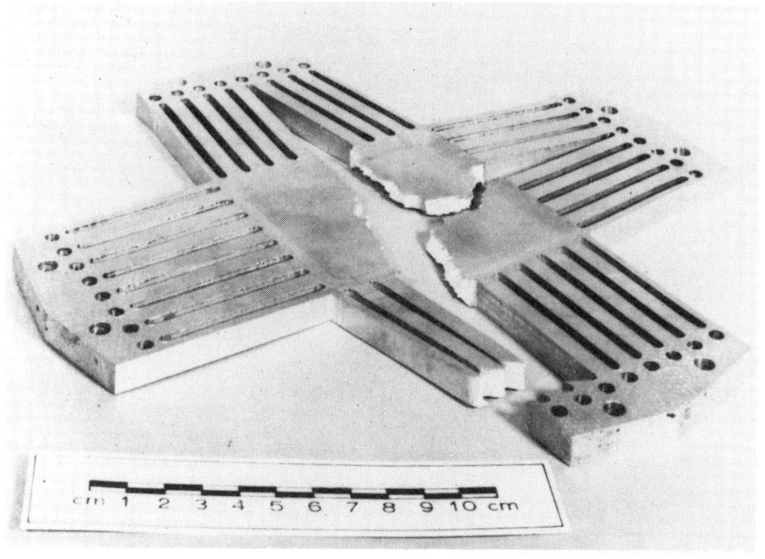

Fig.44.   Uniformly loaded aluminium alloy plate with a central circular hole which failed due to creep under steady load at 210°C.

148

uni-axial creep test to progress from the value of $\omega = 0.7$ to
rupture at $\omega = 1$ is small compared to the duration of the
test.   Similarly, it is assumed that the decrease in the
rupture time of the structure due to the redistribution of the
tangential stresses to zero would not have a significant
effect on the rupture lifetime.

The current location of the zone of rupture can be identified in Fig.43 by the steeply rising section of the curve.
The edge of the zone of rupture moves out slowly from the
edge of the hole and progresses steadily to the region defined
by the vertical lines AB and CD ; after it has passed this
region its rate for progression increases and failure rapidly
ensues (Fig.44).   This style of behaviour can be identified
on the stereoscan view shown in Fig.45, where a zone of low
strain rupture can be identified close to the edge of the hole.
The region corresponds to that defined by $0.5 \leqslant a/r \leqslant 1$ in
Fig.43; the failure surface in the region $0 \leqslant a/r \leqslant 0.5$
shown in Fig.45 has the appearance of a plasticity type of
shear failure.   A mid-thickness micrograph taken from the
region close to the hole in an unfailed plate showed the density of the continuum damage to be circumferentially uniform;
this is to be expected because of the symmetry of the stress
field about the centre of the hole.

Two rupture tests were carried out and in both cases the
lifetimes predicted numerically were close, the maximum error
between experiment and theory being 7%.

Fig.45.   Stereoscan view of the rupture surface
of the plate shown in Fig.44.   The edge of the hole
is at the right-hand border of the figure.

The observation was made that the normalised tangential stress at $a/r = 0.5$ remains approximately constant during the lifetime of the structure.   This observation led (18) to the concept of a representative rupture stress which can be used to determine the lifetime of a complex structure using uni-axial rupture data.

In addition to this study other investigations of axi-symmetric structures have been carried out; in particular investigations as to how the form of the damage rate equations, and the way in which damage is included in the strain rate equation can influence the predicted structural lifetimes. Studies carried out on torsion bars (47) and on rotating discs (48) showed that for steadily loaded structures the precise forms of the constitutive equations are not important provided that the global behaviour is accurately represented.

In the next section the rupture behaviour of non-symmetric plate structure with high stress concentrators will be studied theoretically.

3.5(b) Continuum damage solutions for a uni-axially loaded tension plate containing a circular hole

The structures considered so far have been axi-symmetric; for this type of structure only modest stress concentration factors can be achieved.   In practice high stress concentrations are encountered in asymmetric situations such as in plates containing elliptical holes (39) and cracks.   Here consideration is given first to the situation in which a uniformly stressed uni-axial plate, containing a central circular hole, undergoes steady load creep rupture.   The problem has been studied both theoretically and experimentally for the copper and aluminium alloy discussed in earlier sections.   The plate under consideration is shown in Fig.46. The results of experiments on copper, carried out at $250\,^{\circ}$C, and on an aluminium alloy, carried out at $210\,^{\circ}$C, are compared with uni-axial data in Fig.47.   It may be seen that when the experimental results for the plates are plotted on the basis of the average stress acting on the minimum section, σ_N (= $\sigma_0/(1-a/b)$) then close agreement is obtained with the experimental data.

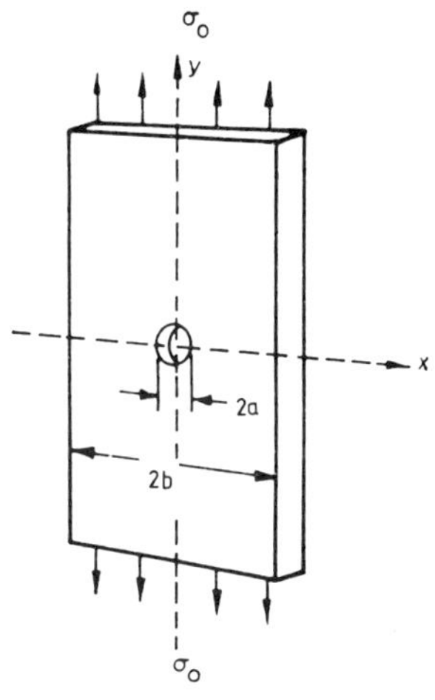

Fig.46.   Uni-axially stressed tension plate containing a central circular hole

150

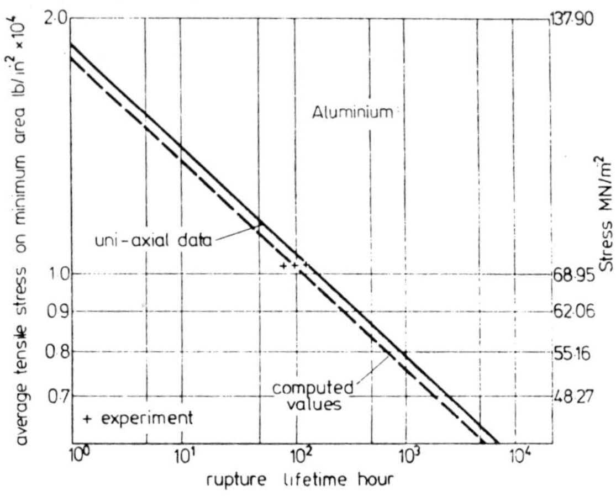

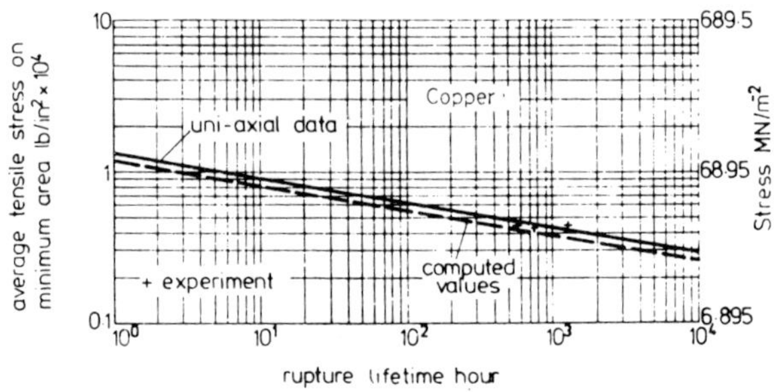

Fig.47.   Comparison of experimental
and computed rupture lifetimes with
uni-axial data for copper and aluminium.

The theoretical results were obtained from a finite element analyses (41) using the rate method in conjunction with the constitutive equations (8) and (19); the detailed procedures are outlined in Section 3.3. The multi-axial rupture behaviour of the materials was described by equation 19; a maximum effective stress, σ_e, criterion, $\alpha=0$, and a maximum principal tension stress criterion, $\alpha=1$, were assumed for the aluminium alloy and copper respectively.   The computed lifetimes are given in Fig.47 and shown to agree well with experimental values.

Computed regions of damage are presented in Fig.48 and compared with the results of a metallographic examination of a copper plate.   Two zones of rupture have been computed; $\omega \geqslant 0.1$, which corresponds to moderately damaged material, and $\omega \geqslant 0.99$, which represents failed material.   Very close agreement can be observed between the damage fields determined theoretically and experimentally.

Similar results have been obtained for aluminium alloy plates, and hence they are not discussed in detail here. However, one significant observation is that the results for the two materials do not reflect the fact that they satisfy different multi-axial rupture criteria.   The reason for this is that the stress fields in the plate are essentially plane

stress and are therefore confined to the tension-tension quadrant where the two rupture criteria almost coincide, cf.Fig.19.

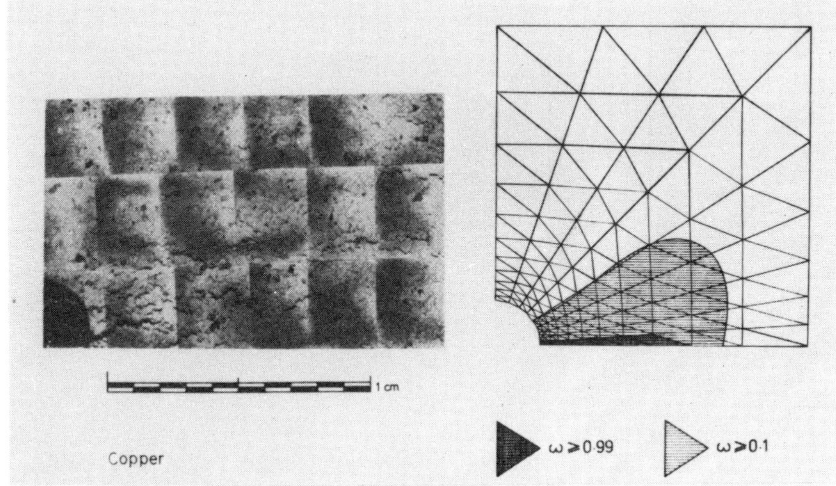

Fig.48.   Comparison of computed regions of damage
with metallographic results.

The most striking result to emerge from this study is that
the effect of the growth of continuum dmage is to annul the
localised high stresses found in the elastic and in the stationary state solutions and to redistribute stresses to values
close to the net section values.

If this result is applicable to structures which contain
more severe stress concentrators then it is of considerable
technological significance since it greatly simplifies the
level of complexity of design calculations.   An experimental
investigation is now described which seeks to answer this
question.

3.5(c) Continuum damage behaviour of uni-axially loaded tension
plates containing holes and slits.

Two structures have been selected to study if the growth
of continuum damage is capable of effectively blunting a sharp
defect or if the defect preserves its sharpness and grows in
size as a sharp crack.   The first structure is a thin plate
containing a sharp slit or crack of length 2a and the second
structure is the one studied in Section 3.5(b) with a hole diameter of 2a; the latter structure has been included for com-
parative purposes.   The structures were manufactured from the
aluminium alloy and copper described earlier and tested at
210°C and 250°C respectively under steady load conditions.
The geometry of the plates is shown in Fig.49 and the value of
a/b = 0.11.

152

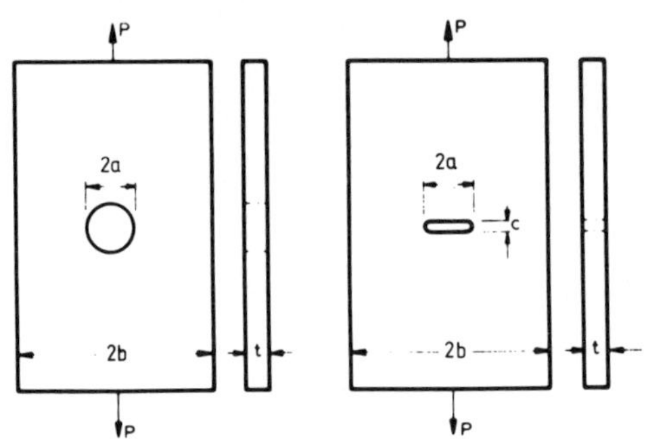

Fig.49.   Plane stress plates with
circular hole and slit
subjected to uni-axial tension.

The results of the tests (49) showed that for both materials the lifetimes of the two structures were indistinguishable when allowance had been made for scatter on the data. In all cases the lifetimes were closely predicted by the net-section stress σ_N ($= \sigma_0/(1-a/b) = P/2bt(1-a/b)$) and uni-axial data.

Metallographic examination of mid-thickness planes (Fig. 50) showed that damage had grown in both structures to form patterns in the region of the stress concentrators which effectively makes the two, initially different, structures appear the same. Clearly if this transformation and the accompanying stress redistribution takes place early in the lifetime then the lifetimes of this particular class of structure can be determined using the net-section stress and uni-axial rupture data.

Since the test results for the aluminium alloy and copper plates were similar no effect of the multi-axial stress rupture criterion has been observed. As stated in Section 3.5(b) this is due to the stress field being plane stress in character. An effect of the multi-axial rupture criterion can be expected in structures where the relative values of σ_1 and σ_e differ. This state of affairs is found in the axi-symmetrically notched bars discussed in Section 2.3(c), cf. Fig.13. The results of a theoretical study of these specimens are now presented and discussed.

3.6 <u>The influence of the multi-axial stress rupture criterion on the growth of continuum damage in axi-symmetrically notched bars.</u>

The structures to be considered in this section are the circular (20) and British Standard (9) notched bars shown in Fig.13.   The elastic and stationary-state (n=5) stress distributions across the minimum section (i.e. for $1 > \bar{r} > 0$, where $\bar{r} = r/a$, at $z = 0$, and z is the axial coordinate measured from the centre of the notch) are presented in Fig.14. It can be seen that, unlike the plane stress plates where σ_1 is similar in magnitude to σ_e , the stationary-state values of σ_1 and σ_e differ considerably.   It may be expected therefore, that the rates of accumulation of creep damage, as

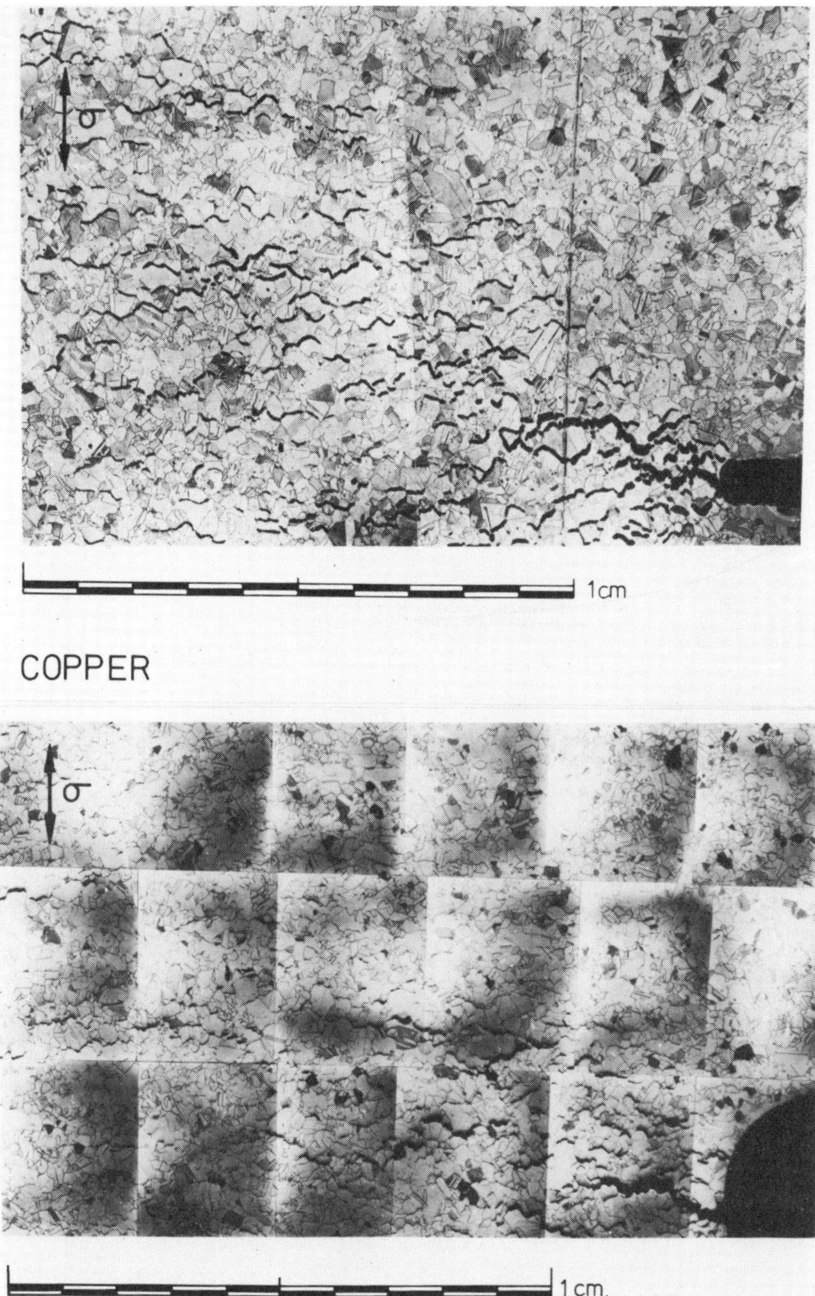

Fig.50.   Mid-thickness micrographs of steadily loaded
uni-axial copper plates immediately before failure.

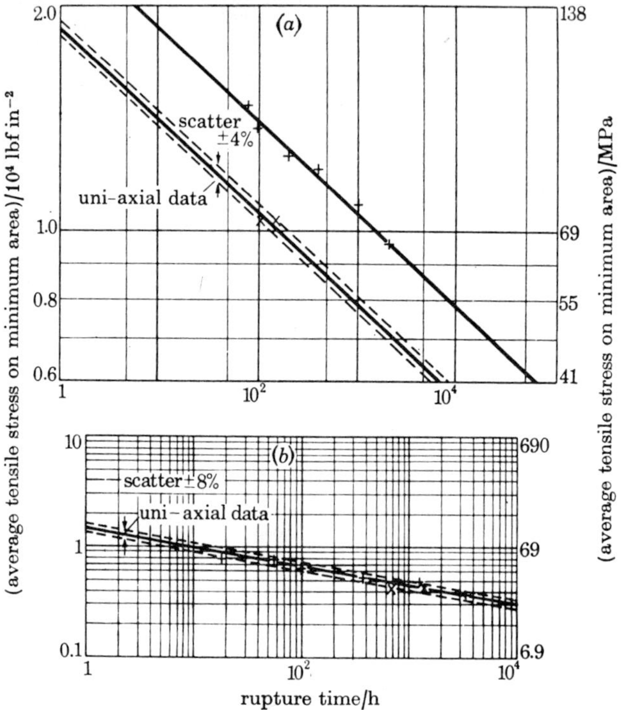

Fig.51.. Comparison of rupture lifetimes for uni-axial
plane stress plates containing circular holes
and axisymmetric circular notched uni-axial tension
specimens, cf. Fig.13(a), with uni-axial rupture
data for (a) an aluminium alloy tested at 210°C
and for (b) copper tested at 250°C.
+, notch specimens; X, plane stress plates

determined by equation (19), and hence the structural lifetimes,will be dependent upon the form of the multi-axial
stress rupture criterion or on the value of α in equation
(19). This effect can be observed in Fig.51 where the rupture lifetimes of circular notched bars and thin plates containing circular holes are compared with uni-axial rupture
data using the average stress acting on the minimum load
bearing section. It can be seen in Fig.51(b) that for copper
the structural lifetimes can be determined by the net section
stress and uni-axial data; for the aluminium alloys Fig.51(a),
the circular notched bars are stronger than uni-axial test
specimens when judged on the basis of the net-section stress.
This is not surprising since aluminium alloys satisfy a maximum σ_e stress rupture criterion, or $\alpha = 0$; examination of

the stationary state stress distribution given in Fig.14 shows that the average value of σ_e is suppressed relative to the average value of the axial component of stress.

In discussions of notch rupture behaviour it is convenient to re-introduce the concept of the normalised representative rupture stress. The performance of a notched or cracked specimen may be expressed in terms of the average stress which acts on the minimum section of the notch $\sigma_N = P/\pi a^2$, compared with the stress, σ_R , in a parallel sided uni-axial creep specimen which has the same lifetime as the notched bar. The normalised representative rupture stress Σ_R is defined by σ_R/σ_N . A specimen which requires a lower uni-axial stress σ_R , than the average stress σ_N is said to be notch strengthening $(\Sigma_R < 1)$ and, conversely a specimen which requires a higher uni-axial stress σ_R , than the average stress σ_N is said to be notch weakening $(\Sigma_R > 1)$. The results for copper bars, given in Fig.51(b), show slight notch strengthening, $\Sigma_R = 0.95$, and those for the aluminium alloy bars show significant notch strengthening, $\Sigma_R = 0.74$. Both the circular and B.S. notches have been studied previously by using an approximate technique based on kinematical determinacy (22) and by using the finite element method of analysis (23). A discussion is now presented of the usage of the latter technique in a study of the behaviour of circular and B.S. notched bars manufactured from 316 stainless steel and tested under steady load conditions at 550°C. However, no multi-axial data is available at this temperature. Instead of performing complex multi-axial tests to determine α an alternative approach has been used. Normalised representative rupture stressses have been determined for both specimens using equations (8) and (19) but for a range of values of α , the results are presented in Fig.52. Since the normalised experimental representative rupture stresses for the circular and the B.S. notched specimens are 0.858 and 0.942 respectively the value of α for 316 stainless steel at 550°C may be determined from Fig.52. The theoretically determined values of the normalised representative rupture stress are 0.873 and 0.936 for the circular and B.S. notched bars respectively, for $\alpha = 0.75$. The fact that these values are close to the corresponding experimental values demonstrates that the theory of continuum damage mechanics is capable of predicting the behaviour of notched bars. The isochronous rupture locus which results is shown in Fig.53 where it is compared with the results of tests (50) carried out on the same material at 600°C; reasonably good agreement may be observed.

3.6(a)   <u>Circular notch</u>

Computed zones of damage, defined by $\omega \geq 0.99$ and $\omega \geq 0.1$, are shown in Fig.54 and are compared with a micrograph taken from a diametral plane of the notch close to

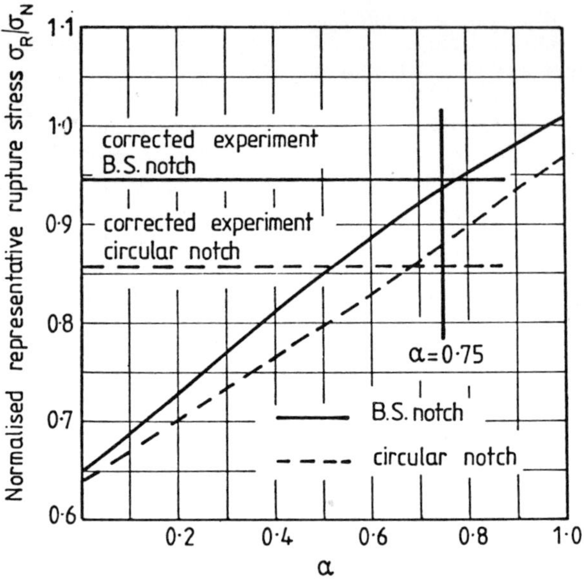

Fig.52.   Variation of normalised representative rupture stress with α for 316 stainless steel.

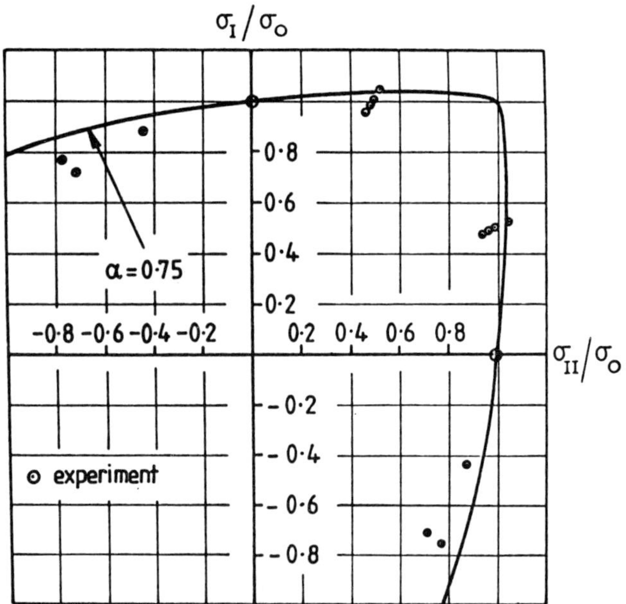

Fig.53.   Comparison of isochronous rupture locus for α=0.75 with experimental results obtained for 316 stainless steel at 600°C, cf. ref. (50).

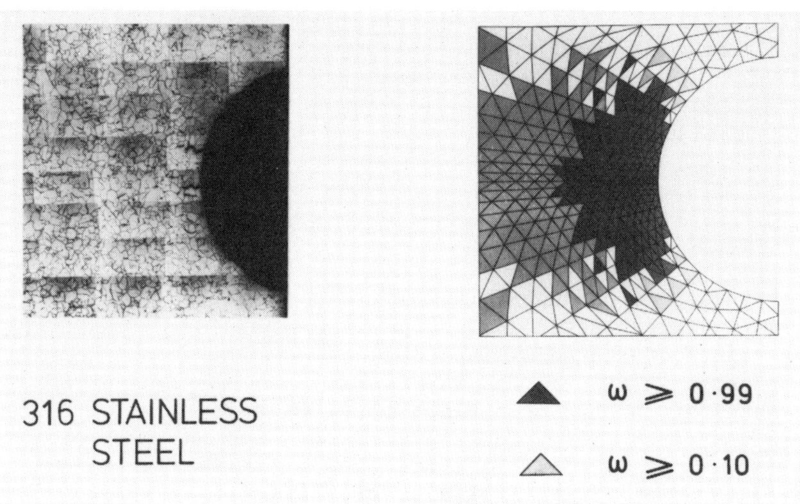

Fig.54.   Comparison of a theoretical prediction of the
distribution of damage in a circular notched bar
with that observed in a micrograph taken from a
diametral plane close to failure.

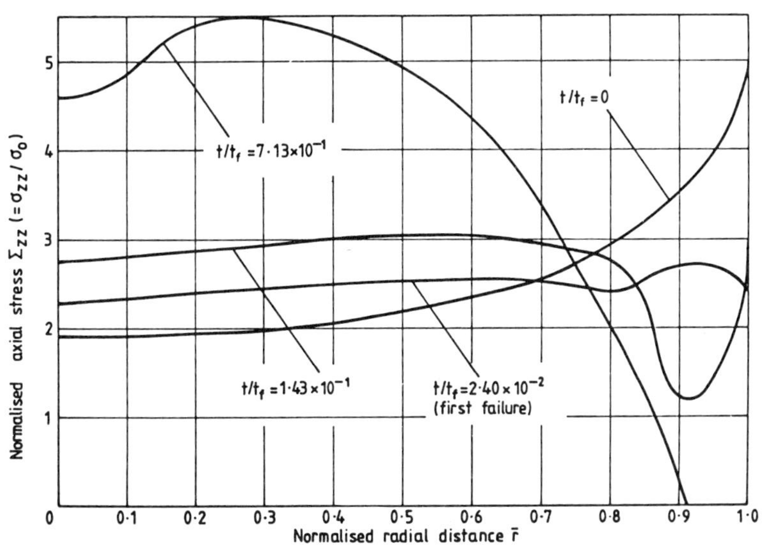

Fig.55.   The change of normalised axial stress with
normalised time (t/t_f) in a circular notch
for 316 stainless steel ($z = 0$).

failure, the specimen centre line coincides with the left
hand margin of the figures.   A region of damage with
$\omega \geq 0.99$ forms at (r=a, z=0) early in the lifetime, it
spreads gradually but the zone remains essentially stable for
the major part of the lifetime.   316 stainless steel is
capable of achieving relatively large creep strains and it
would appear that the stable cracking observed is a consequence of this property.   The micrograph closely reflects
the predicted behaviour.

Values of the normalised stress Σ_{zz} ($=\sigma_{zz}/\sigma_0$, where
$\sigma_0 = P/\pi b^2$) are plotted against normalised radial distance $\bar{r}$
for values of the normalised time t/t_f, where t_f is the
computed time at failure, in Fig.55.   Considerable stress
redistribution has taken place after the first 2.4% of life
when the first failure occurs at the tip of the notch with
$z > 0$.   At 14.3% of life the stress distribution is non-
uniform and at 71.3% of life it has undergone a radical
change.   The stress distributions in this figure show little
resemblance to the equivalent extrapolated stationary-state
distribution, n = 1.74 , obtained by Hayhurst and Henderson
(21).   Consequently the specimen in 316 stainless steel,
unlike those in copper and aluminium, does not behave as a
homogeneously stressed body at the minimum section of the
notch; consequently care must be exercised when using this
specimen as a materials test for 316 stainless steel.

3.6(b)   British Standard notch

Computed zones of damage, defined by $\omega \geq 0.99$ and
$\omega \geq 0.1$, are shown in Fig.56 and are compared with a micrograph taken from a diametral plane of the notch close to failure.   The specimen centre line coincides with the left hand
margin of the figures.   The two regions of computed damage
tend to coincide and to have developed in directions inclined
at approximate angles of 40° to the r-axis.   The micrograph closely reflects the predicted behaviour.

Smooth distributions of the normalised stress Σ_{zz}
along the r-axis for values of the normalised time t/t_f are
shown in Fig.57.   The first crack or failure occurs after
0.0147% of life, after which time the spatial distributions
are similar to the equivalent extrapolated stationary-state
distribution, n = 1.74 obtained by Hayhurst and Henderson
(21).   After 4.43% of life the crack has advanced across the
minimum section to a distance $\bar{r} = 0.92$, thereafter the
crack growth is predominantly away from the r-axis.

3.6(c)   Discussion

Two distinct features emerge on comparison of the results of circular and B.S. notches: firstly, that the

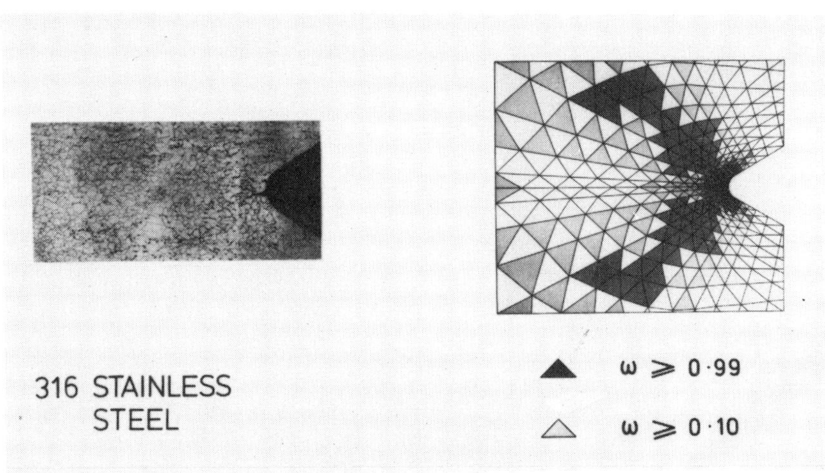

Fig.56.   Comparison of a theoretical prediction of the distribution of damage in a B.S. notched bar with that observed in a micrograph taken from a diametral plane close to failure.

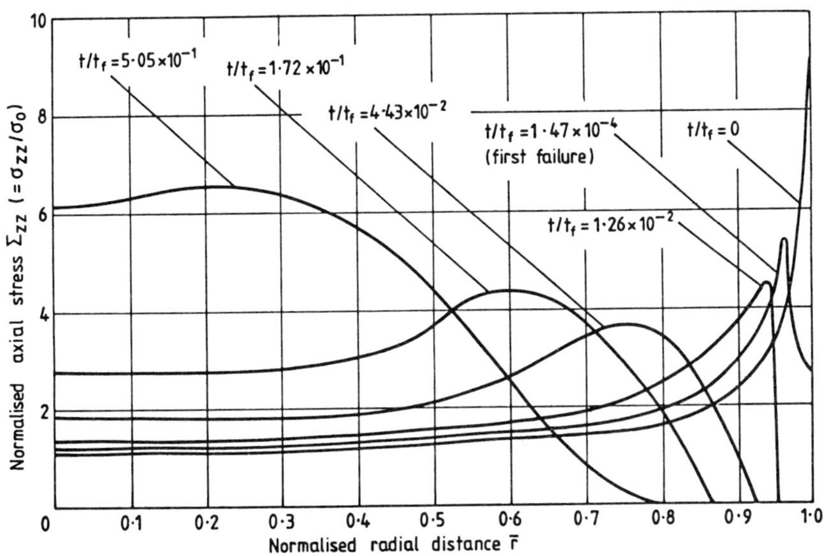

Fig.57.   The change of normalised axial stress with normalised time (t/t_f) in a B.S. notch for 316 stainless steel $(z = 0)$.

circular notch shows characteristics of a specimen which
uniformly damages in the region of the minimum section and in
which failure takes place relatively uniformly towards the
end of life, this behaviour is more pronounced in the copper
and aluminium specimens studied by Hayhurst et al (23);
secondly, that in the case of the B.S. notch the first crack
forms early in the lifetime and propagates relatively slowly
through the specimen.   The latter situation is well removed
from the type of behaviour found in the homogeneously
stressed region of the circular notch; the extent of the regions of damage is confined to the close proximity of the
crack and the behaviour is similar to that for the growth of
a crack due to creep in a slightly damaged continuum.   This
observation suggests that the continuum damage process, used
by the solution procedure to obtain the results of Figs.56
and 57, is probably the mechanism by which well-defined
cracks grow by creep.   This possibility is investigated in
the next section.

3.7   <u>The rôle of cotinuum damage in plane strain creep crack
growth.</u>

Considerable effort has recently been devoted to the
development of techniques to describe the growth of sharp
cracks due to creep (51,52,53).   The majority of the methods
are ad hoc extensions to either linear elastic fracture
mechanics (54) or post yield fracture mechanics (55) to high-
temperature situations.   Although the methods can be used to
relate the results of experiments carried out on similar specimens of the same material, their range of applicability to
different materials, temperatures and structural geometries

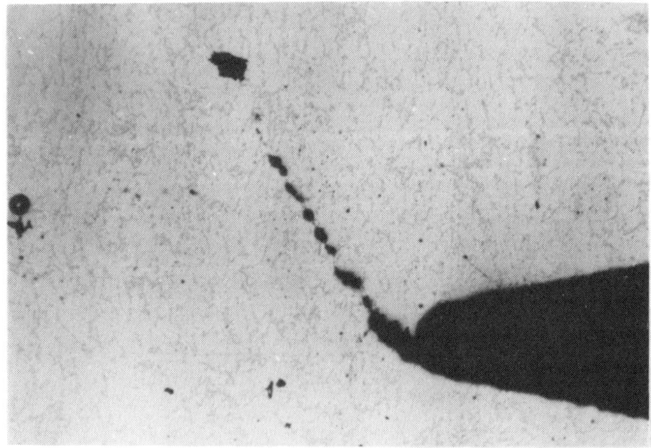

Fig.58.   Crack tip shear damage, close to failure, in an
externally cracked aluminium alloy specimen tested
at 150°C under steady load, (Mag. x 48).

is not clear (56).   For example the available theories
assume that discrete cracks grow in their original plane; it
is well known that under certain conditions cracks grow by
creep on planes which do not coincide with the original crack
plane.   An example of crack growth on a plane inclined at
approximately 55° to the plane of the original crack is
shown in Fig.58 for an aluminium alloy.

The technological aim of a creep crack growth theory is
to relate short-term performance of small laboratory specimens to that of larger in-service components which contain
cracks. Probably the only justifiable way of achieving this
aim is to ensure that the governing physical mechanisms of
the processes of creep crack growth are the same in both
situations.

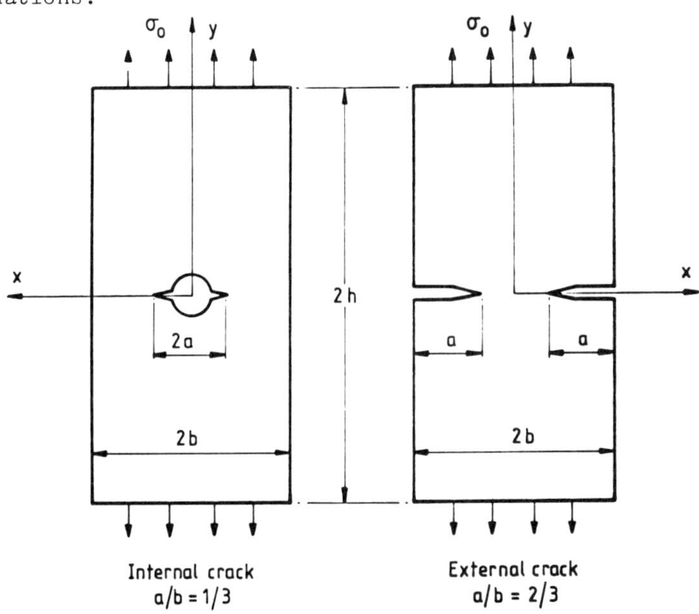

Fig.59.   Definition of co-ordinate system
for internally and externally cracked specimens.

In this section the proposal is investigated that continuum damage mechanics theory can be used to describe the
physical processes responsible for creep crack growth.   The
investigation reported considers both experimentally and theoretically the behaviour of the two specimens shown in Fig.59.
The selection of the specimen geometries and the presentation of the experimental results has been discussed in detail
elsewhere (42,57).   The material considered is the 316
stainless steel, discussed in Section 3.6 and tested at
550 $\pm$ 2°C.   Experimental values of the normalised representative rupture stress, discussed in Section 3.6 and given by
$\Sigma_R = \sigma_R/\sigma_N$, where $\sigma_N = \sigma_0/(1-a/b)$, are 1.25 and 0.93 for the

162

internally and externally cracked specimens respectively. The
internally and externally cracked specimens undergo 25%
weakening and 7% strengthening respectively.

In the theoretical study of the two specimens the
finite element procedures, described earlier, were used to
quantify the growth of damage and to allow the propagation of
a zone of failed material. Numerical techniques for the solution of crack-tip field problems have been developed by
Hayhurst and Brown (58,59) and used to obtain plane strain
stationary-state solutions; the results were shown to be in
close agreement with available solutions. The same constant
strain finite element has been used in this investigation;
the detailed procedures have been discussed elsewhere by
Hayhurst, Brown and Morrison (42). The numerical results
will now be presented and compared with the results of
metallographic examinations carried out on mid-thickness
planes of tested specimens.

Computed values of the normalised representative rupture
stress are 1.30 and 0.91 for the internally and externally
cracked specimens respectively. These results are in close
agreement with the experimental values; the level of agreement is worst for the internally cracked geometry which predicts an additional amount of weakening of 0.05.

Unlike the aluminium specimens tested by Hayhurst,
Brown and Morrison (42) the 316 stainless steel specimens
showed no evidence of the propagation of sharp cracks on
initial loading. However, initial geometry changes were
observed which were shown theoretically not to be significant. Theoretical predictions of crack advancement, defined by $\omega \geq 0.999$, are shown in Fig.60(a), for the externally cracked specimen at 60% of life, and in Fig.61(a), for
the internally cracked specimen almost at failure; the predictions are compared with the distributions of grain boundary damage shown in the micrographs presented in Figs.60(b)
and 61(b). For both specimens the grain boundary damage can
be observed to have formed and linked up producing cracked
regions which are similar in shape and orientation to these
obtained theoretically.

For both specimens the initial crack advance takes
place along the plane of the minimum cross-section, it does
so relatively rapidly since the stresses are typically
greater than the stress $\hat{\sigma}$ at which the log σ - log t_f
uni-axial rupture graph branches into two linear sections.
The high stress portion of the curve has a high value of the
stress index χ , in equation (19). After the crack has
advanced a short distance in this way, the rate of propagation reduces after stress redistribution has lowered the
typical stress levels to below $\hat{\sigma}$. After this event the two
specimens behave differently. The theoretical predictions

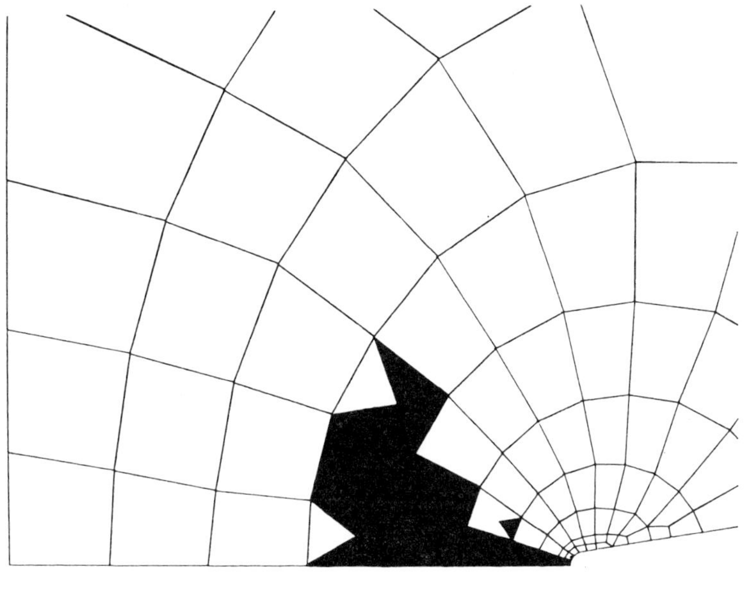

(a)

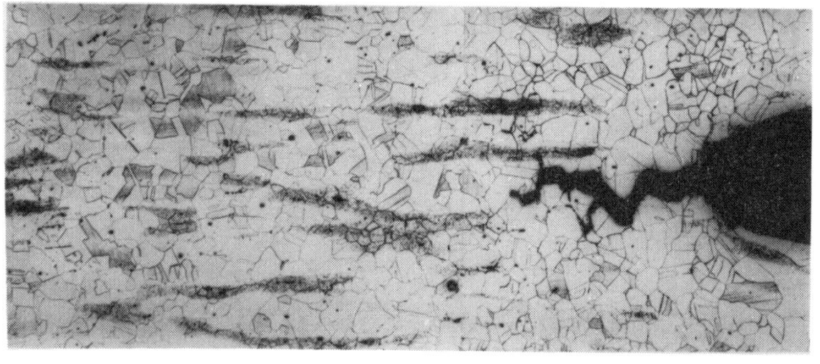

(b)

Fig.60.   Comparison of (a) computed failure path
$\omega \gtrsim 0.999$, with (b) mid-thickness micrograph
taken from the minimum section of an
externally cracked specimen in 316 stainless steel
at $t/t_f = 0.6$.

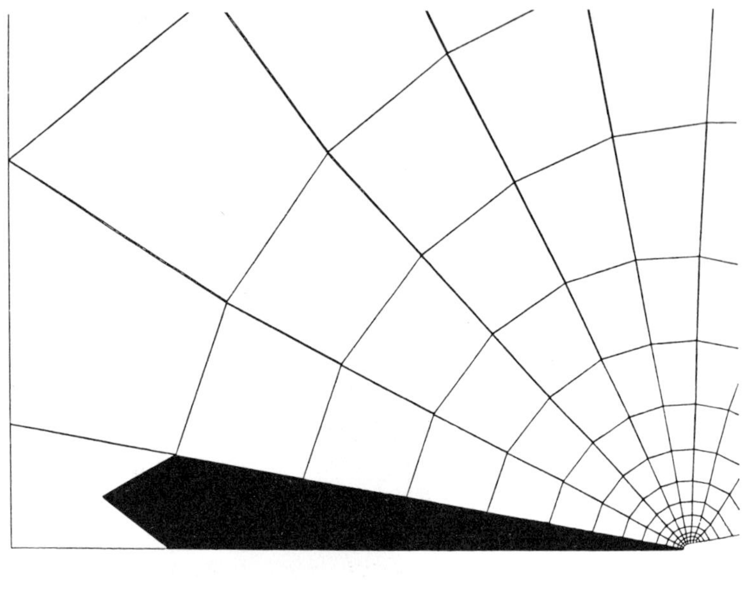

(a)

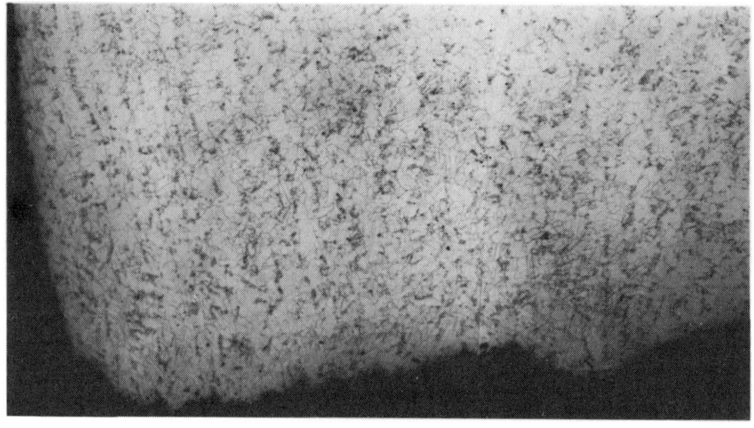

(b)

Fig.61.   Comparison of (a) computed failure path,
$\omega \gtrsim 0.999$, with (b) mid-thickness micrograph
taken from the minimum of an internally cracked
specimen in 316 stainless steel at failure.

for the externally cracked specimen, with σ_N = 341.33 MPa,
show that the plane of crack advancement deviates from the
plane of the minimum section to one which is inclined at an
angle to the x-axis.   Failure takes place by the region
between the current crack front and the y-axis behaving as a
relatively uniformly stressed region; the entire section
fails at the same instant in the manner described by Hayhurst,
Morrison and Brown (57).   This result is typical of theoretical predictions obtained for lower and slightly higher
levels of applied stress.   In the case of the internally
cracked specimens failure is predicted to take place in the
plane of the minimum section and this result is corroborated
by the mid-thickness micrograph given in Fig.61(b).   However, these results were obtained for a value of σ_N = 210.30
MPa and experimental results at higher stresses showed that
the plane of the propagating crack deviates slightly from the
plane of the minimum section as described by Hayhurst,
Morrison and Brown (57).   The theoretical value of Σ_R at
slightly higher stresses were not significantly affected.
Theoretical contour plots of damage and stress states are
now presented.

In Figs.62(a) and 62(b) computed contours of constant
damage are given for the internally and externally cracked
specimens at 60% of life, also shown in the figure is the
manner in which the initial cracks have propagated.   Damage
values ahead of the propagated crack are approximately
ω = 0.7.   In the case of the internally cracked specimen
there is a marked decrease in the value of damage as the
value of x increases.   For the externally cracked specimen
the situation is different; the level of damage across the
minimum section is relatively uniform, varying from ω = 0.807
at the crack to ω = 0.526 at x = 0 .   During the remainder
of the lifetime the crack propagates along a plane inclined
at an angle of $45°$ to the x-axis and failure occurs when the
remaining ligament fails as a uniformly stressed region.

In Fig.63 contours of constant normalised effective
stress, σ_e/σ_0 , are given for both specimens at the same
stage of the lifetime as in Fig.62.   For the internally
cracked specimen the values of σ_e/σ_0 are highest in a zone
between the crack-tip and the edge of the specimen, x = b ,
and above the x-axis.   In the externally cracked specimen
the effective stress is relatively uniform in the region of
the minimum section.   There is not a strong correlation
between the distribution of damage and that of σ_e/σ_0 since
316 stainless steel satisfies a mixed damage growth law,
equation (19), with α = 0.75.

In Fig.64 contours of constant normalised first stress
invariant J_1/σ_0 ($J_1 = \sigma_{ii}$) are given for both specimens at
the same stage in the lifetime as in Figs.62 and 63.   The

166

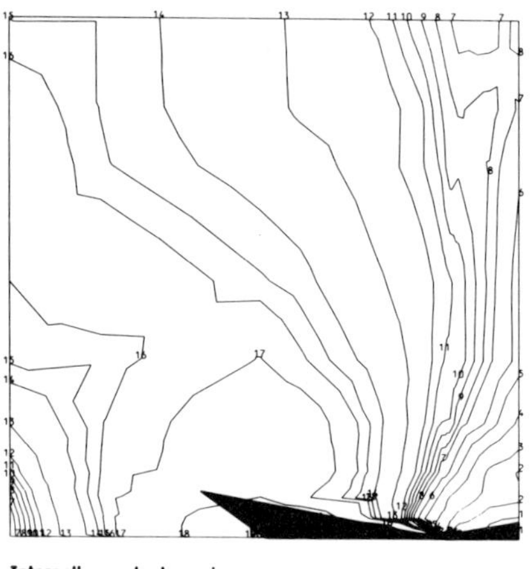

Contour no.	Height	Contour no.	Height
1	0·111	11	0·318
2	0·132	12	0·330
3	0·163	13	0·362
4	0·209	14	0·410
5	0·243	15	0·432
6	0·263	16	0·450
7	0·282	17	0·491
8	0·291	18	0·725
9	0·300	19	0·967
10	0·309	20	0·989

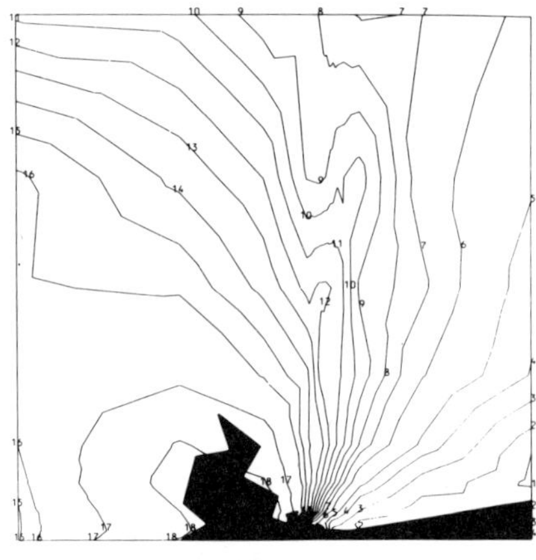

Contour no.	Height	Contour no.	Height
1	0·097	11	0·425
2	0·129	12	0·447
3	0·179	13	0·469
4	0·229	14	0·497
5	0·266	15	0·526
6	0·314	16	0·558
7	0·346	17	0·661
8	0·366	18	0·807
9	0·386	19	0·949
10	0·406	20	0·984

Fig. 62. Comparison of minimum cross-section plots of creep damage for internally and externally cracked specimens at $t/t_f = 0.6$ in 316 stainless steel.

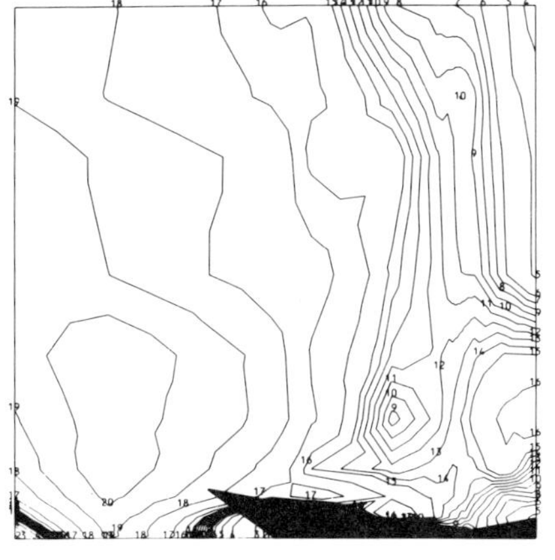

Internally cracked specimen
Base of square region defined by $b \geqslant x \geqslant 2b/9$

(a)

Contour no.	Height	Contact no.	Height
1	0·030	11	0·857
2	0·090	12	0·887
3	0·195	13	0·918
4	0·486	14	0·949
5	0·618	15	0·979
6	0·703	16	1·062
7	0·737	17	1·215
8	0·767	18	1·550
9	0·797	19	1·925
10	0·826	20	2·377

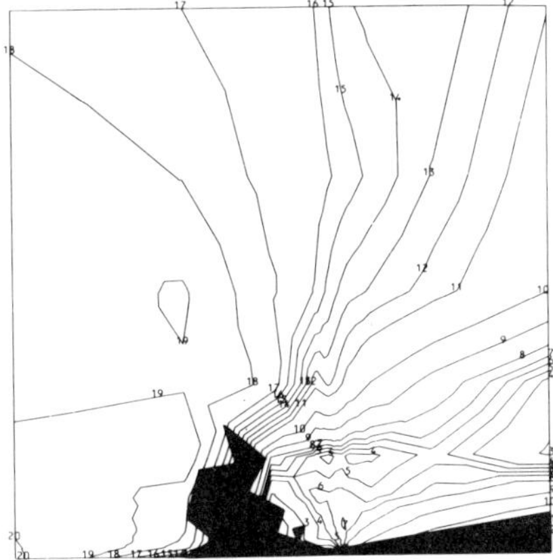

Externally cracked specimen
Base of square region defined by $5b/9 \geqslant x \geqslant 0$

(b)

Contour no.	Height	Contour no.	Height
1	0·046	11	1·159
2	0·139	12	1·298
3	0·279	13	1·437
4	0·390	14	1·576
5	0·432	15	1·735
6	0·475	16	1·908
7	0·518	17	2·131
8	0·582	18	2·418
9	0·721	19	2·747
10	0·902	20	3·560

Fig.63. Comparison of minimum cross-section plots of the normalised effective stress σ_e / σ_0 for internally and externally cracked specimens at $t/t_f = 0.6$ in 316 stainless steel.

168

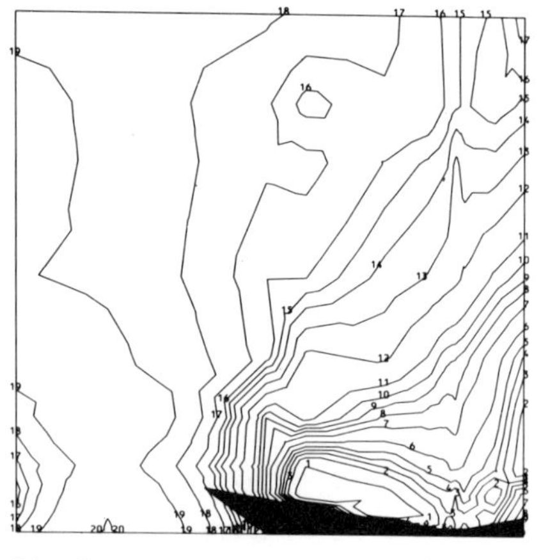

Internally cracked specimen

Base of square region defined by $b \geqslant x \geqslant 2b/9$

(a)

Contour no.	Height	Contour no.	Height
1	-1·621	11	0·097
2	-1·295	12	0·423
3	-1·139	13	0·686
4	-1·016	14	0·921
5	-0·892	15	1·092
6	-0·721	16	1·263
7	-0·482	17	1·607
8	-0·316	18	2·144
9	-0·192	19	3·140
10	-0·068	20	5·926

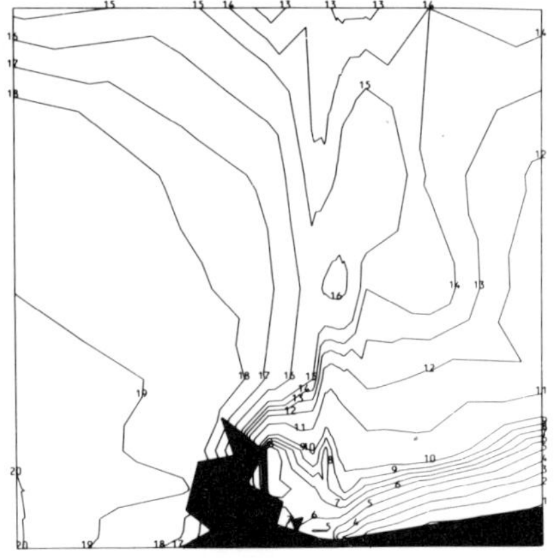

Externally cracked specimen

Base of square region defined by $5b/9 \geqslant x \geqslant 0$

(b)

Contour no.	Height	Contour no.	Height
1	-1·537	11	0·956
2	-1·294	12	1·442
3	-1·051	13	1·742
4	-0·820	14	1·986
5	-0·591	15	2·277
6	-0·365	16	2·942
7	-0·203	17	3·721
8	-0·041	18	4·591
9	0·121	19	7·939
10	0·321	20	10·920

Fig.64. Comparison of minimum cross-section plots
of the normalised first stress invariant J_1/σ_o
for internally and externally cracked specimens
at $t/t_f = 0.6$ in 316 stainless steel.

distributions are similar to those for the maximum principal tension stress and contours for the latter are not presented. In the case of the internally cracked specimen a region of high J_1/σ_0 occurs mid-way between the crack-tip and the edge of the specimen. This region coincides with a zone of high damage and hence the crack propagates into this region, since $\alpha = 0.75$, and the rupture behaviour is dominated by the maximum principal tension stress. In the case of the externally cracked specimen the highest value of J_1/σ_0 is ahead of the crack on a plane inclined at 45° to the x-axis, the region of highest damage is in the same region, cf. Fig.62(b), hence the crack propagates on a plane inclined at approximately 45° to the x-axis.

Observation of Figs.(63) and (64) show that there is no evidence of the stress field having singular characteristics of the type encountered under elastic and under stationary-state creep conditions. The effect of the formation of creep damage is to cause pronounced stress redistribution and the nullification of the stress singularities. The singular character of the strain fields is preserved for much longer than in the case of the stress fields and it is only when the initial crack has been effectively blunted that the strain fields loose their singular character.

3.8 Discussion

In this section an outline has been given of the numerical techniques which have been developed to solve high temperature structural creep problems. It has been shown how the results of stationary-state and of rupture calculations can be checked against results obtained from experiments carried out on model structures. In this way it is possible to check the design methodology of combining constitutive equations with numerical methods to predict the performance of structures. In addition it has been demonstrated how such techniques can be used, together with the theory of continuum damage mechanics, to predict the rupture behaviour and the lifetimes of a broad range of structures, which includes cracked components.

In the case of structures which contain modest stress concentrators, under plane stress conditions, it has been shown that the rôle of continuum damage is to cause considerable stress redistribution and to effectively blunt the region of the stress concentration.

In the axi-symmetrically notched bars, where complex stress-states are found the theory of continuum damage describes the two important features:

(i)   stress redistribution due to damage, and
(ii) the multi-axial stress-rupture criterion of the material.

It has been shown that a description of both these features is necessary to predict the effects of notch strengthening and weakening frequently encountered in notch situations.

In structural components which contain initially sharp cracks it has been shown that the theory of continuum damage is capable of describing the high gradients of strain, stress and damage which are found in close proximity to cracks and of describing the multi-axial stress rupture behaviour of the material in such regions. The approach which has been described here has the advantage that, unlike other theories, it is not necessary to pre-judge the direction of crack growth and to assume crack-tip stress fields which neglect both localised and widespread creep damage.

4.0   DISCUSSION AND CONCLUSIONS

In the first section it has been shown how metals subjected to homogeneous states of stress undergo grain boundary creep damage which is uniformly distributed throughout regions of homogeneous stress. This conclusion has been made from microstructural examinations of specimens tested under uni-axial and bi-axial tension, pure shear, tension and torsion and tri-axial tension stress states, and has led to the usage of the term Continuum Damage. The growth of continuum damage has been independently described by engineering scientists and by materials scientists; the former have used a phenomenological approach, which has been shown to be equivalent to a single damage state variable theory, and the latter have used either an empirical approach or mechanistic models. It has been shown how the approaches of the two groups are equivalent for non-proportional loading conditions. The results of tests show that when the loading conditions become non-proportional it is necessary to understand the directional characteristics of the nucleation and of the growth of grain boundary cavities.

If constitutive equations are to be used to predict the performance of structural components then it is vital that they accurately describe two features:

(i)   the growth of tertiary creep strains;
(ii)   the multi-axial stress rupture criterion of the material

In the second section a discussion has been presented of some of the difficulties which arise in the numerical solution of creep deformation and rupture problems, and it has been shown how numerical methods have been developed for this class of problem. Such numerical methods have been used together with constitutive equations to predict creep deformations and lifetimes in structural components.

It has been shown using simple beams how the growth of continuum damage takes place in a manner which reflects the field character of stress.

In the case of plane stress plates it has been shown how stress redistribution due to tertiary creep has the effect of blunting a geometrical stress concentrator. For axisymmetrically notched bars under tension the stress redistribution due to tertiary creep and the states of multi-axial stress cause different damage growth characteristics according to the multi-axial stress rupture criterion of the material. This material-structural interaction is responsible for the observed notch strengthening and notch weakening behaviour.

The theory of continuum damage has been used to study the growth of cracks due to creep in 316 stainless steel. It has been shown that the constitutive equations and the finite element method of numerical solution can be used to predict the direction of crack growth and to determine the component lifetime. Again the effect of continuum damage is to cause significant stress redistribution and to nullify the elastic and stationary state stress concentration factors. The advantage of using this method to study the growth of cracks due to creep is that no assumptions need be made about the crack growth mechanism; instead, constitutive equations are required which accurately model the material deformation and rupture behaviour at the appropriate stress levels and stress-states. The computational resources required when using this approach are not excessive and the method may become a viable design tool with the development of the next generation of computers.

Continuum damage mechanics is therefore capable of describing the deformation and rupture behaviour of a broad class of structures which range from simple beams to sharp cracks contained in plane strain tension bars.

5.0   REFERENCES

1.   KACHANOV, L.M.   'The Theory of Creep', (English translation edited by KENNEDY, A.J.), Chs. IX, X, National Lending Library, Boston Spa, 1960.

2.   HAYHURST, D.R.   'Isothermal Creep Deformation and Rupture of Structures', Ph.D. Thesis, Cambridge University, 1970.

3.   LECKIE, F.A., and HAYHURST, D.R.   'The Damage Concept in Creep Mechanics', Mech. Res. Comm., $\underline{2}$, 23, 1975.

4.   DYSON, B.F. and McLEAN, D.   'Creep of Nimonic 80 A in Torsion and in Tension', Met. Sci., $\underline{2}$, 37, 1977.

5.   GREENWOOD, G.   'Creep Life and Ductility', Int. Congress
     on Metals, Cambridge, 1973.   Microstructure and the
     Design of Alloys, 2, 91, 1973.

6.   DYSON, B.F.   'Constrained Cavity Growth, its use in
     Quantifying Recent Creep Fracture Experiments', Canadian
     Met. Quart. 18, 31, 1979.

7.   COCKS, A.C.F. and ASHBY, M.F.   'On Creep Fracture by
     Void Growth',   Prog. in Materials Science, 27, 3-4,
     189, 1982.

8.   GLEN, J., JOHNSON, R.F., MAY, M.J. and SWEETMAN, D.
     'Elevated-temperature Tensile Creep and Rupture Proper-
     ties of Various Carbon Steels', Proc. Joint Conf. on
     High-temperature Properties of Steels, British Iron and
     Steel Research Association and Iron and Steel Institute,
     Eastbourne, 1966.

9.   British Standards 1969 no.3500 'Methods for Creep
     Rupture Testing of Metals'.

10. HAYHURST, D.R.   'The Effects of Test Variable on Scatter
     in High-temperature Tensile Creep-Rupture Data', Int. J.
     Mech. Sci., 16, 829, 1974.

11. HOFF, N.J.   'The Necking and Rupture of Rods, Subjected
     to Constant Tensile Loads', J. Appl. Mech., 20, 105, 1953.

12. NEMY, A.S., and RHINES, F.N.   'On the Origin of Tertiary
     Creep in an Aluminium Alloy',   Trans. A.I.M.E. (Met.Soc.),
     215, 992, 1959.

13. BOETTNER, R.C. and ROBERTSON, W.D.   'Study of Growth of
     Voids in Copper During Creep Process by Measurement of
     Accompanying Change in Density', Trans. Metall. Soc.
     A.I.M.E., 221, 613, 1961.

14. HAYHURST, D.R.   'Creep Rupture under Multi-axial States
     of Stress', J. Mech. Phys. Solids, 20, 381, 1972.

15. JOHNSON, A.E., HENDERSON, J. and KHAN, B.   'Complex-
     Stress Creep, Relaxation and Fracture of Metallic Alloys',
     Ch.4, H.M.S.O., London 1962.

16. MÖNCH, E., and GALSTER, D.   'A Method for Producing a
     Defined Uniform Bi-axial Tensile Stress Field',
     British Jnl. Appl. Physics, 14, 810, 1963.

17. PENNY, R.K. and HAYHURST, D.R. 'The Deformations and Stresses in a Stretched Thin Plate Containing a Hole During Stress Redistribution Caused by Creep', Int. J. Mech. Sci., 11, 23, 1969.

18. HAYHURST, D.R. 'Stress Redistribution and Rupture Due to Creep in a Uniformly Stretched Thin Plate Containing a Circular Hole', J. Appl. Mech., 1, 40, 244, 1973.

19. HAYHURST,D.R. and STORÅKERS, B. 'Creep Rupture of the Andrade Shear Disk', Proc.R.Soc.London. A.,349, 369, 1976.

20. BRIDGMAN, P.W. 'Studies in Large Plastic Flow and Fracture', New York: McGraw-Hill, 1952.

21. HAYHURST, D.R., and HENDERSON, J.T. 'Creep Stress Redistribution in Notched Bars', Int. J. Mech. Sci. 19, 133, 1977.

22. HAYHURST, D.R., LECKIE, F.A. and MORRISON, C.J. 'Creep Rupture of Notched Bars', Proc. R. Soc. Lond. A. 360, 243, 1978.

23. HAYHURST, D.R., DIMMER, P.R. and MORRISON, C.J. 'Development of Continuum Damage in the Creep Rupture of Notched Bars', Leicester University, Department of Engineering, Report No.83-2, 1983.

24. DYSON, B.F. and LOVEDAY, M.S. 'Creep Fracture in Nimonic 80 A under Tri-axial Tensile Stressing', Creep in Structures, 1980, (I.U.T.A.M. Symposium, Leicester, U.K.) (ed. A.R.S.Ponter & D.R.Hayhurst) 406, Berlin: Springer-Verlag, 1981.

25. NEEDHAM, N.G. and GLADMAN, T. 'Effect of Stress State on the Processes Controlling Creep Fracture in a 2.25%Cr 1% M_O Steel', Proc. Int. Conf. on Eng. Aspects of Creep, 15-19 Sept. 1980, Sheffield, 1 paper C190/80, p.49, London: I.Mech.E. 1980.

26. HAYHURST, D.R., LECKIE, F.A. and HENDERSON, J.T. 'Design of Notched Bars for Creep-Rupture Testing under Tri-axial Stresses', Int. J. Mech. Sci., 19, 147, 1977.

27. ODQVIST, F.K.G. 'Mathematical Theory of Creep and Creep Rupture', (2nd Ed.), Ch.12, Oxford: Clarendon Press, 1974.

28. LECKIE, F.A. and PONTER, A.R.S. 'On the State Variable Description of Creeping Materials', Ing. Arch., 43, 158, 1974.

29. RABOTNOV, Y.N. 'Creep Problems in Structural Members', Amsterdam: North Holland Publishing Company 1969.

30. KACHANOV, L.M.   'Time of the Fracture Process under
     Creep Conditions', IZV. Akad. Nauk. SSSR O.T.N. Teckh.
     Nauk. $\underline{8}$, 26, 1958.

31. SDOBYREV, V.P.   'Long-term Strength of Alloy, EI-437B
     Under Complex Stresses', IZV. Akad Nauk. SSR, Otd.Tekh.
     Nauk. $\underline{4}$, 92, 1958.

32. TRAMPCYZNSKI, W.A., MORRISON, C.J. and TOPLISS, W.   'A
     Tension-Torsion Creep-Rupture Testing Machine', Jnl. of
     Strain Analysis, $\underline{15}$, 3, 151, 1980.

33. TRAMPCZYNSKI, W.A., HAYHURST, D.R. and LECKIE, F.A.
     'Creep Rupture of Copper and Aluminium under Non-Propor-
     tional Loading', J. Mech. Phys. Solids, $\underline{29}$, 5/6, 353, 1981.

34. ASHBY, M.F. and RAJ, M.   'Creep Fracture',   Proceedings
     of the Conference on Mechanics and Physics of Metals,
     Cambridge: The Metals Society, Institute of Physics, 1975.

35. LECKIE, F.A. and HAYHURST, D.R.   'Constitutive Equations
     for Creep Rupture', Acta Metallurgica, $\underline{25}$, 1059, 1977.

36. LECKIE, F.A. and HAYHURST, D.R.   'Creep Rupture of
     Structures', Proc. R. Soc. Lond. A., $\underline{340}$, 323, 1974.

37. DYSON, B.F. and RODGERS, M.J.   'Prestrain, Cavitation
     and Creep Ductility', Metal. Sci., $\underline{8}$, 261, 1974.

38. HAYHURST, D.R., TRAMPCZYNSKI, W.A. and LECKIE, F.A.
     'Creep Rupture Under Non-proportional Loading', Acta
     Metall., $\underline{28}$, 1171, 1980.

39. HAYHURST, D.R.   'A Time Hardening Transient Creep Solu-
     tion for Steadily Loaded Uni-axial Tension Panels
     Containing Circular and Elliptical Holes under Conditions
     of Plane Stress', Int. J. Mech. Sci., $\underline{14}$, 885, 1972.

40. HAYHURST, D.R. and KRZECZKOWSKI, A.J.   'Numerical Solu-
     tion of Creep Problems', Comp. Meth. in Appl. Mech. and
     Eng., $\underline{20}$, 151, 1979.

41. HAYHURST, D.R., DIMMER, P.R. and CHERNUKA, M.W.   'Esti-
     mates of the Creep Rupture Lifetime of Structures Using
     the Finite Element Method', J. Mech. Phys. Solids, $\underline{23}$,
     335, 1975.

42. HAYHURST, D.R., BROWN, P.R. and MORRISON, C.J.   'The
     Rôle of Continuum Damage in Creep Crack Growth', Leicester
     University, Department of Engineering, Report No.83-4,
     1983.

43. BUDIANSKY, B.   'A Reassessment of Deformation Theories of Plasticity', J. Appl. Mech., $\underline{81}$, 159, 1959.

44. BUDIANSKY, B. and MANGASARIAN, O.L.   'Plastic Stress Concentration at a Circular Hole in an Infinite Sheet Subjected to Equal Bi-axial Tension', J. Appl. Mech., $\underline{82}$, 59, 1960.

45. HAYHURST, D.R. and PENNY, R.K.   'A Long Narrow Foil Strain Gauge for Special Purposes',   Strain, $\underline{5}$, 1, 34, 1969.

46. HAYHURST, D.R.   'Estimates of the Creep Rupture Lives of Structures Subjected to Cyclic Loading', Int. J. Mech. Sci., $\underline{18}$, 75, 1976.

47. HAYHURST, D.R. and LECKIE, F.A.   'The Effect of Creep Constitutive and Damage Relationships Upon the Rupture Time of a Solid Circular Torsion Bar', J. Mech. Phys. Solids, $\underline{21}$, 431, 1973.

48. HAYHURST, D.R.   'The Prediction of Creep-Rupture Times of Rotating Discs Using Bi-axial Damage Relationships', J. Appl. Mech. $\underline{40}$, 915, 1973.

49. HAYHURST, D.R., MORRISON, C.J. and LECKIE, F.A.   'The Effect of Stress Concentrations on the Creep Rupture of Tension Panels', J. Appl. Mech., $\underline{42}$, 613, 1975.

50. CHUBB, E.J. and BOLTON, C.J.   'Stress State Dependence of Creep Deformation and Fracture in AISI Type 316 Stainless Steel', Proc. Int. Conf. on Eng. Aspects of Creep, 15-19 Sept. 1980, Sheffield, 1, Paper C201/80, p.39, London: I.Mech.E.

51. NEATE, G.J. and SIVERNS, M.J.   'The Application of Fracture Mechanics to Creep Crack Growth', Conference on Creep and Fatigue in Elev. Tem. applics.   Philadelphia: ASME/I.Mech.E. 1973.

52. RIEDEL, H.   'The Extension of a Macroscopic Crack at Elevated Temperature by the Growth and Coalescence of Microvoids', (IUTAM Symposium, Leicester, U.K.) ed. A.R.S.Ponter and D.R.Hayhurst, Berlin: Springer-Verlag 1981.

53. OHTANI, R.   'Finite Element Analysis and Experimental Investigation on Creep Crack Propagation', (IUTAM Symposium, Leicester, U.K.) ed. A.R.S.Ponter and D.R.Hayhurst, Berlin: Springer-Verlag, 1981.

54.   WEBSTER, G.A.   'The Application of Fracture Mechanics to Creep Cracking', Proc. Conf. Mech. and Phys. of Fracture, January (Cambridge, England), paper 18, London: Institute of Physics, 1975.

55.   RICE, J.R.   'A Path Independent Integral and the Approximate Analysis of Strain Concentration by Notches and Cracks', Jnl. Appl. Mech., $\underline{35}$, 379, 1968.

56.   AINSWORTH, R.A.   'Some Observations on Creep Crack Growth', Int. Jnl. of Fracture, $\underline{18}$, 26, 1982.

57.   HAYHURST, D.R., MORRISON, C.J. and BROWN, P.R.   'Creep Crack Growth', Creep in Structures, 1980 (IUTAM Symposium, Leicester, U.K.) ed. A.R.S.Ponter and D.R.Hayhurst, Berlin: Springer-Verlag, p.564, 1981.

58.   BROWN, P.R., and HAYHURST, D.R.   'Using the Schwarz-Christoffel Transformation in Mesh Generation for the Solution of Two Dimensional Problems', Computers in Mechanical Engineering, Pub: ASME, $\underline{1}$, 1, 73, 1982.

59.   HAYHURST, D.R. and BROWN, P.R.   'The Use of Finite Element Creep Solutions to Obtain J Integrals for Plane Strain Cracked Members', University of Leicester, Department of Engineering, Report No.83-1, 1983.

6.0 ACKNOWLEDGEMENT

The author acknowledges the provision of Fig.9 by Dr. D.A. Kelly.

Chapter 4

CONSTITUTIVE EQUATIONS IN CREEP-FRACTURE DAMAGE

J.L. CHABOCHE

OFFICE NATIONAL D'ETUDES ET DE RECHERCHES AEROSPATIALES
29 avenue de la Division Leclerc BP 72 92322 Châtillon Cedex

A class of viscoplastic constitutive equations is discussed for plastic behavior under monotonic and cyclic loadings, based on internal variable concepts and a thermodynamic framework. They include isotropic and non-linear kinematic hardening, inelastic strain memorization, recovery and possibilities of aging effects. Applications are presented for some refractory superalloys and the 316 stainless steel.

Damaging creep processes are then described by means of the Continuum Damage Mechanics, introducing damage "measures" and damage constitutive equations in the tensile case as well as in the three-dimensional general case. Different choices for taking into account the anisotropy of damage are discussed, as well as the influence of damage on the viscoplastic behavior introduced through the effective stress concept. Examples are given for the superalloy IN 100.

The concluding remarks explain the present limits of the approach and the future research needed to improve the macroscopic description of actual engineering materials, including anisotropic hardening and damage, out-of-phase loadings, aging effects, etc.

1. INTRODUCTION

Predicting the lifetime of a structure or, more precisely, the crack initiation, entails three stages :

- determination of the stress-strain constitutive laws
  for the materials under various time, temperature,
  work-hardening conditions, etc.

178

- Computation of the stresses and strains in the structure knowing the applied loads, the temperature field
  and their variations. This calculation is carried
  out in the framework of Continuum Mechanics, for
  example with the finite element technique, using the
  constitutive laws for the material.

- Prediction of the damage of the structure or, more
  precisely, of the time or number of cycles before
  macroscopic cracking initiates. This prediction uses
  the damage laws of the material and the stresses and
  strains calculated in the previous step.

For the engineer, then, the first schematization of the
behavior of the material consists of considering the strain
and damage processes differently. In Low-Cycle fatigue, for
example, the number of cycles before crack initiation is
calculated from the characteristics of the stabilized cycle,
most often at mid-life independently of the development of
damage.

The distinction between damage and strain is sometimes
a subtle one that can even be considered subjective. It involves the following considerations :

- The isolated strain process may involve irreversible
  strains (plastic strains and creep), but we may still
  return to the initial state (zero strain), e.g. by
  applying an opposite load. For the engineer, the strain
  is then only a change of shape.

- Of course the accumulation of strains, and in particular the microscopic localization of strains in slip
  bands (fatigue) or intergranular cavities (creep), leads
  to a progressive damaging of the material. The damage
  is distinguished from the strain by the fact that it
  is truly irreversible within the normal loading and
  temperature ranges. Only a very high temperature treatment at high pressure can suppress the cavities resulting from creep.

- The creep test (Fig. 1) clearly illustrates the succes-
  sion of these processes. Work-hardening during the
  initial creep phase (at decreasing rate), then the
  progressive appearance of the damage, an unstable process, occurring during the secondary and tertiary
  phases.

The strain and damage aspects of the response of the material are treated separately in this section. The conventional
constitutive equations are reviewed and more sophisticated models are developed to describe the cyclic effects as well as
the processes of time recovery, aging or various effects of

work-hardening memorization. The type 316 stainless steel is used as an example of these possibilities. The damage is defined in terms of effective stress. The generalization of these ideas to cover three-dimensional anisotropy is discussed in the framework of two theories.

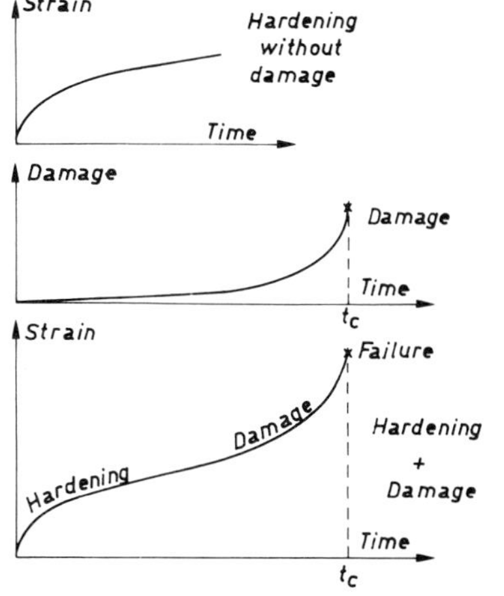

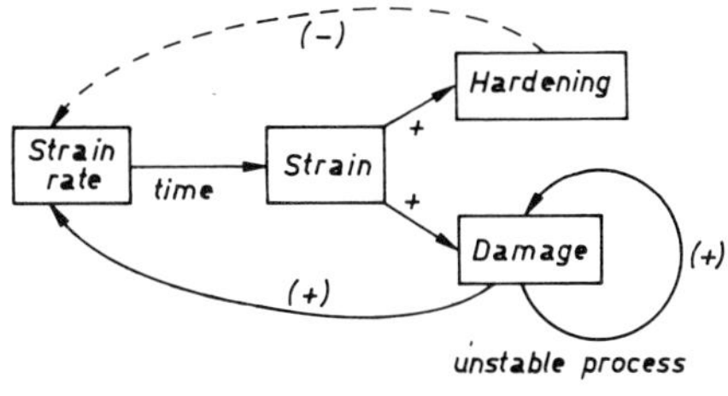

Fig. 1 - Schematic illustration of the superposition of hardening and damage processes in the creep test.

The models developed for both strain and damage fit into a general thermodynamic framework [1, 2, 3] that uses internal variables to express the irreversible processes. Let us summarize this theory briefly for small quasi-static transformations, denoting the temperature by T, the thermoelastic strain tensor by $\mathcal{E}_e$ and the family of internal variables, describing the present state of the element of volume, by α_j .

Two potentials are used : a thermodynamic potential such as the free energy :

$$\Psi = \Psi(\mathcal{E}_e,\ T,\ \alpha_j)\tag{1}$$

where the thermoelastic strain $\mathcal{E}_e$ comes from the partitioning of total strain into elastic and plastic strain : $\mathcal{E} = \mathcal{E}_e + \mathcal{E}_p$. The second potential governs the dissipation :

$$\varphi = \varphi(\dot{\mathcal{E}}_e,\ \dot{\alpha}_j\ ;\ \mathcal{E}_e,\ T,\ \alpha_j)\tag{2}$$

The variables $\mathcal{E}_e$, T and α_j affecting this potential are considered as parameters. The Clausius-Duhem inequality, which is an expression of the second principle of thermodynamics, leads to :

$$\sigma = \rho\,\frac{\partial\Psi}{\partial\mathcal{E}_e} \qquad \mathcal{S} = -\frac{\partial\Psi}{\partial T} \qquad A_j = \rho\,\frac{\partial\Psi}{\partial\alpha_j}\tag{3}$$

where σ is the stress tensor, $\mathcal{S}$ is the entropy, ρ the density and A_j the thermodynamic forces associated with the in-

180

ternal variables α_j . The dissipation is then expressed as :

$$\mathcal{D} = \sigma : \dot{\varepsilon}_p - A_j \, \dot{\alpha}_j - \frac{1}{T} \, \vec{q} \cdot \overrightarrow{\mathrm{grad}} \, T \geqslant 0 \tag{4}$$

This must be positive because of the second principle. This is automatically true if we adopt the framework of the normality hypothesis, generalized from the potential φ^*, obtained from φ by a Legendre-Fenschel transform on the variables $\dot{\varepsilon}_e$ and $\dot{\alpha}_j$:

$$\varphi^* = \varphi^* (\sigma , A_j \, ; \, \dot{\varepsilon}_e, \, T \, , \, \dot{\alpha}_j) \tag{5}$$

For a theory with time (or rate) effects, the generalized normality is expressed as :

$$\dot{\varepsilon}_p = \frac{\partial \varphi^*}{\partial \sigma} \qquad \dot{\alpha}_j = - \frac{\partial \varphi^*}{\partial A_j} \tag{6}$$

It can be seen that the intrinsic dissipation is necessarily positive if φ^* is convex, positive and zero at the origin $(\sigma = A_j = 0)$:

$$\mathcal{D}_i = \sigma : \dot{\varepsilon}_p - A_j \dot{\alpha}_j = \sigma : \frac{\partial \varphi^*}{\partial \sigma} + A_j \frac{\partial \varphi^*}{\partial A_j} \geqslant A_j \geqslant 0 \tag{7}$$

This theory is given here only for reference. The models developed in the following paragraphs are compatible with these hypotheses but, for the sake of simplicity, the proofs will not be given.

2. MODELLING OF THE STRESS-STRAIN BEHAVIOR

2.1 Strain-hardening

Inelastic strain in a material almost always causes a work-hardening effect. This can be perceived microscopically, such as after a plastic flow under tensile stress. The apparent elastic limit is increased and the material exhibits greater strength against subsequent flow. The decrease in the strain rate during the primary phase of the creep test is also a result of work-hardening. An increase in the inelastic strain can thus be associated with an increase in the density of the dislocations, which become less mobile as they pile up on obstacles or form cells.

A natural hardening parameter is thus the cumulative plastic strain p, defined by :

$$\dot{p} = \sqrt{\frac{2}{3} \, \dot{\varepsilon}_{p_{ij}} \, \dot{\varepsilon}_{p_{ij}}} \tag{8}$$

The hypothesis of strain-hardening (SH) is justified in comparison with the hypothesis of time-hardening (TH) by observations in creep tests at two levels. The figure 2 schematically indicates the creep at the second level σ_2 under the two hypotheses, when $\sigma_2 > \sigma_1$. The experimental data [4, 5] match the first hypothesis most frequently, as the creep at the second level is rather more rapid than for the SH curve.

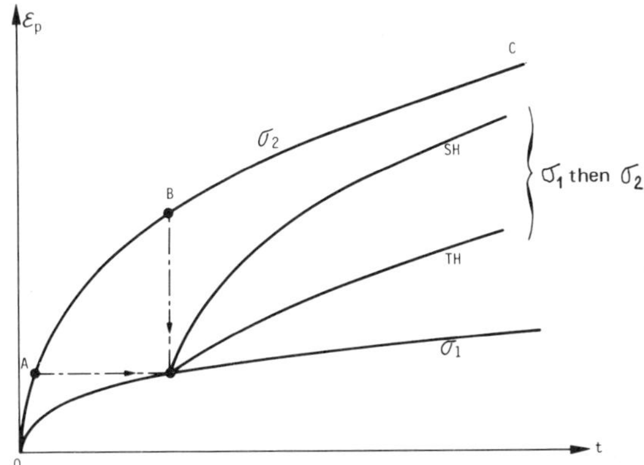

Fig. 2 - Time- and strain-hardening, visualized by two-level creep tests (schematic)

These basic observations have often led to the hypothesis of a one-dimensional equation of state expressed as :

$$F(\sigma, \varepsilon_p, \dot{\varepsilon}_p) = 0 \tag{9}$$

A product of power functions yields good results over a large domain in the primary creep as well as in the tensile and relaxation tests [4, 6].

$$\sigma = K \varepsilon_p^{1/m} \dot{\varepsilon}_p^{1/n} \tag{10}$$

This three-coefficient relationship can be extended easily to the three-dimensional case using an isotropic hardening hypothesis. With the Von-Mises criterion, we let

$$J(\sigma) = \sqrt{\frac{3}{2} \sigma : \sigma} = K\, p^{1/m}\, \dot{p}^{1/n} \tag{11}$$

where $\dot{p}$ is determined from (8) and $\sigma' = \sigma - \frac{1}{3} T_r(\sigma)\mathbb{1}$ is the deviator of the stress tensor. The viscoplastic strains take place under zero volume change, which gives, under normality hypothesis :

$$\dot{\varepsilon}_p = \frac{3}{2}\, \dot{p}\, \frac{\sigma'}{J_2(\sigma)} \tag{12}$$

The elementary theory of viscoplasticity with isotropic hardening is valid for noncyclic quasi-proportional loading. Under other conditions, internal stresses must be introduced.

2.2 <u>The different internal stresses</u>

The idea of internal stress is not a new one. It was introduced in the thirties by Orowan to provide a better explanation of the observed microscopic behavior, particularly in polycrystals and in materials containing several phases. In these materials, the exponent N of the secondary creep law for example (Norton's law) is very high :

$$\dot{\mathcal{E}}_p = (\sigma/\Lambda)^N \tag{13}$$

Introducing an internal stress σ_i makes it possible, among other things, to reduce this exponent in a large measure, cutting it down to 2, 3 or 4 as predicted by the physical theory. It can be measured through the secondary phase of creep by successive step unloadings [7, 8, 9]. We have :

$$\dot{\mathcal{E}}_p = \left(\frac{\sigma - \sigma_i}{K}\right)^n \tag{14}$$

The internal stress can be scalar or tensorial depending on the type of interaction considered. Four types of internal stresses can be introduced at a microscopic level. These correspond to an additive decomposition of the applied stress. The three-dimensional equation would be written :

$$J\left(\sigma - X\right) - k - R - R^* - K\,\dot{p}^{1/n} = 0 \tag{15}$$

where J is the chosen scalar invariant, J_2 for example. For a pure tensile load, this would be simplified to read :

$$\sigma = X + k + R + R^* + K\,\dot{p}^{1/n} = \sigma_i + K\,\dot{p}^{1/n} \tag{16}$$

where :

- X is a second order tensor, called the kinematic stress, back stress or rest stress, expressing interactions at long distances : intergranular stresses, caused by plastic strains that are nonhomogenous from one grain to the next, interactions between dislocations and precipitates, as shown by precise calculations on the scale of the precipitates [10].

- k is the isotropic (or scalar) Orowan stress corresponding to the initial elastic limit of the material and depending, among other things, on the volume fraction of precipitates and on the initial density of the dislocations. The value of this flow threshold k also depends on the size of the precipitates, because of

the different mechanisms (bypass or shearing) by which
the dislocations cross the precipitates. Let us note
that a microscopic model [11, 12] has been developed
to reflect the variations in these internal stresses
caused by changes of temperature (partial dissolution
of the medium-size precipitates, followed by repreci-
pitation of a finer size phase). These models, succes-
fully applied to the alloy IN100, will not be treated
in this section.

- R is the Orowan stress variation caused by the plastic
  deformation. It is directly related to the increase
  in the density of dislocations, but may also depend
  on the configuration of the dislocations -such as the
  creation of dislocation cells, the size and wall thickness of these cells, etc.

- R^{*}, sometimes called R_{sol} [13] or the drag stress,
  describes the hardening by the atoms or particles in
  solution that reduce the dislocation velocity by a
  drag effect.

- The last term, $\sigma_{v} = K \, \dot{p}^{1/n}$ is the viscous stress itself
  (viscous friction). A power function may be used as a
  first approximation of this term. Of course the equation (15) can be rewritten in the usual form[*]:

$$\dot{p} = \left\langle \frac{J(\sigma - X) - k - R - R^{*}}{K} \right\rangle^{n} \qquad (17)$$

Many models have been developed to describe the variations in the internal stresses , R, R . Certain of these
are simple, such as the model used by Lagneborg [14] for the
internal stress R, while other are more complexe [13, 15, 16,
17]. In the following paragraphs, we will cover the non-linear
hardening models based on a principle of evanescent strain
path memory

2.3 <u>Law of viscoplasticity - limit case of plasticity</u>

The viscoplasticity equations are deduced from a viscoplastic potential of the form [18] :

$$\Omega = \frac{K}{n+1} \left\langle \frac{J(\sigma - X) - k - R - R^{*}}{K} \right\rangle^{n+1} \qquad (18)$$

Using the expression (17) for the cumulative strain rate $\dot{p}$,
we find through the normality hypothesis :

(*) The brackets $\langle \rangle$ designate the positive part. $\langle u \rangle = u$ if $u \geqslant 0$
and $\langle u \rangle = 0$ if $u < 0$.

$$\dot{\mathcal{E}}_p = \frac{\partial \Omega}{\partial \sigma} = \frac{3}{2} \, \dot{p} \, \frac{\sigma' - X'}{J(\sigma - X)} \tag{19}$$

Here $J(\sigma - X)$ designates for example the second invariant of the deviator $\sigma' - X'$ of $\sigma - X$ for a Von-Mises type material.

This formulation uses an elastic limit given by $k_e = k + R + R^*$. The figure 3 shows schematically the present range of elasticity, centered on X , of radius k_e, in the deviator plane of the stress space. The surfaces of equal dissipation (or of equal plastic strain rate $\dot{p}$) are deduced homothetically.

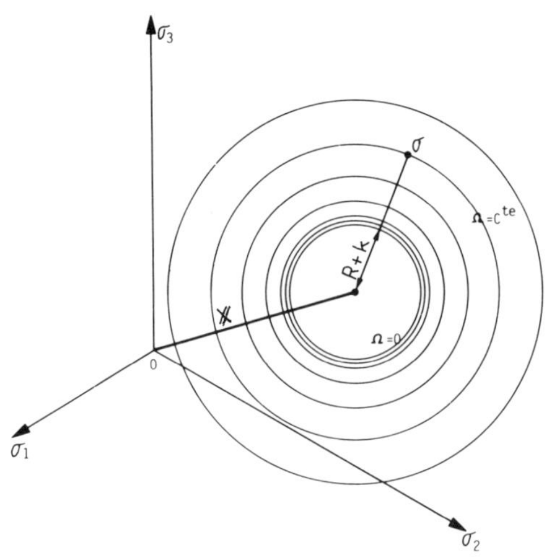

Fig. 3 - Elastic domain and viscoplastic equipotentials

In the limiting case of a very slightly viscous material (with n very large) the theory must be replaced with a time-independent theory of plasticity that we will not develop here. It brings in a threshold surface, a load-unload criterion and a plastic multiplier determined by the consistency condition. Let us simply note that all of the equidissipation surfaces are then identical to the elastic limit surface.

2.4 Non-linear kinematic hardening

First let $R = R^* = 0$. The only hardening effect is kinematic hardening corresponding to a translational movement of the surfaces in the stress space. This hardening effect is preponderant in many actual situations [19] and is essential in order to correctly describe the cyclic behavior. The simplest kinematic model is the linear one developed by Prager [20], which consists of assuming :

$$\cancel{X} = \frac{2}{3} \ c \ \mathcal{E}_p \qquad \text{or} \qquad d\cancel{X} = \frac{2}{3} \ c \ d\mathcal{E}_p \qquad (20)$$

This linear kinematic model has two disadvantages. Firstly, the shape of the stress-strain relation is badly expressed in the limiting case of time-independent plasticity. Secondly, the ratchetting (at a given stress) or relaxation effects of the mean stress (at a given cyclic strain) cannot be described by this model. The figure 4a shows that stabilization starts at the first cycle.

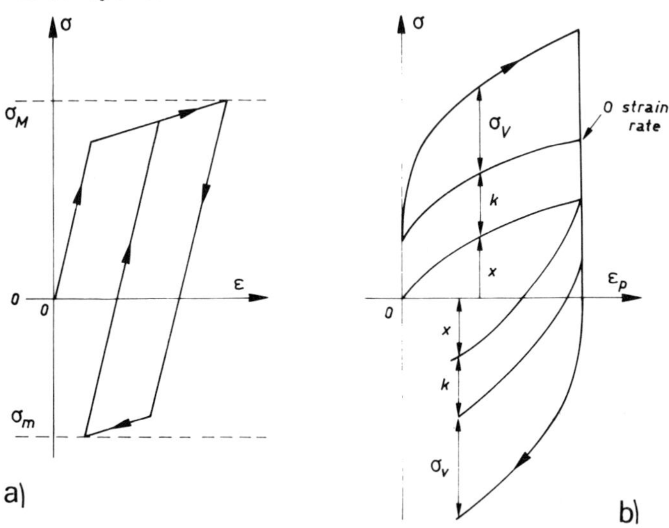

Fig. 4 - A) Immediate accomodation of the linear kinematic model (time-independent scheme)

- B) Non linear kinematic model : stress decomposition

A major improvement can be made by introducing an evanescent plastic strain path memory effect with a recall term [21, 16, 18] :

$$d\cancel{X} = \frac{2}{3} \ c \ a \ d\mathcal{E}_p - c \ \cancel{X} \ dp \qquad\qquad \cancel{X}(0) = 0 \qquad (21)$$

Let us note clearly the essential difference between the two terms of plastic strain $d\mathcal{E}_p$ and dp. Under tension-compression, for example, we have

$$dX = c \ a \ d\mathcal{E}_p - c \ X \ \left| d\mathcal{E}_p \right| \qquad (22)$$

The non-linearity introduced by the recall term is thus not the same during a flow under tension as it is under compression. The relation between X and $\mathcal{E}_p$ is non-unique and the concavity of the stress-strain curves is correctly reproduced even in the limiting case of plasticity (Fig. 4b). During each half-cycle, begining at $\mathcal{E}_{p_o}$, X_o the kinematic model (22)

186
is explicitly integrated as

$$X(\mathcal{E}_p) = \nu a + (X_0 - \nu a)\exp(-c(\mathcal{E}_p - \mathcal{E}_{p_0})) \qquad (23)$$

and the stress is then expressed as

$$\sigma = X(\mathcal{E}_p) + \nu k + \nu K \dot{p}^{1/n} \qquad (24)$$

where $\nu = \pm 1$ indicates the direction of the plastic flow.
The figure 4b schematically shows this stress decomposition.

The kinematic model gives the shape of the hysteresis
loops, but also gives the relations between the amplitudes in
the stabilized cycle (the stabilization here is very fast for
symmetrical periodic loading). The cyclic curves are expressed
as :

$$\frac{\Delta\sigma}{2} = a \; \mathrm{Th}(c\,\frac{\Delta\mathcal{E}_p}{2}) + k + K \dot{p}^{1/n} \qquad (25)$$

For the viscoplastic material, these depend on the strain rate.

This model of viscoplasticity, which uses only five temperature-dependent coefficients n, K, k, a, c, already yields
a very good approximation of the cyclic behavior, as shown by
the case of the alloy IN100 in the figures 5 and 6. Over a
rather limited range of strain (< 0.5 %) the model correctly
reproduces the cyclic curves, the hysteresis loops and the
effect of the loading rate, or of the hold time in the cyclic
creep tests.

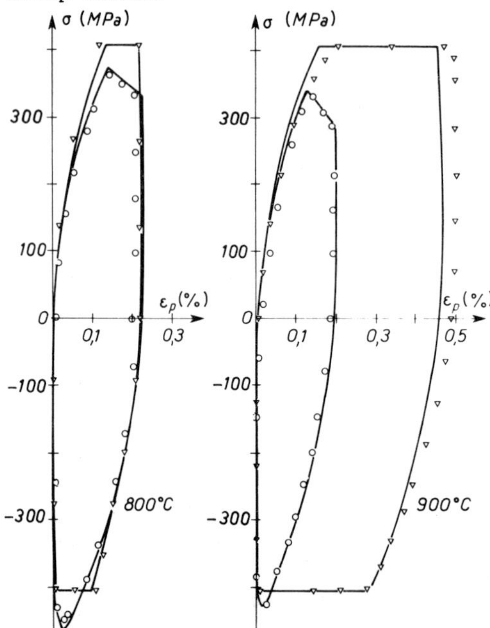

Fig. 5 – Application of
the model of cyclic vis-
coplasticity to the
alloy IN100 : stabilized
hysteresis loops.

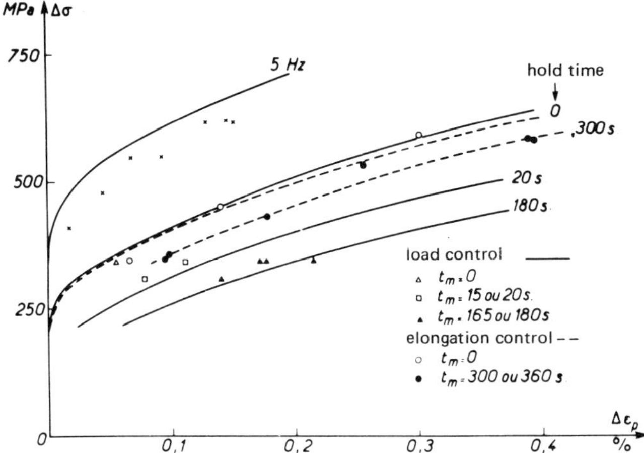

Fig. 6 - Modelling of the cyclic curves for the alloy IN100 at 1000°C (model with five coefficients : n = 8, K = 443, a = 140, C = 690 and k = 10)

Let us note that the non-linear kinematic model allows cycle stabilization only if the loading is symmetrical (at zero average stress). It thus describes qualitatively the ratchetting effect (non zero σ_{moy}) just as well in tension-compression as, for example, under cyclic torsion superposed to a constant tensile stress, and the mean stress relaxation effects under controlled-strain loading (non zero ε_{moy}).

The strain domain of validitiy is widened, and the quantitative description of the ratchetting effects is improved, when several models of the same type are superposed as follows [22] :

$$X = \sum_{i=1}^{m} X_i$$

$$dX_i = \frac{2}{3} c_i a_i d\varepsilon_p - c_i X_i \, dp$$

(26)

Fig. 7 - Plastic stress-strain curve obtained by superposition of three kinematic variables

One of these models can be linear, for example :

$$d\mathbf{X}_m = \frac{2}{3} \, c \, d\mathcal{E}_p \tag{27}$$

The integration of the hardening model remains explicit under proportional loading. The figure 7 illustrates the modelling possibilities.

2.5 Isotropic hardening

The isotropic hardening allows a better description of the monotonic hardening curve, but especially the cyclic variations, i.e. cyclic hardening or softening. A simpler form, similar to the kinematic law, is :

$$dR = b \left(R_s - R \right) dp \qquad R(0) = 0 \tag{28}$$

This internal stress increases or decreases as a function of the cumulative plastic strain p. After a certain number of cycles, the number of which is inversely proportional to the strain range, it stabilizes at the value R_S (necessarily, or else the only stabilized cycle possible would be elastic). We integrate to get :

$$R(p) = R_s \left(1 - \exp(-bp) \right) \tag{29}$$

The coefficients b and R_S are temperature-dependent. Under tension-compression, the stress is now expressed in the form

$$\sigma = X(\mathcal{E}_p) + \nu k + \nu R(p) + \nu K \, \dot{p}^{1/n} \tag{30}$$

For a given strain range (reversed conditions for simplicity), we have at each cycle

$$\sigma_M = X_M(\Delta\mathcal{E}_p) + k + R(p) + K \, \dot{\mathcal{E}}_{pM}^{1/n} \tag{31}$$

where $\dot{\mathcal{E}}_{pM}$ is the plastic strain rate for the maximum stress. In this loading case, $\dot{\mathcal{E}}_{pM}$ is approximately the same as the applied total strain rate. In any case, this quantity can be considered approximately the same at each cycle. Furthermore, X_M is very close to $\alpha \, Th \, (c \, \Delta\mathcal{E}_p/2)$. By applying relation (31) at each cycle, at the stabilized cycle and at the first cycle, considering the above approximations, we have :

$$\frac{\sigma_M - \sigma_{M_0}}{\sigma_{M_s} - \sigma_{M_0}} \simeq \frac{R}{R_s} = 1 - \exp(-bp) \simeq 1 - \exp(-2b \, \Delta\mathcal{E}_p \, N) \tag{32}$$

where σ_{M_s} and σ_{M_0} are the peak stresses in the stabilized cycle and in the first cycle. The figure 8 shows, for the 316 stainless steel (23), that the variation in the above quantity depends on the cumulative plastic strain, independently of the strain range, and that the relation (32) provides a good approximation.

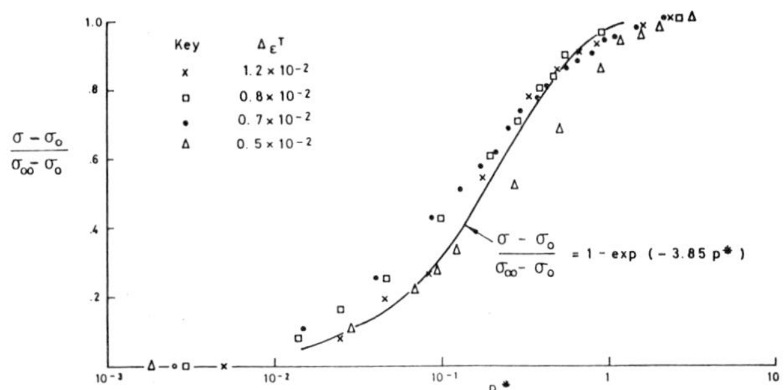

Fig. 8 - Cyclic hardening model : steel 316 at 600°C, from Goodall

Let us note that the only hardening variable that is not constant in stabilized cyclic conditions is the kinematic variable. The figure 9 shows then why a creep test, carried out on the stabilized cycle during the loading (A), causes much greater creep than if the test is carried out during the unloading (B), as this was observed on the 316 steel [23] and on the Hastelloy alloy [24].

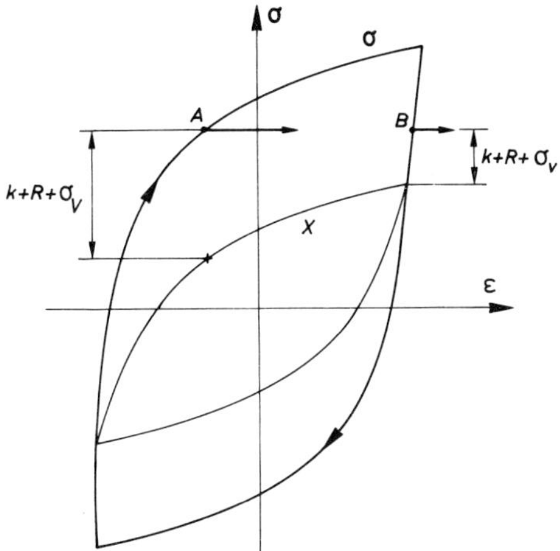

Fig. 9 - Observation by creep tests of various states of hardening during the stabilized cycle (schematic)

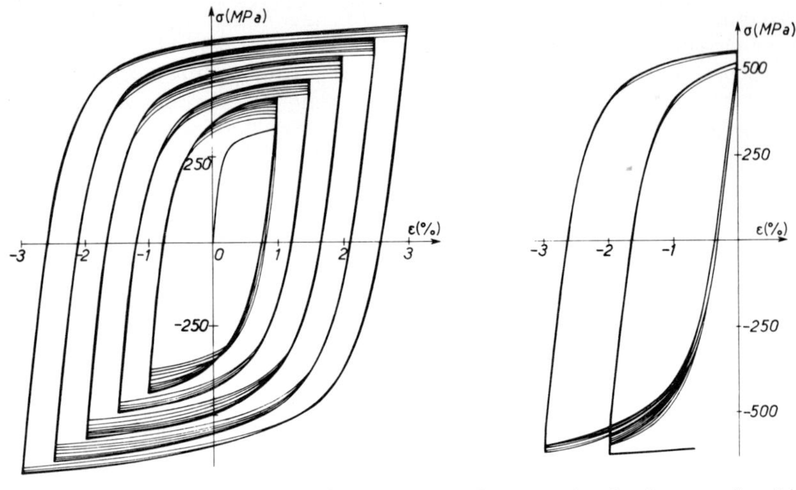

a — Test with increasing levels.　　　　b — Experimental cycles after precycling of ± 3%.

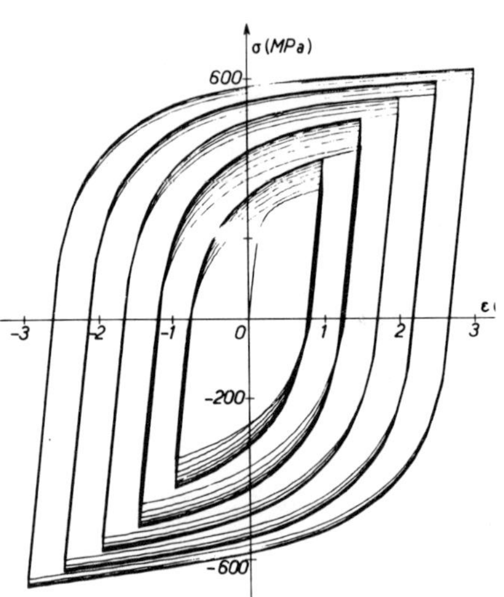

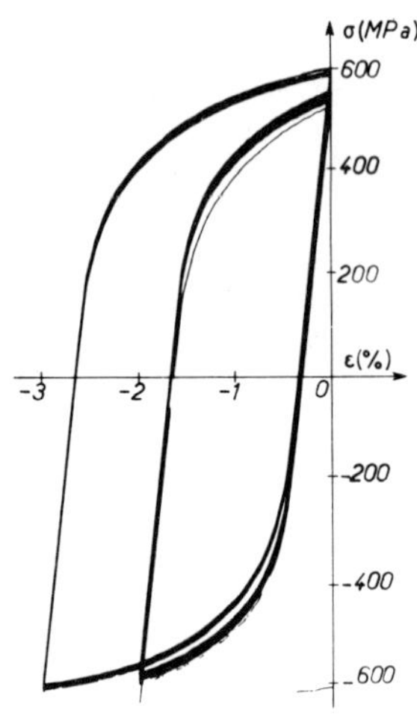

c — Calculation of cycles «a».　　　　d — Calculation of cycles «b».

Fig. 10 - Controlled-strain cyclic tests on the steel 316L
at 20°C : experiments and models

2.6 Maximum plastic strain memorization

The isotropic hardening thus describes the cyclic hardening, but a single value R_S of the variable R is obtained
under stabilized conditions regardless of the strain history.
Experiments show that this single value is not obtained, at
least for certain materials. For the steel 316 L, for example,
at ambient temperature :

- After saturation of cyclic hardening ($R = R_{s1}$) at a given strain level ($\pm$ 1 %, for example), an increase in the amplitude of the cyclic straining ($\pm$ 1.5 %) increases the cyclic hardening (Fig. 10 a) for several cycles until saturation is reached again. This can be manifested only if $R_{s1.5} > R_{s1}$.

- After cycling at the highest level ($\pm$ 3 %) the hardening obtained (R_{s3}) is at least partially memorized. The stabilized cycle at the $\pm$ 1 % level has a much higher stress range than when there is no precycling (compare the figures 10 a and 10 b).

- The cyclic curves (relations between ΔJ and $\Delta \mathcal{E}$) differe greatly, depending on whether a new test specimen is used for each level or if an incremental method is used (Fig. 11).

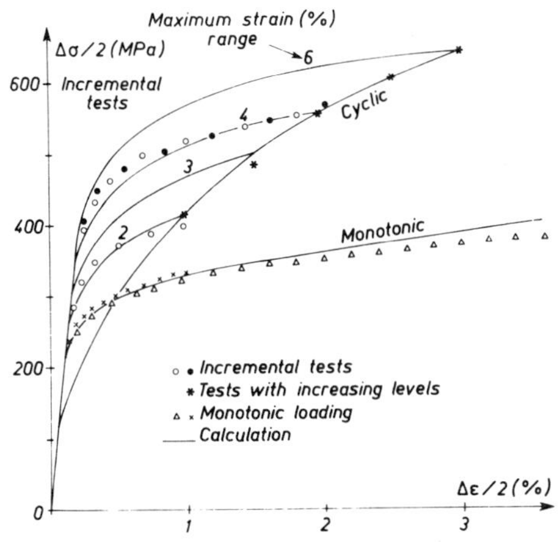

Fig. 11 - Monotonic and cyclic curves for the steel 316L at 20°C : experiments and models

A model was proposed to express the observed dependence between R_S and the maximum plastic strain amplitude [25]. This model consists of introducing a new hardening variable which memorizes the amplitude of the maximum plastic strain. On the physical level, microstructural analyses show that this variable can be associated with the presence of dislocation cells, and expresses both their size and the thickness of the walls [26].

The general formulation consists of introducing a non-hardening surface in the plastic strain space (Fig. 12) :

$$F = \frac{2}{3} J(\mathcal{E}_p - \zeta) - q \leqslant 0 \qquad (33)$$

192

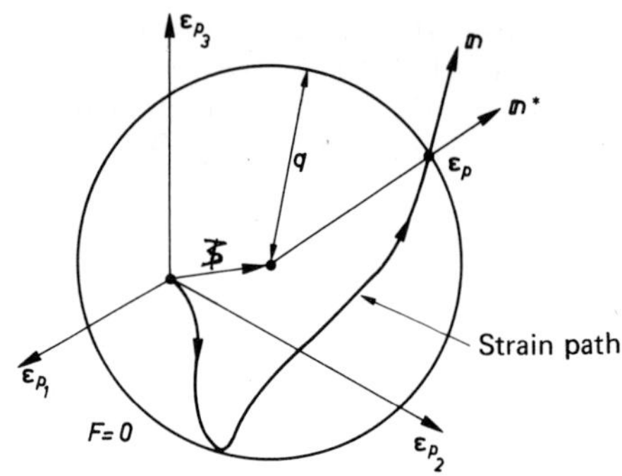

Fig. 12 - Definition of variables describing the maximum
plastic strain range memorization

As long as the strain remains within this surface, we assume
that dq = 0. Generally we let (*) :

$$dq = \frac{1}{2}\, H(F)\ (\mathbf{n}:\mathbf{n}^*)\, dp \tag{34}$$

where $\mathbf{n}$ and $\mathbf{n}^*$ are respectively the unit normals to equi-
potential Ω and surface F = 0, at σ and $\mathcal{E}_P$, in stress
and plastic strain spaces. The consistency condition (dF = 0
if F = 0) associated with the threshold surface F $\leqslant$ 0 determines the equations defining the variation of ξ :

$$d\xi = \frac{1}{2}\, H(F)\ (\mathbf{n}:\mathbf{n}^*)\, \mathbf{n}^*\, dp \tag{35}$$

The cyclic hardening (isotropic hardening) law is then
modified to take the value of q into account. This is expressed for example, as :

$$dR = b\, (\, Q - R\,)\, dp \tag{36}$$

$$dQ = 2\mu(\, A - Q\,)\, dq = \mu\, H(F)\, (\, A - Q\,)\, (\mathbf{n}:\mathbf{n}^*)\, dp \tag{37}$$

For tension-compression, q is $\Delta\mathcal{E}_{p\,max}/2$, the maximum plastic strain range. The relation (37) integrates and the value
at cyclic hardening saturation becomes :

$$Q(q) = Q\left(\frac{\Delta\mathcal{E}_p}{2}{}_{max}\right) = A + (\, Q_0 - A\,)\, \exp(-\mu\Delta\mathcal{E}_p) \tag{38}$$

This complete memorization model expresses rather well
the effects observed on the 316 steel. Associated with the
non-linear kinematic hardening, it can be used to reproduce

(*) H is the Heaviside function: H(u)=0 if u<0 ; H(u)=1 if u$\geqslant$0

at the same time the monotonic tension curve (Fig. 11), the
successive cycles with several levels of saturation (Figs. 10 a
and 10 c), the memorization at the $\pm$ 1 % level of the effect
induced by the cycling at $\pm$ 3 % (Figs. 10 b and 10 d), the
normal cyclic curve obtained with one test per level, and the
differences between cyclic curves obtained by incremental
methods (Fig. 11), in particular, the effect of the maximum
amplitude of these particular cyclic loadings. (The calculated
curves were obtained under the limiting case of time indepen-
dant plasticity).

A modification suggested by Ohno [27] introduces a certain progressivity in the memory effect ; a single cycle is
thus differentiated from several successive cycles at the
same level. This modification consists of replacing (34) by :

$$dq = \eta \, H(F) \, (\underline{n} : \underline{n}^*) \, dp \qquad (39)$$

giving a value between 0 and 0.5 (η = 0.08, for example).
Certain cyclings are improved, in particular when the mean
strain is changed together with the strain range.

The memorization thus expressed is perfect insofar as it
cannot diminish. After monotonic or cyclic hardening, testing
at a lower strain level shows a slow evanescence manifested
over a large number of cycles. The case of cold-worked materials undergoing cyclic softening effects is well-known. To
express such an evanescence the preceeding model can be modified to :

$$dq = \eta \, H(F) \, (\underline{n} : \underline{n}^*) \, dp - \xi \, q \, dp \qquad (40)$$

The results over several cycles will not be greatly affected
if the coefficient ξ is small enough in comparison with η .
On the other hand, if H(F) = 0 over a large number of cycles,
the memorization will gradually evanesce.

2.7 <u>Time recovery</u>

Recovery effects appear at high temperature. By recovery,
we mean a time induced decrease in the hardening effects induced by prior strain. By dislocations climbing, for example,
the thermal agitation induces a progressive annihilation of
the defects. The effects of recrystallization can also be
observed. On the microscopic level, not only a partial recovery
of strain is observed, but also a decrease in the level of
work-hardening.

The non-linear kinematic model can express a partial
recovery of the viscoplastic strain, at least if the threshold
k + R is not too high. In effect, the strain rate at zero
stress may be negative : according to (17, 18), after a tensile creep test and unloading, we have :

194

$$\dot{\mathcal{E}}_p = \left\langle \frac{|-X| - k - R}{K} \right\rangle^n \; \text{Sign}(-X) \tag{41}$$

The recovery -even partial- of the initial structure,
i.e. the decrease in the work-hardening, in particular the
isotropic hardening, is possible only by introducing a recovery term in the equation defining the variation of the internal variables. We would have for example :

$$d X = \frac{2}{3} \; c \; a \; d\mathcal{E}_p \; - c X \, dp \; - \Gamma'(X) \, X \, dt \tag{42}$$

$$dR = b \, (\, Q - R \,) \, dp \; - \gamma \, (\, R - Q_r \,) \, dt \tag{43}$$

where the function Γ and the coefficients γ and Q_r are
highly dependent on the temperature.

The recovery effects have major consequences on the
response of the materials in time-dependent loadings. For
example, they lead to a non-unique internal stress in the
secondary creep stage. Considering only the kinematic part,
the variation of which is preponderant during a monotonic
loading, the internal stress X_s in the secondary phase of
tensile creep is such that :

$$c \, (a - X_s) \left\langle \frac{\sigma - X_s - k}{K} \right\rangle^n \; - \Gamma'(X_s) \, X_s \; = 0 \tag{44}$$

The figure 13 shows the dependence between the stabilized
value of the internal stress X_s and the external applied
stress. Measurements of internal stress by partial unloading
indicates similar tendencies [7].

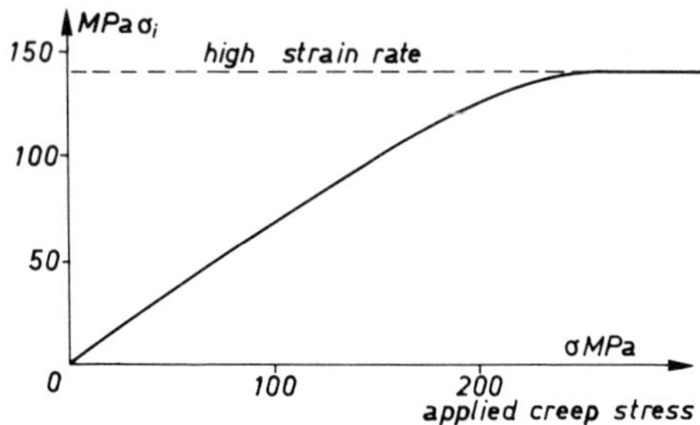

Fig. 13 - Stabilized internal stress in secondary creep
(nonlinear kinematic model)

The time recovery terms also make it possible to express
the effects observed in creep at two levels : acceleration at
the second level when $\sigma_2 > \sigma_1$ and creep delay when $\sigma_2 < \sigma_1$.
The figure 14 illustrates the test at increasing levels. At

the beginning of the second level, the internal stress is
$\sigma_{i_1} = X_1 + R_1 + k$ instead of the normal value at the level
$\sigma_{i_2} = X_2 + R_2 + k$. Since $\sigma_{i_1} < \sigma_{i_2}$, the start of creep at the
second level occurs at a higher rate than that predicted by
the strain hardening hypothesis (model without recovery).
This matches much experimental data [4, 6]. The reverse effect
occurs when $\sigma_2 < \sigma_1$: after a stress decrease, the strain rate
is lower, or even 0.

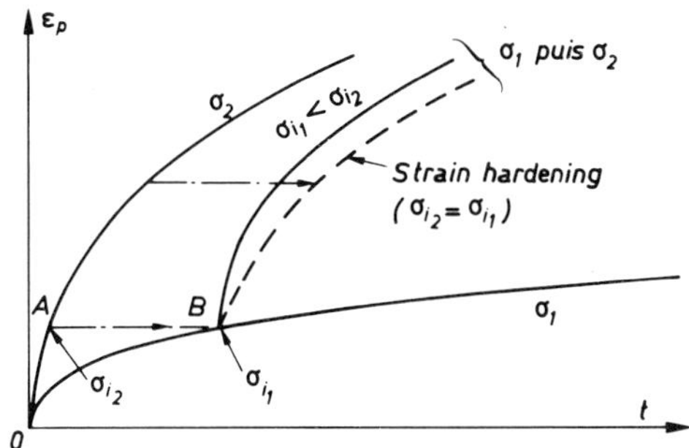

Fig. 14 - Two-level creep (schematic) : inclusion of the
recovery effect

The recovery term in the isotropic hardening equation
(43) brings a time effect into the cyclic tests containing
dwell periods. For a fatigue-relaxation test, the variation
of R at each cycle can be approached by :

$$dR = \left[2b \, (Q - R) \Delta\varepsilon_p - \gamma (R - Q_r) \, t_m \right] dN \qquad (45)$$

where $\Delta\varepsilon_p$ is the plastic strain range (approximately constant
for a controlled-strain test) and tm is the relaxation time.
We obtain a stabilized cycle (dR = 0) for :

$$R = R_s = \frac{2b \, Q \, \Delta\varepsilon_p + \gamma \, Q_r \, t_m}{2b \, \Delta\varepsilon_p + \gamma \, t_m} \qquad (46)$$

If $0 < Q_r < Q$, the value of R_s decreases as the hold time increases (for the same strain range). The maximum stress varies
similarly. This is observed experimentally, e.g. for the type
316 stainless steel (Fig. 15) [23].

Integrating equation (45) gives the variation of R before
stabilization. $\Delta\varepsilon_p$ and the maximum value of X are approximately constant. We then have :

$$\frac{\sigma_M - \sigma_{M_0}}{\sigma_{M_s} - \sigma_{M_0}} = \frac{R}{R_s} = 1 - \exp(-(2b\Delta\varepsilon_p + \gamma t_m)) \qquad (47)$$

This variation depends on the dwell time, which agrees with
the experimental data on the steel 316 [23] (Fig. 16).

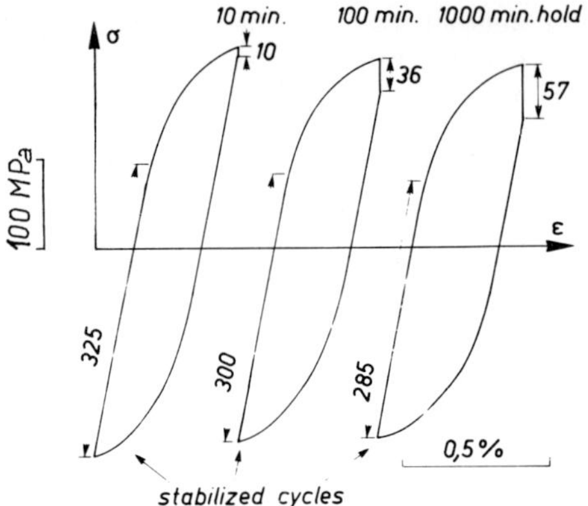

Fig. 15 - Effect of the hold time on the stress range in the stabilized cycle and on the size of the apparent elastic range for steeel 316 at 600°C, from Goodall

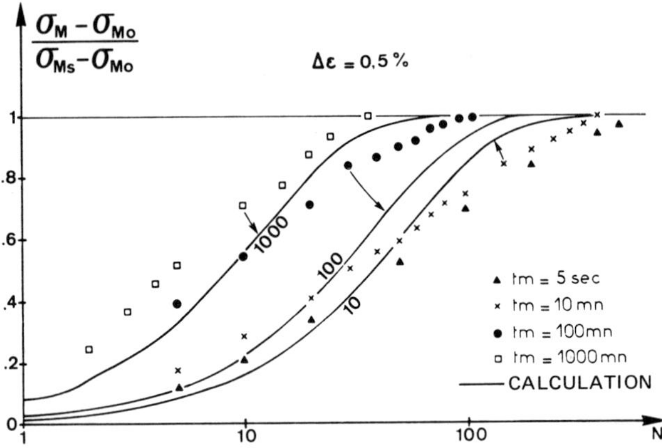

Fig. 16 - Description of the effect of the hold time on the cyclic hardening rate of the steel 316 at 600°C : model with recovery of the isotropic hardening

2.8 Aging effects

Aging effects are time effects related to modifications in the microstructure. The most common case is the precipitation, either after overheatings [11, 12], or during the working of the material at service temperature (carbide precipitations in certain stainless steels [28]). Models have been developed to express the first effect, but these can also been applied to the second case.

An initial aging most often increases the resistance to the flow as the creep test data show for the 304 steel [29], (Fig. 17). This effect can be described by combination of strain hardening and time hardening such as the equation (43).

Since Ro = 0, by setting Qr < Q, the initial aging causes an increase in R and therefore a hardening effect.

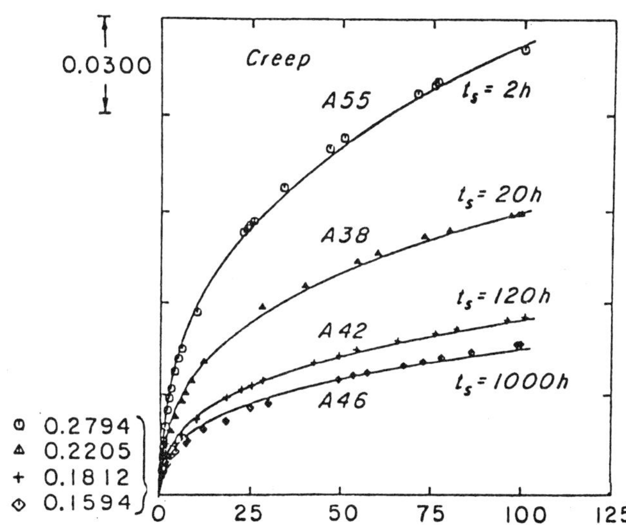

Fig. 17 - Effect of an initial aging on the creep and the recovery of the steel 304 at 593°C, from Findley [29]

The dynamic aging effects require a different model. Little work has been done on this subject up to now ; yet we know that this microstructural instability occurs in the medium temperature range (300°C to 600°C for the type 316 stainless steels) and at medium strain rates. This mechanism is responsible for the Portevin-Le Chatellier effect and depends very much on the strain rate. Fairly accurate microstructural studies [30] show that the dislocations are slowed down by the drag effect of the atoms in solution. When the rate increases the dislocations are freed, with a dynamic effect and a major localization of the strain. The movement of dislocations is then limited only by the mechanism of friction. The figure 18 schematically illustrates these catasstrophic type effect of dynamic instability.

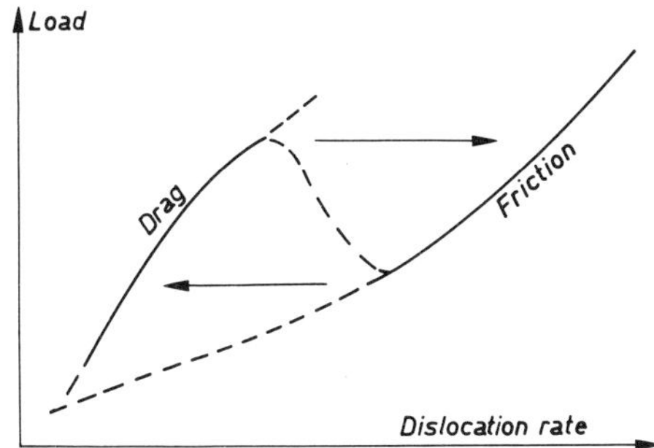

Fig. 18 - The two mechanisms influencing the dynamic aging effect.

198

The macroscopic model associated with this phenomenon
requires the introduction of an additional internal stress R^* ,
or the drag stress. The preceeding figure shows that R^* must
be highly dependent on the rate. We then have, as in the model
developed by Miller [13],

$$\dot{p} = \left\langle \frac{|\sigma - X| - k - R - R^*}{K} \right\rangle^n \qquad (48)$$

$$R^* = R^*(\dot{p})$$

and the relation (42) and (43) for the other variables. In
fact, it is more practical to eliminate R^* between the two
relations :

$$|\sigma - X| - k - R = R(p) + K\,\dot{p}^{1/n} \qquad (49)$$

We get an equation in the form

$$\dot{p} = G\,(|\sigma - X| - k - R) \qquad (50)$$

in which the expression for the function G depends on the form
of the function R^*.

The dynamic aging explains the following fact, observed
on many steels such as the stainless steel 316 (Fig. 19) [31] :
creep tests carried out in the intermediate temperature range
show a large strain under transient loading, followed by very
small strain even after 10,000 h of creep. The rate of loading
is high (applied in 1 min) and is carried out in the domain of
the friction effect. As soon as the creep phase begins the
rate decreases (work-hardening effect) and the drag effect
takes over (trapping the dislocations), resulting in an increase in R . As the applied stress is constant, the rate decreases very quickely, which explains the low creep strains.

The two types of aging -static and dynamic- can lead to
effects that counter the normal rate effects. In a certain
range, an increase in the rate corresponds to a decrease in
the stress. The figure 20 illustrates this effect. In certain
cases, the apparent independence of the rate corresponds to
the combination of two counteracting time effects : the normal
viscosity and the aging.

This remark, and the preceding one, tend to support an
approach different from the classical one. The classical
approach decomposes the inelastic strain into the sum of a
plastic strain with a creep strain, and brings in two types
of hardening independently. On the other hand, we believe it
is possible to express the same effects by the viscoplastic
strain alone, by separating rapid and slow hardening using a
combination of strain hardening and time hardening, possibly
including dynamic aging.

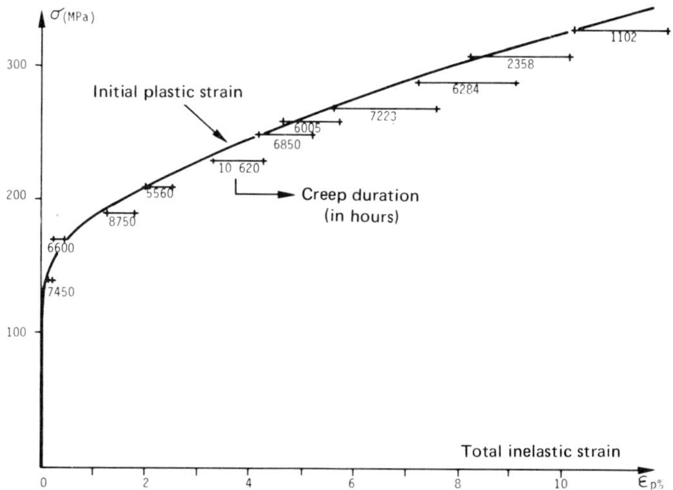

Fig. 19 - Initial strain and creep strain in the steel 316L at 550°C.

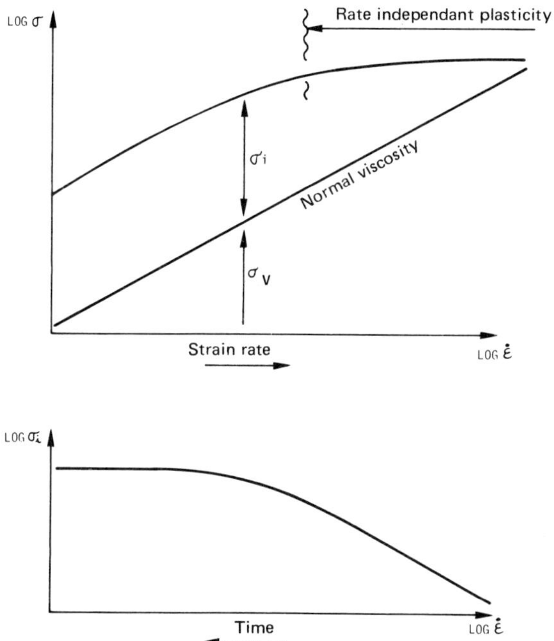

Fig. 20 - Superposition of aging effects on the viscosity effect

3. DAMAGE LAWS

3.1 Ideas of damage and crack initiation

During the strain process, by localization and dislocation accumulation effects, irreversible defects appear that
can be considered as damage : intercrystalline cavities in
creep, persistent slip bands in fatigue, etc. These damages
are the cause of the rupture of the volume element of material
or, more precisely, of crack initiation.

The purpose of damage theory is to describe the processes
involved prior to the initiation of a macroscopic crack.
Defining crack initiation is one of the first difficulties.
It depends on the scale on which the phenomenon is observed.
Under fatigue conditions, for example, microscopic cracks
start to appear very soon, often on the surface of the test
specimen ; and a local definition of crack initiation, used
by metallurgists, is the change from stage I to stage II, i.e.
a crack size of some one to several grains. The engineer uses
a more macroscopic definition when a crack is clearly preponderant (when it has unloaded the neighboring microcracks) and
has already progressed sufficiently. Its geometry can then be
approximated in the framework of Fracture Mechanics (Fig. 21).
The Damage Mechanics discussed in this section is a tool used
to deal with the defects in the intermediate domain where
Fracture Mechanics is no longer applicable, while remaining
within the framework of Continuum Mechanics.

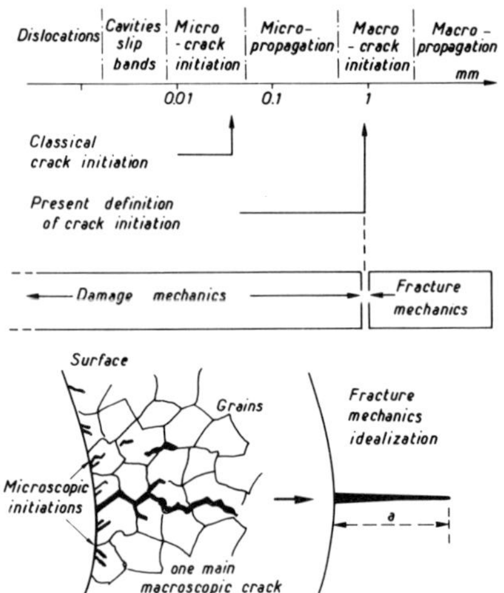

Fig. 21 - Schematic illustration of a macroscopic crack
initiation concept

The engineer also often refers to the remaining lifetime
to express the idea of damage. Damage exists as soon as the
remaining life under a given loading is less than the nominal

life under the same loading. This is the definition applied for calculating the potential life of a component. Often used to express two-level test results (linear or non-linear cumulative rules), this definition has the disadvantage of leading to a partial indetermination of the quantitative value of damage, as the figure 22 illustrates.

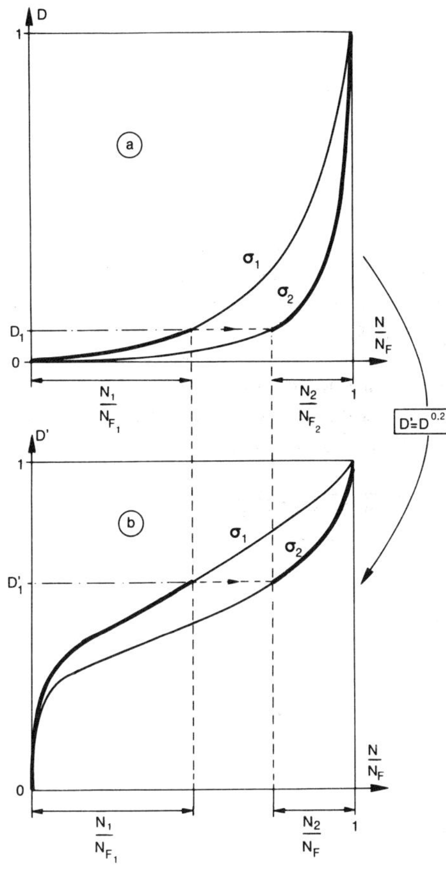

Fig. 22 - Nonlinear accumulation and indeterminate value of the damage defined by the remaining life.

To eliminate this indetermination, we can use either microstructural defect density measurements, more global measurements involving the variation of physical quantities (resistivity [22], density [33]) or strain resistance characteristics [34]. This last approach, based on the concept of effective stress, is the one we will develop in the following paragraphs.

3.2 Net stress and effective stress

The existence of decohesions in a volume element of material, for example intercrystalline cavities under creep conditions, leads to a reduction in the strength section, which means an increase in the effective stress. There are three types of stress increases :

202

- the true stress $\sigma_v \simeq (1 + \mathcal{E})$, which takes into account the geometric reduction in the sectional area due to the macroscopic deformation

- the net stress $\sigma^* = \sigma_v \, S/S^* = \sigma_v / (1 - \Omega)$, where the present geometrical section is reduced to $S^* = S(1 - \Omega)$, where ΩS is the mean area of decohesion

- the effective stress $\tilde{\sigma} = \sigma_v \, S/\tilde{S} = \sigma_v / (1 - D)$, which takes into account the local stress concentrations and the interactions between defects.

Continuum Mechanics can be used in the two latter cases because of the theories of homogenization : geometrical homogenization for the net stress and mechanical behavior homogenization for the effective stress. This effective stress $\tilde{\sigma}$ is thus defined as the stress that would have to be applied to the undamaged volume element for it to attain the same macroscopic strain $\mathcal{E}$ at the damaged volume element subjected to the true stress σ_v (Fig. 23).

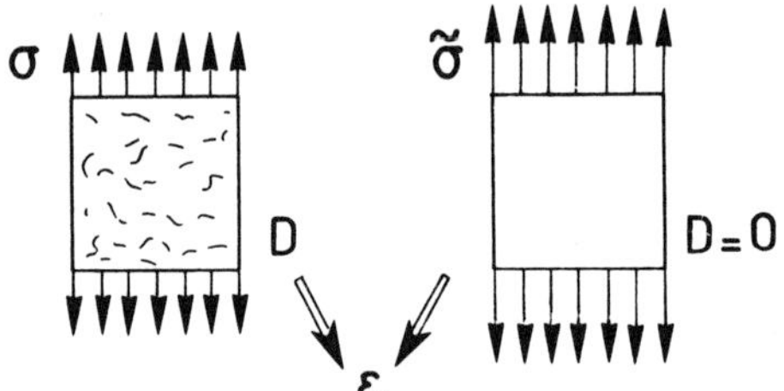

Fig. 23 — Effective stress concept.

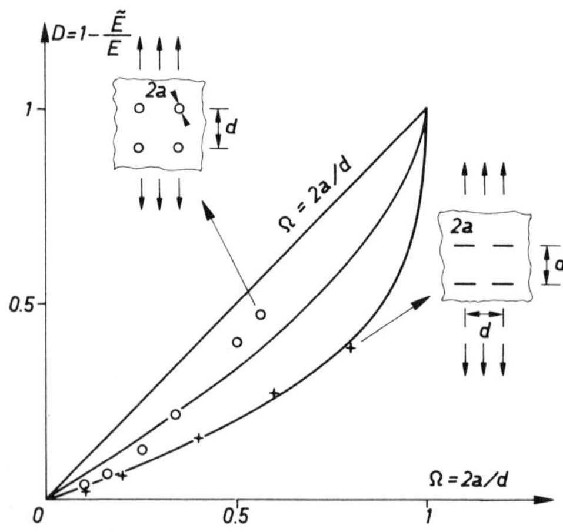

Fig. 24 — Damages defined by the net cross-section and by the effective cross-section. Results of calculations on perforated or cracked plates.

The relationship between the net and effective stresses is not simple, or at least not linear, as the comparisons in the figure 24 show. These were obtained by calculations for perforated plates [35] and for cracks perpendicular to the direction of tensile stress [36]. In what follows σ denotes the true stress.

3.3 One-dimensional damage models

Many damage models exist to calculate, among other things, the evolution of the damage and the lifetime under creep, fatigue and combined fatigue-creep conditions. The most widely known law is the one introduced by Kachanov to describe creep. In the field where fatigue applies, we will mention more particularly the works of Mayia and Majumdar [37], Krempl [38], Bui-Quoc [39], Chrzanowski [40], Lemaitre and Chaboche [34, 41]. Certain of these are based on strain parameters, for creep as well as for fatigue [37, 42] as it is the case for the strain range partitioning method [43]. On the other hand, other are applied using the stresses. Another classification of t hese models depends on whether or not it is possible to describe the non-linear cumulative effects of the damages.

These methods cannot be detailed here. We shall simply summarize Kachanov's law and outline the approach developed by Lemaitre and Chaboche.

a) Kachanov-Rabotnov law for creep

We suppose that the damage D_C varies between 0, for the unstressed material, and 1 for the rupture of the volume element. Kachanov's law [44], generalized by Rabotnov [45], is written as follows for pure tensile stress :

$$dD_C = \left(\frac{\sigma}{A}\right)^r (1 - D_C)^{-k} dt \tag{51}$$

where r, k and A are material -and temperature- dependent coefficients, determined partially by the constant stress (true stress) creep tests for which the integration from 0 to 1 gives the rupture time [*]

$$t_c = \frac{1}{k+1} \left(\frac{\sigma}{A}\right)^{-r} \tag{52}$$

while the evolution of the damage in this constant stress test is given by :

$$D_c = 1 - (1 - t/t_c)^{1/k+1} \tag{53}$$

---

[*]To identify the coefficients of the models, refer to chapter III and to references [46] and [47].

204

The relation (51) can be improved and the cumulative non-linear effects (two-level tests) can be described by considering a dependence between the exponent k and the applied stress [41].

Rabotnov introduced the concept of effective stress in the framework of the secondary creep law (Norton's law) :

$$\dot{\varepsilon}_p = \left[\frac{\sigma}{\Lambda(1 - D_c)} \right]^n \tag{54}$$

Replacing D_c with (53), the integration for σ = Cte yields

$$\varepsilon_p = \varepsilon_{p_R} \left[1 - \left(1 - \frac{t}{t_c}\right)^{\frac{k-n+1}{k+1}} \right] \tag{55}$$

where ε_{p_R} is given by :

$$\varepsilon_{p_R} = \frac{k+1}{k-n+1} \left(\frac{\sigma}{\Lambda}\right)^n t_c \tag{56}$$

This matches quite closely the tertiary creep curves and the ductility at the rupture point, as the figure 25 shows it for the alloy IN100 [48] (the law used also takes into account the primary creep, by a kinematic hardening variable).

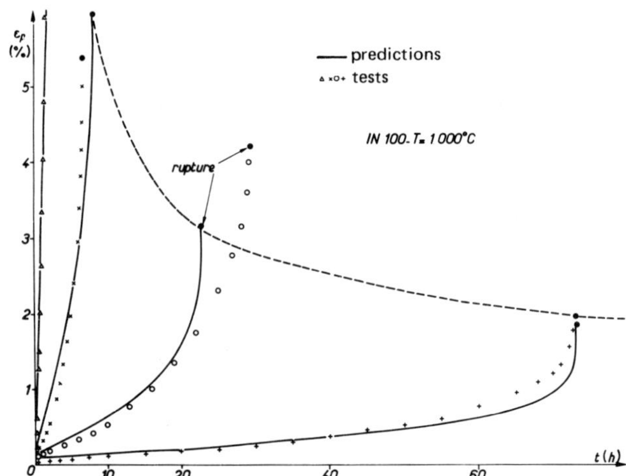

Fig. 25 - Prediction of the tertiary creep and of the rupture strain by strain-damage coupling - IN 100 alloy - 1000°C

b) <u>Fatigue damage law</u>

The fatigue damage D_F is introduced by reference to the remaining lifetime. It also varies from 0 to 1. Many models have been developed to express the effects of non-linear accumulation, which are large under fatigue conditions. The nonlinear cumulative effect is illustrated in the figure 22-a, which shows two hypothetical curves for the variation of the damage, corresponding to two levels of loading. Note that the number of cycles is referred to the corresponding number of cycles to failure and that the effect of non-linearity of the accumulated damages comes from the non-uniqueness of the curves $D_F = f(N/N_f)$. When these curves are independent of the loading, we return to the linear cumulative rule :

$$\frac{N_1}{N_{F_1}} + \frac{N_2}{N_{F_2}} = 1 \tag{57}$$

To obtain the non-uniqueness, we can either introduce a separation between the periods of incipient cracking and micropropagation, with different laws [49] or use an equation describing the evolution of the damage, in which the damage D_F and the loading are inseparable.

$$dD_F = f(\sigma_M, \bar{\sigma}, D_F)\, dN \tag{58}$$

This second method was used by several researchers [39] in the range of high cycle fatigue. The following equation, developed by the author [50], provides a good model of the cumulative non-linear effects :

$$dD_F = \left[1 - (1 - D_F)^{\beta+1}\right]^{\alpha(\sigma_M, \bar{\sigma})} \left[\frac{\sigma_M - \bar{\sigma}}{M(\bar{\sigma})\,(1 - D_F)}\right]^{\beta} dN \tag{59}$$

σ_M and $\bar{\sigma}$ are the maximum stress and mean stress during the cycle. The function α is selected in a form that makes it possible to include the fatigue limits and the static rupture limit :

$$\alpha(\sigma_M, \bar{\sigma}) = 1 - a\,\frac{\sigma_M - \sigma_1(\bar{\sigma})}{\sigma_u - \sigma_M}$$

$$\sigma_1(\bar{\sigma}) = \bar{\sigma} + \sigma_{1_0}(1 - b\bar{\sigma}) \tag{60}$$

$$M(\bar{\sigma}) = M_0(1 - b\bar{\sigma})$$

where σ_{l_0} and σ_u are, respectively, the fatigue limit under reversed stressing and the static rupture stress. β, M_0, b, a are the characteristic coefficients of the material.

This equation, expressed in terms of stresses, includes all of the Wöhler curves (Fig. 26). In effect, after integrating from 0 to 1 we have :

$$N_F = \frac{\sigma_u - \sigma_M}{a\,(\beta+1)\langle\sigma_M - \sigma_1(\bar{\sigma})\rangle}\left[\frac{\sigma_M - \bar{\sigma}}{M(\bar{\sigma})}\right]^{-\beta} \tag{61}$$

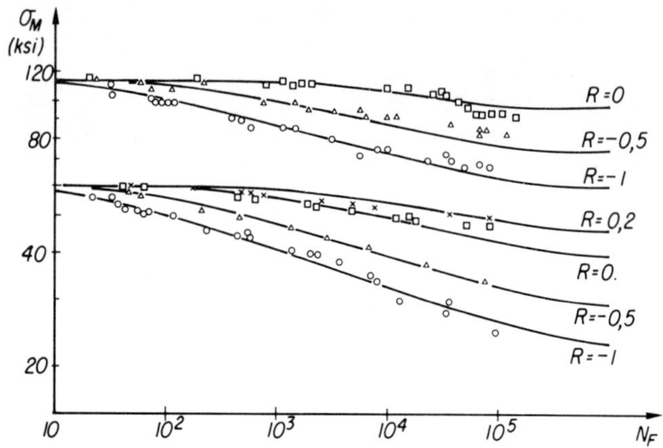

Fig. 26 - Modelling of Wöhler curves for steels A201 and A517 at room temperature

The variables D_F and σ_M, $\bar{\sigma}$ are inseparable because of the exponent α. For a two-level test, for example, the integration of the model (59) yields the lifetime remaining at the second level (under reversed loading) :

$$\frac{N_2}{N_{F_2}} = 1 - \left(\frac{N_1}{N_{F_1}}\right)^{\eta} \tag{62}$$

$$\eta = \frac{1-\alpha_2}{1-\alpha_1} = \frac{\sigma_{M2}-\sigma_{1_o}}{\sigma_{M1}-\sigma_{1_o}} \cdot \frac{\sigma_u-\sigma_{M1}}{\sigma_u-\sigma_{M2}} = \frac{N_{F1}}{N_{F2}}\left(\frac{\sigma_{M1}}{\sigma_{M2}}\right)^{\beta} \tag{63}$$

where N_1 and N_2 are the numbers of cycles run at the level σ_{M_1} and σ_{M_2} and N_{F1}, N_{F2} are the reference number of cycles to failure. This model very closely describes many test results [50].

Note that the relation (62) has been confirmed by other author [51, 52] by different approaches, with only the exponent η being different :

$$\eta = \left(\frac{N_{F1}}{N_{F2}}\right)^{0.4} \quad \text{or} \quad \eta = \frac{\sigma_{M2}-\sigma_{1_o}}{\sigma_{M1}-\sigma_{1_o}} \tag{64}$$

Let us further note that the law can be expressed as a function of the strains by using the cyclic curve of the material. We then obtain an equation in the form

$$dD_F = g\left(\Delta\varepsilon, \bar{\sigma}, D_F\right) dN \tag{65}$$

including the total strain range and the mean stress.

c) <u>Fatigue-creep interaction</u>

This is the case where the cyclic loading at high temperature brings out the influence of time (low frequency, hold times, etc.). The fatigue and creep damages may develop simultaneously and interact more or less strongly. We consider here that all of the time effects are contained in the creep term.

The model of Mayia and Majumdar [37] expresses such an interaction. These authors considered two types of damages : creep cavities represented by c and fatigue microcracking represented by a. The laws governing the variationsof c and a depend on the plastic strains, determinated in pure creep and pure fatigue. When the two processes exist simultaneously, the model introduces the effect of the creep cavities on the growth rate of the fatigue cracks. The complementary effect, that is the acceleration in the growth of the cavities due to the stress concentrations induced by the cracks, is not introduced.

In the approach used by Lemaître and Chaboche, on the other hand, the two phenomena accelerate each other. The creep damage growth law depends on the fatigue damage and vice versa :

$$dD_c = f_c (\sigma, D_c, D_F) \, dt$$

$$dD_F = f_F (\sigma_M, \bar{\sigma}, D_F, D_c) \, dN$$

(66)

The base laws, pure fatigue and creep, were determined on the basis of the effective stress concept. The measurement of the tertiary creep provided an indirect measurement of the creep damage D_c ; and an analogous method for the fatigue [50] made it possible to remove the ambiguity of the coefficient a (only the product $a \, M_c^{-\beta}$ can be defined by the Wöhler curve). The figure 27 gives an example of such measurements for the alloy IN100. The effective stress concept thus makes it possible to connect the creep and fatigue theory and it is then natural to consider that D_c and D_F add together in the two equations (66) as :

$$dD_c = f_c (\sigma, D_c + D_F) \, dt$$

$$dD_F = f_F (\sigma_M, \bar{\sigma}, D_c + D_F) \, dN$$

(67)

The complete model can thus be written in the form :

$$dD = f_c (\sigma, D) \, dt + f_F (\sigma_M, \bar{\sigma}, D) \, dN \qquad (68)$$

The creep and fatigue damages are of a different physical nature, but with D defined in this way it represents the overall effect of these defects on the strength of the material.

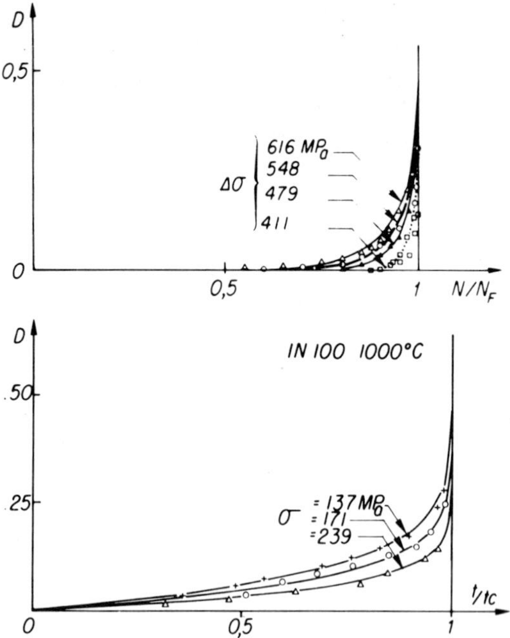

Fig. 27 - Damage measurements by the effective stress
concept and the corresponding model. Cases of creep and
fatigue on the alloy IN100 at 1000°C

With the functions f_C and f_F being determined separately
by tensile creep tests and fatigue tests at high frequency,
the theory predicts the cases of loading at low frequency, or
loading including dwell periods. From equations (51) and (59),
we get :

$$dD = \left[\frac{(1 - D)^{-k}}{(k+1)\,N_c} + \frac{\left[1 - (1 - D)^{\beta+1}\right]^{\alpha}}{(1-\alpha)\,(\beta+1)\,(1 - D)\,N_F} \right] dN \qquad (69)$$

where N_C and N_F are the reference number of cycles to failure
that we get, respectively, by neglecting the effect of fatigue
and the effect of creep.

We note that the two terms in this equation make it impos-
sible to isolate D on the left side. The result is a non-linear
interaction related to the difference between the creep and
fatigue damage kinetics. The figure 28 illustrates the high
non-linearity of the interaction for the alloy IN100 in
cyclic creep tests.

Let us note further that the application of the effective
stress concept makes it possible to describe the cyclic viscoplastic behavior of the damage material [53] by substituting
$\tilde{\sigma} = \sigma/(1-D)$ for σ in the viscoplastic potential (18). The

figures 29 and 30 show the case of tension-compression on the alloy IN100 at 1000°C. The three phases -cyclic softening, stabilization and acceleration due to damage- are well described for controlled load tests, and the controlled strain cycles are correctly described in the damage phase (the very last cycles are not described because of the closing of the cracks during compression).

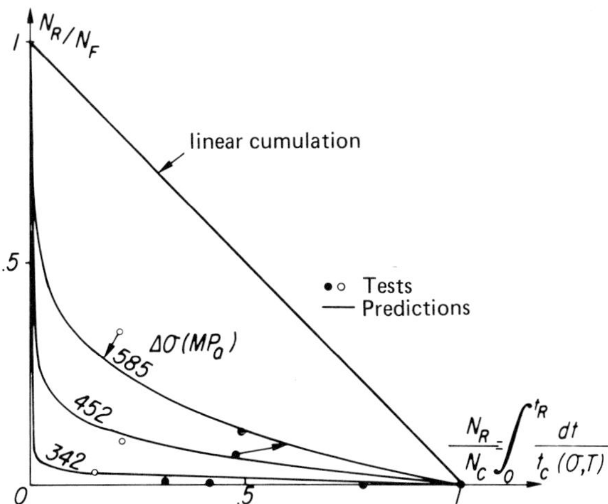

Fig. 28 - Prediction of the nonlinear creep-fatigue interaction alloy IN100 at 1000°C

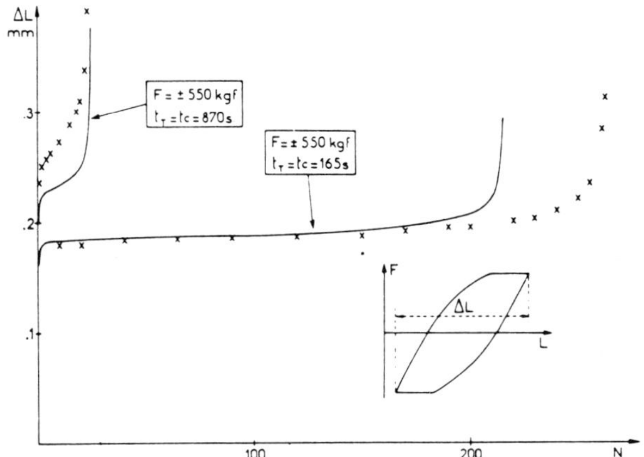

Fig. 29 - Prediction of the damage effect on the variation in the elongation in two controlled-load tests, IN100 at 1000°C

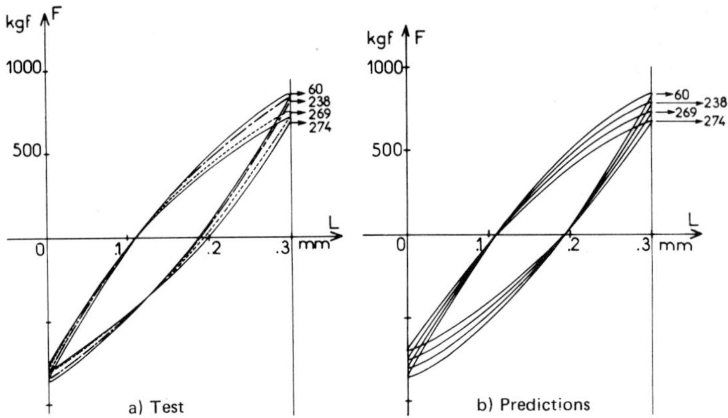

Fig. 30 - Prediction of the damage effect on the force-elongation cycles in a viscoplastic fatigue test with controlled elongation, IN100 at 1000°C

3.4 Multiaxial damage

The three-dimensional generalization of the damage law poses two problems :

- the determination of the multiaxial criteria, i.e. the isodamage surface (in the stress space, for example) representing equal lifetimes. This aspect is important for proportional loading. It is covered in chapter III and in paragraph (a) below.

- The anisotropy of the damage appearing under non-proportional loading. The microstructural observations show that the creep defects can be distributed and orientated in planes perpendicular to the maximum principal stress.

Including the anisotropy of the damage entails major difficulties. The initial formulation in which the damage variable remains scalar but affects only the main stress component [4,54] cannot be universally applied to any desired loading. Tensorial variables must then be introduced. Two creep theories are thus summarized below to complement the theories developed in chapter (III) of this book.

In both theories the following aspects must be considered successively :

- Definition of the damage tensor, from which a possible means of measurement can be deduced

- effect of the present damage on the strain behavior

- the damage evolution law.

a) <u>Multiaxial criteria</u>

We shall only address the simplest case here, i.e.
creep. An additional complication arises for fatigue because
of the simultaneous effects of an "amplitude" parameter and
of a "mean value" parameter. Among the general criteria that
can be used for any given loading, we can mention the
criteria of Sines [55], Crossland, Dan Van [56] for high-
cycle fatigue, and the criteria developed by Miller [57] or
Pineau [58] for low-cycle fatigue.

The effect of multiaxiality in creep is usually
described using the idea of isochronous surfaces, which are
loci of the constant stress state that lead to rupture at
the same time. For an initially isotropic material, these
surfaces can be defined by three intervals :

- the octahedral shear stress J_2 (σ) related to the
effects of shear,

- the hydrostatic stress J_1 (σ) = T_r (σ) = $\sigma : \mathbf{1}$
which greatly affects the growth of the cavities,

- the maximum principal stress J_0 (σ) = σ_1 which
opens the microcracks and causes them to grow.

Following Hayhurst's method, these three invariants can
be combined into

$$\chi(\sigma) = \alpha\, J_0(\sigma) + \beta\, J_1(\sigma) + (1-\alpha-\beta)\, J_2(\sigma) \tag{70}$$

where α, β are coefficients dependent on the material, and
temperature. The time to failure under a fixed multiaxial
stress is expressed as

$$t_R = \frac{1}{k+1}\left\langle \frac{\chi(\sigma)}{A} \right\rangle^{-r} \tag{71}$$

which generalizes the simple tension relation (52).

Note the use of the symbol $\langle\rangle$, which implies an
infinite time to rupture when $\chi(\sigma)$ is negative. This is
true, for example, under pure compression for a material
like copper.

212

b) <u>The anisotropic theory of Murakami and Ohno [60,61]</u>

    This theory uses the intercrystalline microcavities
directly to define the damage variable through the reduction
in the strength section (where the geometric section is
reduced by the shape of the cavities). The resulting stress
is thus a direct generalization of the net stress introduced
in paragraph 3.2. The variable describing the present state
of damage of the element of volume is the symmetrical second-
rank tensor[*] :

$$\Omega = \frac{3}{S_g(V)} \sum_{k=1}^{N} \int_V \boldsymbol{\nu}^{(k)} \otimes \boldsymbol{\nu}^{(k)} \ dS_g^{(k)} \tag{72}$$

where $dS_g^{(k)}$ is the area of the grain boundary occupied by
the kth cavity, and $\boldsymbol{\nu}(k)$ is the vector normal to the
boundary. $S_g(V)$ is the total area taken up by the boundaries
in the volume element.

    We can insert the principal values Ω_j of the tensor
and the principal corresponding direction $\boldsymbol{\nu}_j$ to get :

$$\Omega = \sum_{j=1}^{3} \Omega_j \ \boldsymbol{\nu}_j \otimes \boldsymbol{\nu}_j \qquad\qquad S\,\mathbf{t} = \boldsymbol{\sigma}.(S\boldsymbol{\nu}) \tag{73}$$

Ω_j represents the surface density of the cavities on the
plane perpendicular to $\boldsymbol{\nu}_j$. This brings a decrease in the
strength, characterized by an increase in the stress tensor
as shown in the figure 31a, taken from [61] . Under these
conditions, the force vector acting on the surface $S\boldsymbol{\nu}$ is
$S\mathbf{t} = \boldsymbol{\sigma}.(S\boldsymbol{\nu})$, where $\mathbf{t}$ is the corresponding stress vector, $\boldsymbol{\sigma}$
is the Cauchy tensor acting on the considered volume
element. The effective surface element $A^* B^* C^*$ is subjected
to the same force vector and we have (fig. 31b)

$$S^* \mathbf{t}^* = S\,\mathbf{t} \tag{74}$$

$$\boldsymbol{\sigma}^*.(S^*\boldsymbol{\nu}^*) = \boldsymbol{\sigma}.(S\boldsymbol{\nu}) = \boldsymbol{\sigma}.(\mathbf{1} - \Omega)^{-1}.(S^*\boldsymbol{\nu}^*)$$

whence the not-necessarily-symmetrical net stress tensor
$\boldsymbol{\sigma}^* = \boldsymbol{\sigma}.(\mathbf{1} - \Omega)^{-1}$. In practice we use the symmetrical part
of this tensor which will also be denoted by $\boldsymbol{\sigma}^*$:

$$\boldsymbol{\sigma}^* = \frac{1}{2} (\boldsymbol{\sigma}.\boldsymbol{\Phi} + \boldsymbol{\Phi}.\boldsymbol{\sigma}) \qquad\qquad \boldsymbol{\Phi} = (\mathbf{1} - \Omega)^{-1} \tag{75}$$

---

[*]In this section $\otimes$ denotes the tensor product of two
vectors, the resulting quantity being a second order tensor.

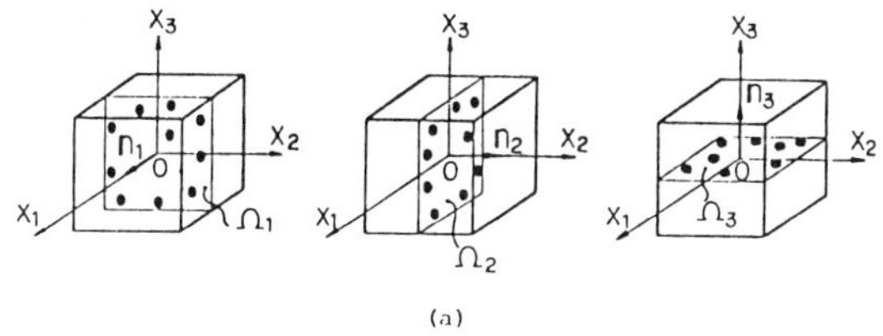

(a)

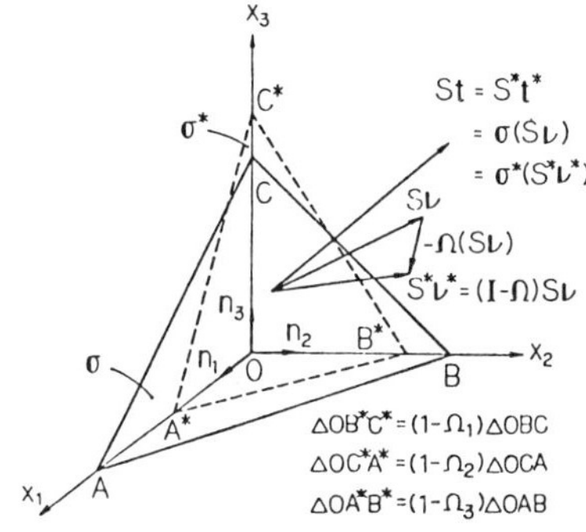

(b)

Fig. 31 - The net stress concept :
a) Reduction of the cross-section on the principal planes of the damage tensor

b) Reduction of the cross-section on any given plane

The damage evolution law, i.e. the law for the tensor is expressed for creep in the form

$$\dot{\Omega} = \left\langle \frac{\chi(\sigma)}{A} \right\rangle^{r} \left[\gamma \, \mathbb{1} + (1-\gamma) \, \nu^{(1)} \otimes \nu^{(1)} \right] \tag{76}$$

where $\chi(\sigma^{*})$ is the invariant describing the isochronous surfaces (see above), with $\nu^{(1)}$ being the direction of the maximum principal stress. A, r and γ are the coefficients dependent on the material. The latter adjusts the degree of anisotropy of the damage evolution.

The net stress tensor is thus employed to express the evolution of the damage variable. On the other hand, the effect of the present damage on the strain behavior (elastic, plastic, creep) can only be introduced through a fourth-rank tensor as we will see in the next paragraph. Murakami and Ohno use an effective stress on the form

$$\tilde{S} = \frac{1}{2} \left[\Gamma : \sigma + (\Gamma : \sigma)^{T} \right] \tag{77}$$

214

where the asymmetrical fourth-rank tensor Γ is deduced from the tensor Ω, or rather from $\Phi = (1-\Omega)^{-1}$ by a general expression deduced from the representation theory of Rivlin and Ericksen [63] using a linear combination of fourth-rank tensors formed by the tensor Φ and the unit tensor. Note that this expression contains second-, third- and fourth-order terms in Φ .

According to the effective stress concept, the law of viscoplasticity of the damaged material is then obtained by substituting the effective stress $\widetilde{S}$ for σ . For the special case where only the isotropic hardening is taken into account, the authors use expressions of the form

$$\dot{p} = \left[\frac{J(\widetilde{S})}{K}\right]^{n} p^{-n/m}$$

$$\dot{\mathcal{E}}_p = \frac{3}{2} \dot{p} \frac{\widetilde{S}'}{J(\widetilde{S})} \tag{78}$$

where $\widetilde{S}'$ is the deviator of $\widetilde{S}$.

For simple tension the equations (76) (78) reduce to

$$\dot{\Omega} = \left[\frac{\sigma}{A (1-\Omega)}\right]^{r} \tag{79}$$

$$\dot{\mathcal{E}}_p = \left[\frac{\sigma}{K (1-\Omega)}\right]^{n} p^{-n/m} \tag{80}$$

If we integrate the damage equation (79) for constant stress creep we get :

$$\Omega = 1 - (1 - t/t_c)^{1/r+1} \tag{81}$$

This theory is general. It can be applied to any given loading, even non-proportional. It is very attractive for two reasons :

- the internal variable expressing the state of damage of the volume element is only a second-rank symmetrical tensor (six components).

- the theory has been firmly supported by a serie of experiments on plates with multiple perforations

[62,64]. This confirmation is indicated in paragraph
(d).

Three disadvantages should nevertheless be pointed out:

- The elasticity law of the damaged medium is not
described explicitly.

- A large number of coefficients is used in defining
the effective stress included in the law of
viscoplasticity.

- No reference is made to a dissipative theory, by which
the equation of viscoplasticity (78) would be slightly
modified as we will see below.

c) <u>The anisotropic theory of Chaboche</u> [55]

To avoid the above mentioned disadvantages, we
introduce the tensorial damage variable through the
effective stress concept and the effective behavior of the
damaged material. The principle is very simple. It
generalizes the elementary idea as reviewed in paragraph
3.2.

The effective stress tensor $\widetilde{\sigma}$ is the one that must
be applied to the undamaged volume element to obtain the
same strain tensor as the one produced by the stress σ
applied to the damaged volume element (from the rest actual
state defined by σ =0).

In the case of the linear elastic material the law of
the undamaged material is written

$$\sigma = \Lambda : \mathcal{E}_e \tag{82}$$

while the behavior of the damaged material is expressed

$$\sigma = \widetilde{\Lambda} : \mathcal{E}_e \tag{83}$$

By definition, the effective stress is obtained

216

$$\widetilde{\sigma} = \Lambda : \mathcal{E}_e = \Lambda : \widetilde{\Lambda}^{-1} : \sigma \qquad (84)$$

The fourth-rank tensor $\mathbb{M} = \Lambda : \widetilde{\Lambda}^{-1}$ can be set in a form that generalizes the initial theories of Kachanov and Rabotnov, $\mathbb{M} = (\mathbb{1} - \mathbb{D})^{-1}$ such that the effective stress can be expressed[**]

$$\widetilde{\sigma} = (\mathbb{1} - \mathbb{D})^{-1} : \sigma \qquad (85)$$

where $\mathbb{D}$ is an asymmetrical fourth-rank damage tensor. It can be measured throught the elastic behaviors measured in the present state and in the initial (undamaged) state :

$$\mathbb{D} = \mathbb{1} - \widetilde{\Lambda} : \Lambda^{-1} \qquad (86)$$

Such an operator $\mathbb{D}$ can represent any observed behavior for the damaged elastic material. We have

$$\sigma = \widetilde{\Lambda} : \mathcal{E}_e = (\mathbb{1} - \mathbb{D}) : \Lambda : \mathcal{E}_e \qquad (87)$$

and the symmetry of the operators Λ and $\widetilde{\Lambda}$ is complied with ($\Lambda_{ijkl} = \Lambda_{klij}$). The definition adopted for the damage tensor and for the effective stress $\widetilde{\sigma}$ thus matches the results of the homogenization theories [66], which show that, for a periodic space distribution of the defects,

$$\widetilde{\Lambda}_{ijkl} = \left[\Delta_{ijpq} - (1 - \frac{V^*}{V}) \Delta_{ijpq} - \frac{1}{V} \int_{V^*} b_{ijpq} \, dV \right] \Lambda_{pqkl} \qquad (88)$$

where V is the apparent volume of material, V^* is the solid volume (V minus the volume of empty space) and b_{ijp} is a field of elementary elastic solutions. Let us note that slightly different formulations have ben developed, using only a second order damage tensor [67] .

[**]The tensor $\mathbb{1}$ is the fourth-rank unit tensor whose component can be denoted as $\Delta_{ijkl} = \delta_{ij}\,\delta_{kl}$

Thermodynamic aspects and viscoplastic flow law

Anisotropic damage can be introduced for elasticity and viscoplasticity through a general thermodynamic framework. If we limit ourselves here to the isothermal case, the reversible part of the free energy depends on $\mathcal{E}_e$ and on the damage $\mathbb{D}$:

$$\rho \Psi = \rho \Psi_e(\mathcal{E}_e, \mathbb{D}) + \rho \Psi_p(\alpha_j, \dots)$$

$$\rho \Psi_e = \frac{1}{2}(\mathbb{1} - \mathbb{D}) : \Lambda : \mathcal{E}_e : \mathcal{E}_e \tag{89}$$

From the classic demonstration we have

$$\sigma = \rho \frac{\partial \Psi}{\partial \mathcal{E}_e} = (\mathbb{1} - \mathbb{D}) : \Lambda : \mathcal{E}_e \tag{90}$$

and the thermodynamic force associated with the tensor $\mathbb{D}$ is the fourth-rank tensor[*]

$$\mathcal{Y} = \rho \frac{\partial \Psi}{\partial \mathbb{D}} = -\frac{1}{2}(\Lambda : \mathcal{E}_e) \otimes \mathcal{E}_e \tag{91}$$

Still in the isothermal case, the second principle requires that the intrinsic dissipation be positive. From (7) we find

$$\sigma : \dot{\mathcal{E}}_p - A_j \dot{\alpha}_j - \mathcal{Y} :: \dot{\mathbb{D}} \geqslant 0 \tag{92}$$

For viscoplastic flow we use a potential dependent on the effective stress :

$$\varphi^* = \varphi^*(\tilde{\sigma}, A_j ; \alpha_j)$$

Without going into the details, the normality assumption then leads to an expression that is slightly different from (78). In the case with isotropic hardening the viscoplastic potential is written, for example,

$$\varphi^* = \frac{K}{n+1}\left[\frac{J(\tilde{\sigma})}{K}\right]^{n+1} p^{-n/m} \tag{93}$$

---

[*]In this section $\otimes$ denotes the tensorial product ot two second order tensors, resulting in a fourth order one, and :: is the corresponding contracted product (over four indices).

218

and the law of viscoplastic flow is

$$\dot{\mathcal{E}}_p = \frac{\partial \varphi^*}{\partial \sigma} = \frac{3}{2}\left[\frac{J(\widetilde{\sigma})}{K}\right]^n \; p^{-1/n} \; (1-D)^{-1} : \frac{\widetilde{\sigma}'}{J(\widetilde{\sigma})} \qquad (94)$$

In simple tension, this equation reduces to

$$\dot{\mathcal{E}}_p = \frac{1}{1-D}\left[\frac{\sigma}{K\,(1-D)}\right]^n \; p^{-n/m} \qquad (95)$$

In comparison with (80) the damage term now has an exponent n + 1 instead of n. This is because the effective stress replaces the applied stress in the potential and not in the flow law itself. Because of the anisotropic damage we note that the plastic flow is no longer necessarily at constant volume.

<u>Creep damage law</u>

This law is established in two steps :

- the relation selected to describe the simple tensile case provides the law for the scalar variable D representing the volume density of the defects.

- the directional development of the damage is taken into account by an anisotropy tensor $\mathbb{Q}$. We suppose [65], as do Murakami and Ohno [64], that the anisotropy ratios remain fixed under proportional loading, which is expressed by

$$\dot{\mathbb{D}} = \mathbb{Q} \; \dot{D} \qquad (96)$$

where $\mathbb{Q}$ depends on the material and on the principal directions of the effective stress tensor $\widetilde{\sigma}$.

In the general case $\mathbb{Q}$ is introduced by rotating a tensor $\mathbb{Q}^*$, that only depends on the material[*]. To make identification less difficult we can define $\mathbb{Q}^*$ with a linear combination of the isotropic case and of a particular case

------

[*]To simplify, we note that this tensor $\mathbb{Q}^*$ does not depend on the present state, but is a fixed tensor.

of anisotropy, e.g. the one corresponding to planar microcracks, all parallel, developing perpendicular to the direction of the maximum principal stress :

$$\mathbb{Q}^* = \gamma \mathbb{1} + (1-\gamma)\,\mathbb{G} \qquad (97)$$

The anisotropy tensor $\mathbb{G}$ is obtained from elastic analysis results [36]. In matrix form, corresponding to stress components 11, 22, 33, 23, 31, 12, we can say

$$\mathbb{G} = \begin{bmatrix} 1 & 0 & 0 & & & \\ \nu/_{1-\nu} & 0 & 0 & & & \\ \nu/_{1-\nu} & 0 & 0 & & & \\ & & & 0 & & \\ & & & & a & \\ & & & & & a \end{bmatrix} \qquad (98)$$

where ν is the Poisson contraction ratio (a and ν depending on the material).

We consider that the scalar damage is an invariant of $\mathbb{D}$, for example $D = T_r\,(\mathbb{D})/T_r\,(\mathbb{Q})$ and its law is given by identification with the simple tensile case, attempting to describe as above the isochronous surfaces in the stress space. To do this, a function such as the following must be used :

$$\chi(\widetilde{\sigma},\,D) = \alpha\,J_0(\widetilde{\sigma}) + \frac{\beta}{1+2B}\,J_1(\widetilde{\sigma}) + \frac{1-\alpha-\beta}{1-B}\,J_2(\widetilde{\sigma}) \qquad (99)$$

where J_0, J_1, J_2 are the three invariants already defined and $A = \dfrac{\nu}{1-\nu} \cdot \dfrac{(1-\gamma)D}{1-\gamma D}$. The damage law is expressed for the principal axes of $\widetilde{\sigma}$:

$$\dot{\mathbb{D}} = \left[\gamma \mathbb{1} + (1-\gamma)\,\mathbb{G} \right]\dot{D}$$

$$\dot{D} = \left\langle \frac{\chi(\widetilde{\sigma},\,D)}{A} \right\rangle^k \left[\frac{\chi(\sigma)}{A} \right]^{r-k} \qquad (100)$$

The expression (99) is selected to produce the Rabotnov

220

law in the simple tensile case. We then have $\chi = \tilde{\sigma} = \dfrac{\sigma}{1-D}$ and :

$$\dot{D} = \left(\frac{\sigma}{A}\right)^{r} (1 - D)^{-k}$$

(101)

When this equation is integrated it yields (53) for the damage variation.

d) <u>Comparison of the two anisotropic theories</u>

In the two theories described above the problems posed by the multiaxial creep loadings are solved in two steps :

  - definition of a stress invariant for the creep rupture under constant multiaxial stressing. It describes the surfaces of equal time to rupture.

  - Definition of a creep damage law with fixed anisotropy with respect to the principal directions. This anisotropy does not depend on time, but only on the material. The nonlinearity of the variation in the damage is described by the selection of the equation used in the simple tensile case.

The essential differences between these two theories stem from the definition of the damage variable (second-rank or fourth-rank tensor), the selection of the effective stress used in the damage law (net stress σ^{*} or effective stress $\tilde{\sigma}$), and then from the way the stress-strain behavior is described. The table 1 summarizes the differences and similarities, in particular for the form of the equations used in the simple tensile case. Note in particular the idea of equivalence that is used in the damage law for the theory of Murakami and Ohno, and in the constitutive law for the theory of Chaboche.

Not many experiments can validate these theories, in particular under multiaxial loading and in creep. Experiments carried out on plates having multiple perforations give rather precise validation [35,64,68], for which the present distribution of the damage (simplified by the perforations) is well-known. The quantitative and directional aspects can thus be studied at the same time as in [64] . The quality and precision of the tensile experiments given in [64] thus offer :

Table 1 — Comparison of anisotropy theories of Murakami-Ohno and Chaboche

TABLE I		Theory of Murakami-Ohno	Theory of Chaboche
Definitions	Damage Definition	through the net section : $\Omega = 1 - S^*/S$	through the effective behavior : $D = 1 - \widetilde{E}/E$
	Effective Stress	net stress: $\sigma^* = \dfrac{\sigma}{1-\Omega}$	effective stress : $\widetilde{\sigma} = \dfrac{\sigma}{1-D}$
	Equivalence on the damage law ?	yes: $\dot{\Omega}(\sigma,\Omega) = \dot{\Omega}(\sigma^*,0)$	no: $\dot{D}(\sigma,D) \neq \dot{D}(\widetilde{\sigma},0)$
	Equivalence on the viscoplastic law ?	no : $\dot{\varepsilon}_p(\sigma,\Omega) \neq \dot{\varepsilon}_p(\sigma^*,0)$	yes : $\dot{\varepsilon}_p(\sigma,\Omega) = \dot{\varepsilon}_p(\widetilde{\sigma},0)$
Simplified equations	Damage law	$\dot{\Omega} = \left[\dfrac{\sigma}{A(1-\Omega)}\right]^r$	$\dot{D} = \dfrac{(\sigma/A)^r}{(1-D)^k}$
	Damage evolution during creep (σ = cte)	$\Omega = 1-\left(1-\dfrac{t}{t_c}\right)^{1/r+1}$	$D = 1-\left(1-\dfrac{t}{t_c}\right)^{1/k+1}$
	Constitutive equation (secondary and tertiary creep)	$\dot{\varepsilon}_p = B\left(\dfrac{\sigma}{1-c\,\Omega}\right)^n$	$\dot{\varepsilon}_p = \dfrac{B}{1-D}\left(\dfrac{\sigma}{1-D}\right)^n$
Generalization	Damage tensors	Second order tensor : $\underline{\Omega}$	Fourth order asymmetrical tensor : $\mathbb{D}$
	Effective stress tensors	$\sigma^* = \dfrac{1}{2}\left[\sigma \cdot \Phi + \Phi \cdot \sigma\right]$ $\Phi = (\mathbb{1}-\Omega)^{-1}$ $\mathbb{S} = \dfrac{1}{2}\left[\Gamma:\sigma+(\Gamma:\sigma)^T\right]$ $\Gamma = \Gamma(\Omega)$	$\widetilde{\sigma} = (\mathbb{1}-\mathbb{D})^{-1}:\sigma$

1) the justification of the net stress concept through the idea of equivalency at rupture (Fig. 32). ϕ'_{11} represents the ratio between the stress at rupture for the unperforated plate and the stress at rupture for the perforated plates. The calculated curve corresponds to $\phi'_{11} = 1/(1-\tilde{\Omega})$.

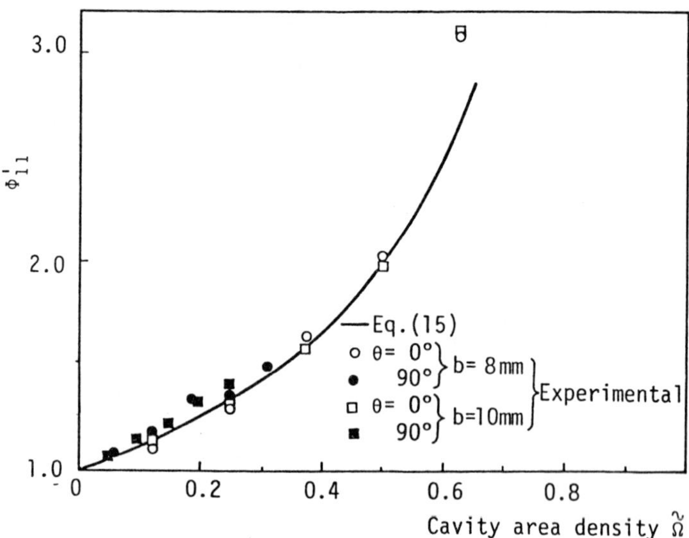

Fi. 32 - Relation between the effective stress, measured by the rupture, and the reduction in area

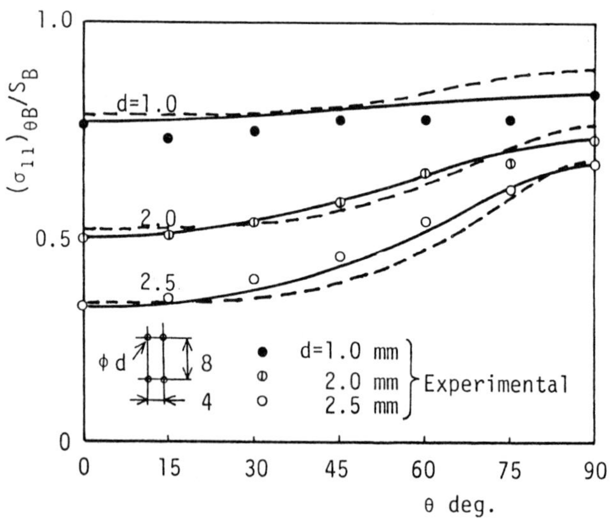

Fig. 33 - Relation between the measured and calculated rupture stress and the direction of uniaxial tension. ———— theory of Murakami and Ohno. ----- theory of Chaboche. S_B is the rupture stress of the unperforated plate

2) Verification of the same concept as a function of
the direction of the tensile stress with respect to the
principal axes of the drilling pattern (Fig. 33) (the
plain lines correspond to the theory of Murakami).

3) The determination of the tensor Γ as a function
of the tensor Ω through the plastic strain behavior
(Fig. 34).

The second theory presented also applies to the same
type of experiment. Here, the fourth-rank tensor $\mathbb{D}$ is
determined from measurements of the plastic strain (Fig. 34).
The form selected for $\mathbb{D}$ derives from the application of the
damage law (100) by selecting $\nu = 0.33$:

$$
\mathbb{D} = \begin{bmatrix}
D & 0 & 0 & & & \\
.5(1-\gamma)D & \gamma D & 0 & & & \\
.5(1-\gamma)D & 0 & \gamma D & & & \\
& & & \gamma D & & \\
& & & & (\gamma+(1-\gamma)a)D & \\
& & & & & (\gamma+(1-\gamma)a)D
\end{bmatrix}
\tag{102}
$$

where a and γ are unknown coefficients as well as the
scalar value of damage D. For a given drilling pattern the
values are determined using a similar procedure as in [64],
through the plastic response, under the hypothesis of a Von-
Mises material. a and γ are considered as constants,
independent of the hole diameter.

For the rupture, as in [64] we suppose that only the
maximum principal stress acts ($\alpha = 1$, $\beta = 0$). Relation (100)
and an equivalence between the perforated plate (σ, D) and
the undamaged one ($\tilde{\sigma}$, 0) lead to :

$$
\frac{\sigma_{11\theta}}{\sigma_0} = \left[\frac{\tilde{\sigma}_{1max}(\sigma_{11\theta})}{\sigma_{11\theta}} \right]^{-k/r}
\tag{103}
$$

The dotted lines in Fig. 33 and 34 are obtained using
this theory, with $\gamma = 0.5$, a = 0.84, k = r. The comparison
shows that experimental results of [64] can be described
with this second theory, for the rupture as well as for the
plastic behavior. A better correlation could easily be
obtained by considering that γ, a, k depend on the hole
diameter.

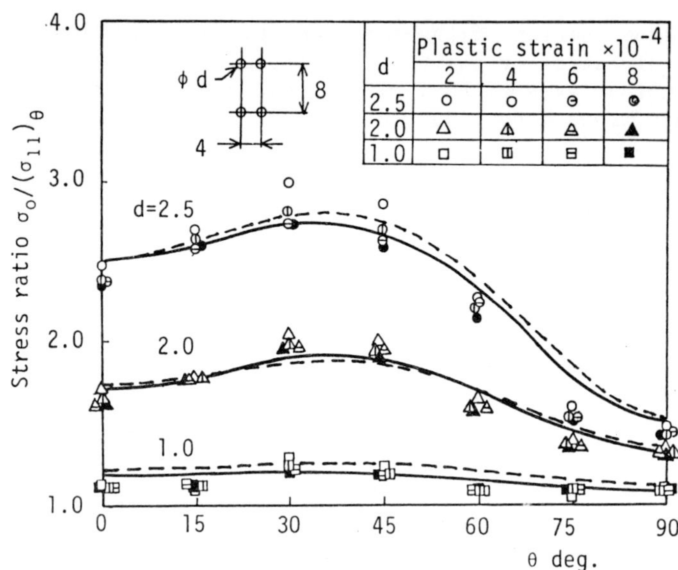

Fig. 34 - Effect of the orientation of the perforations on the uniaxial plastic strain. ———— theory of Murakami and Ohno. ------- theory of Chaboche. σ_0 is the stress corresponding to the same plastic strain on the nonperforated plate

The two definitions of damage used in these two theories can be related together, for example for the tensile creep case. Comparison of (53) and (81) shows that

$$D = 1 - (1 - \Omega)^{r+1/k+1} \tag{104}$$

The difference between the two concepts is illustrated in Fig. 24 where the continuous lines are calculated by means of (104) with $r+1/k+1=0.6$ in the circular hole case, 0.33 in the case of parallel cracks.

Let us also note that the two theories can be used to calculate the tertiary tensile creep, in particular the strain at rupture. The relations used by Murakami integrate explicitly only in the particular case where [61] :

$$\dot{\varepsilon}_p = \left[\frac{\sigma}{\Lambda(1 - c\Omega)}\right]^n \tag{105}$$

where c is a constant, and when r = n - 2. After calculation we have :

$$\mathcal{E}_p = \mathcal{E}_{p_R} \left[1 - \frac{1 - t/t_c}{\left[1-c + c(1- t/t_c)^{1/n-1}\right]^{n-1}} \right] \qquad (106)$$

where the rupture strain is

$$\mathcal{E}_{p_R} = \left(\frac{\sigma}{\Lambda}\right)^n \frac{t_c}{1-c} \qquad (107)$$

The theory developed by Chaboche integrates explicitly for tensile creep

$$\mathcal{E}_p = \mathcal{E}_{p_R} \left[1 - \left(1 - \frac{t}{t_c}\right)^{\frac{k-n}{k+1}} \right] \qquad (108)$$

where the rupture strain is

$$\mathcal{E}_{p_R} = \left(\frac{\sigma}{\Lambda}\right)^n t_c \frac{k+1}{k-n} \qquad (109)$$

We then clearly see the link existing between the two approaches and the relation between the phenomenological coefficients : $c \simeq n+1/k+1$.

4. CONCLUSION

The prediction of the lifetime of structures requires a knowledge of the constitutive equations of materials and of crack initiation equations. These two aspects have been discussed, with an emphasis on viscoplasticity, whether monotonic or cyclic, and on creep damage. For each aspect we first reviewed the most conventional and simplest laws, then recent developments and various improvements, etc., that are still not currently used in applications.

These new possibilities for describing the behavior of materials are certain to become more and more widely used as computing tools are improved and as costs go down. The sophistication of these laws, suggested for the purpose of

226

making them more representative, is fully justified with a
view to obtaining more reliable structural design results in
the field of plasticity, creep or rupture, which are in any
case long and costly.

At first, the constitutive and damage laws will be
applied successively, uncoupled, with the structural
analysis using the first type of law to provide data for the
second type of model. The figure 35, drawn from reference
[62], illustrates this method as well as the more general
method in which the three aspects are treated simultaneously.
This latter method, potentially usable by means of Continuum
Damage Mechanics, can be used for certain applications in
which the strain-damage coupling aspect is preponderant. It
can be used more systematically only if there is a major
increase in the power of the computers.

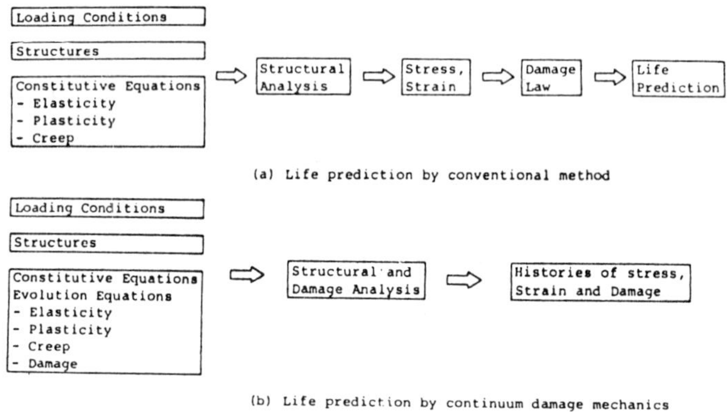

Fig. 35 - Lifetime prediction methods

The possibilities for the equations as discussed in the
present chapter are already very large. They are summarized
below and certain insufficiencies are pointed out briefly.

For Strain behavior

The monotonic and cyclic plasticity can be correctly
described by superposition of the isotropic and kinematic
hardening models, describing among other things the
Bauschinger effects, the cyclic hardening or softening, the
stabilization and the hardening memorization.

The same hardening theories can be used to generate

viscoplasticity equations that give a correct description of the creep, the relaxation and the rate effects, whether monotonic or cyclic.

The effects of recovery from hardening, at high temperature, are easy to model. On the other hand, inclusion of microstructural instability effects, if possible in certain cases, needs more thorough study. The same is true for static or dynamic aging effects.

Among the inadequately solved problems are :

- the ratchet effects when cyclic loading is superimposed on an approximately constant (mean stress) primary loading. The models described in this chapter must be at least redetermined, possibly modified, when the flows are very low at each cycle.

- the multiaxial effects under nonproportional loading. The kinematic hardening makes it possible to include a first type of anisotropy (translation of the elastic domain). However, certain experiments show second-order effects by rotation or change of shape of the elastic domain. Another aspect is that of the cyclic multiaxial out-of-phase loadings that have undergone little study up to now. On materials such as copper [69] and stainless steel [70], we observe the effects of additional cyclic hardening not described by the the classical approaches.

For Damage

Phenomenological models for the variation of damage exist and are well-identified for uniaxial loading (plasticity, fatigue, creep). They are based on an internal damage variable defined, for example, by means of an effective stress concept making it possible to continue considering the medium as continuous and homogeneous, in the framework of what we can call Damage Mechanics.

These models describe the evolution of the damage between the initial state and the crack initiation or rupture state, including the cumulative effects of the damage under variable loads or by superposition of various processes (fatigue and creep, for example).

The field that is now the widest open is that of
multiaxial loadings, in particular nonproportional loadings.
The two theories developed in this section to describe the
effects of the creep damage anisotropy yield similar
possibilities, but should be evaluated more thoroughly in
comparison with experimental data. The definition of the
fourth-rank tensor $\mathbb{D}$ fits easily into a thermodynamic
framework and describes the elastic behavior of the damaged
(anisotropic) medium. The theory of Murakami is more complex
in use and in identification for the description of the
behavior. Furthermore, the measurement of the tensor $\mathbb{D}$, and
thus the identification of its law of variation, is easier
using the effective stress concept, whereas the second-rank
tensor introduced by Murakami et al. requires measurements
of defects on the microstructural level. Thus the difficulty
associated with the large number of variables to be used in
the general case (36 instead of 6) is compensated partially
by more ease in identification.

The anisotropic damage laws make it possible to
describe the behavior of the damaged material more correctly,
but especially the damage evolution and thus variations in
the rupture conditions when the loading is not proportional.
A significant example is that of the two-level creep, e.g.
tensile followed by torsional, in which the time to rupture
is greater than the reference time to rupture under tension
for the same equivalent stress (isotropic case). The figure
36 illustrates the predictions that can be made. Depending on
the coefficients, the theory of Chaboche rather matches the
theory of Murakami [60] or the initial theory of Kachanov
[71].

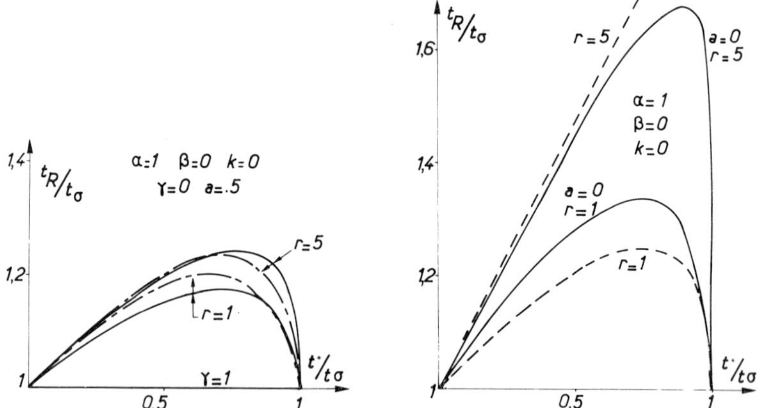

Fig. 36 - Tension then torsion : effect of the tension time
on the total rupture time

    a) —————— theory of Chaboche (a = 0.5),-------theory
of Murakami

    b) —————— theory of Chaboche (a = 0), ------- theory
of Kachanov

Among the phenomena still poorly reflected in the current models, we may mention :

- the effects of interaction with the environment. These effects can be introduced indirectly in fatigue mode by using, as in [72], a separation between the crack initiation and propagation phases.

- the evolution of the creep damage under compression (for lack of significant tests) and the possibilities of damage decrease ($\dot{D}<0$) that may occur in these phases of compression by the diminution of the cavities.

- the separation during tertiary creep of the effects related to a recovery of the hardening from the damage effects. These two phenomena increase the creep rate and they must be separated to apply the concept of effective stress correctly.

In summary, the equations describing the behavior and damage of metals and alloys offer large possibilities. Additional validations and improvements are to be expected from research undertaken in the fields of multiaxial loading, interactions with the microstructural modifications and with the environment.

REFERENCES

1. GERMAIN, P.. - Cours de Mécanique des Milieux continus, Tome 1, Masson, Paris, 1973.

2. SIDOROFF, F. - On the formulation of plasticity and viscoplasticity with internal variables, Arch. Mech., Poland, Vol. 27, 5-6, 1975, p. 807-819.

3. HALPHEN, B., NGUYEN, S. - Sur les matériaux standard généralisés, J. Mécanique, Vol. 14, n° 1, 1975, p. 39-63.

4. RABOTNOV, Y.N. - Creep problems in structural members, North Holland Publ. Comp., 1969.

5. LARSSON, B., STORAKERS, B. - A state variable interpretation of some rate-dependent inelastic properties of steel, J. of Engng Materials and Technology, Vol. 100, 1978, p. 395-401.

230

6.   LEMAITRE, J. - Sur la détermination des lois de comportement des matériaux élasto-viscoplastiques Publication ONERA n° 135 (1971).

7.   HARRISON, G.F., EVANS, W.J. - A creep deformation map for Nimonic 90 - Its construction, interpretation and implications for life prediction, Int. Conf. on Engineering aspects of Creep, Sheffields, 1980.

8.   OYTANA, C., DELOBELLE, P., MERMET, A. - Constitutive equations study in biaxial stress experiments, J. of Engng Mat. and Technology, Trans. ASME, Vol. 104 n° 1, 1982.

9.   CULIE, J.P., CAILLETAUD, G., and LASALMONIE, A. - Internal stress in viscoplasticity : comparison between mechanical and microscopic approaches - La Recherche Aérospatiale, 1982-2.

10. CARRY, C., STRUDEL, J.L. - Apparent and effective creep parameters in single crystals of a nickel base superalloy - secondary creep, Acta Metallurgica, Vol. 26, 1978, p. 859-870.

11. CAILLETAUD, G., CHABOCHE, J.L. - Macroscopic description of the microstructural changes induced by varying temperature. Example of IN 100 behaviour, ICM 3, Cambridge, 1979.

12. CAILLETAUD, G., CULIE, J.P., KACZMAREK, H. - Mechanical description of viscoplastic and damage behaviour in presence of microstructural instabilities induced by variable temperature, 3rd IUTAM Symp. on Creep in Structures, Leicester, 1980.

13. SCHMIDT, C.G., MILLER, A.K. - A unified phenomenological model for non-elastic deformation of type 316 Stainless steel - Part I : development of the model and calculation of the material constants - Res. Mechanica, Vol. 3, 1981, p. 109-129.

14. LAGNEBORG, R. - A modified recovery-creep model and its evaluation, Met. Sci. J., Vol. 6, 1972, p. 127-133.

15. MERZER, A., BODNER, S.R. - Analytical formulation of a rate and temperature dependent stress-strain relation, J. of Engng. Materials and Technology, Trans. ASME, Vol. 101, 1979, p. 254-257.

16. MALININ, N.N., KHADJINSKY, G.M. - Theory of creep with anisotropic hardening, Int. J. Mech. Sci., Vol. 14, 1972, p. 235-246.

17. LEE, D., ZAVERL, F. Jr. - A generalized strain rate
     dependent constitutive equation for anisotropic metals,
     Acta Metallurgica, Vol. 29, 1978, p. 1771-1780.

18. CHABOCHE, J.L. - Viscoplastic constitutive equations
     for the description of cyclic and anisotropic behaviour
     of metals, Bull. de l'Acad. Polonaise des Sciences,
     Série Sc. et Tech., Vol. 25, n° 1, 1977, p. 33-42.

19. MORETON, D.N., MOFFAT, D.G., PARKINSON, D.B. - The
     yield surface behaviour of pressure vessel steels, J. of
     strain analysis, Vol. 16, n° 2, 1981, p. 127-136.

20. PRAGER, W. - Recent developments in the mathematical
     theory of plasticity, J. Appl. Phys., Vol. 20, n° 3,
     1949, p. 235-241.

21. ARMSTRONG, P.J., FREDERICK, C.O. - A mathematical
     representation of the multiaxial Bauschinger effect,
     C.E.G.B. report RD/B/N 731, 1966.

22. CHABOCHE, J.L., ROUSSELIER, G. - On the plastic and
     viscoplastic constitutive equations based on the
     internal variables concept, 3rd Int. Seminar "Inelastic
     Analysis and Life Prediction in High Temperature
     Environment" SMIRT-6, Paris, 1981.

23. GOODALL, I.W., HALES, R., WALTERS, D.J. - On
     constitutive relations and failure criteria of an
     austenitic steel under cyclic loading at elevated
     temperature, IUTAM Symp., Leicester, 1980, C.E.G.B.,
     Report RB/B/N 4916.

24. WALKER, K.P., - Research and development program for
     non-linear structural modeling with advanced time-
     temperature dependent constitutive relationships,
     Rapport PWA-5700-50, NASA CR-165533, 1981.

25. CHABOCHE, J.L., DANG-VAN K., CORDIER G. - Modelization
     of the strain memory effect on the cyclic hardening of
     316 stainless steel, SMIRT-5, Division L, Berlin, 1979.

26. BOLLINGER, E., CAILLETAUD, G., CHABOCHE, J.L.,
     DUFAILLY, J., KACZMAREK, H., LIEURADE, H.P., MARQUIS,
     D., RIBES, A. - Interaction de l'écrouissage et de
     l'endommagement sous sollicitations complexes dans un
     acier inoxydable austénique, rapport DDSTI, 1982.

27. OHNO, N. - A constitutive model of cyclic plasticity
     with a non-hardening strain region, to be published in
     ASME Journal of Applied Mechanics.

232

28. ROBERT, G., SILVENT, A. - 4th International Conference on the strength of Metals and Alloys. Nancy, Août 1976.

29. CHO, U.W., FINDLEY, W.N. - Creep and creep recovery of 304 stainless steel at low stresses with effects of aging on creep and plastic strains, ASME J. of Applied Mechanics, Vol. 48, 1981, p. 785.

30. STRUDEL, J.L. - Interactions des dislocations avec des impuretés mobiles, "Dislocation et déformation plastique", Ecole d'Eté Yravals, 1979, Les Editions de Physique, p. 199-222.

31. Etude des caractéristiques mécaniques à chaud d'une tôle en acier inoxydable austénitique 18.12 au mobybdène à très bas carbone et azote contrôlé, nuance ICL 167 SPH de Creusot-Loire, EdF, Département Etude des Matériaux, HT/PV D. 465 MAT/T. 43 du 11 Juillet 1980.

32. CAILLETAUD, G., POLICELLA, H., BAUDIN, G. - Mesure de déformation et d'endommagement par méthode électrique, La Recherche Aérospatiale, n° 1980-1.

33. PIATTI, G., BERNASCONI, G., COZZARELLI, F.A. - Damage equations for creep rupture in steels, Paper L11/4, SMIRT-5 Conference, Berlin, 1979.

34. LEMAITRE, J., CHABOCHE, J.L. - Aspect phénoménologique de la rupture par endommagement, J. Mécanique Appliquée, Vol. 2, n° 3, 1978, p. 317-365.

35. CORDEBOIS, J.P. - Comportement et résistance des milieux métalliques multiperforés. Thèse 3è cycle, laboratoire de Mécanique et Technologie, Paris VI, 1976.

36. DELAMETER, W.R., HERMANN, G. - Weakening of elastic solids by doubly-periodic arrays of cracks, "Topics in Applied Continuum Mechanics", Springer Verlag, 1974.

37. MAJUMDAR, S., MAIYA, P.S. - A mechanistic model for time-dependent fatigue, ASME, J. of Engng. Materials and Technology, 1980, Vol. 102, p. 159.

38. OSTERGREN, W.J., KREMPL, E. - A uniaxial damage accumulation law for time-varying loading including creep-fatigue interaction, ASME, J. of Pressure Vessel Technology, Vol. 101, 1979, p. 118.

39. BUI-QUOC T., DUBUC, J., BAZERGUI, A., BIRON, A. - Cumulative fatigue damage under stress controlled conditions, J. of Basic Engng., Trans. ASME, 1971, p. 691-698.

40. CHRZANOWSKI, M. - Use of the damage concept in describing creep-fatigue interaction under prescribed stress, Int. J. Mech. Sci., Vol. 18, 1976, p. 69-73.

41. LEMAITRE, J., CHABOCHE, J.L. - A non-linear model of creep-fatigue damage cumulation and interaction, in "Mechanics of Visco-Plastic Media and Bodies", Ed. Jan Hult-Springer, 1975, p. 291-301.

42. PLUMTREE, A., LEMAITRE, J. - Application of damage concepts to predict creep-fatigue failures, ASME. Pressure Vessel and Piping Conf., Montreal, June 1978, Paper n° 78-PLP-26, Trans. of the ASME, Vol. 101 (1979), pp. 284-292.

43. MANSON, S.S., HALFORD, G.R., HIRSCHBERG, M.H. - Creep-fatigue analysis by strain range partitioning - Symp. on Design for elevated temperature environment, ASME, NASA TMX 67838, 1971.

44. KACHANOV, L.M. - Time of the rupture process under creep conditions, Izv. Akad. Nauk. SSR OTd Tekh. Nauk, n° 8, 1958.

45. RABOTNOV, Y.N. - Creep rupture, 12th Int. Congress of Applied Mechanics, Stanford, 1968.

46. STORAKERS, B. - Finite creep of a circular membrane under hydrostatic pressure, Acta Polytechnica scandinavica, Mech. Engng Series, N° 44, 1969.

47 CAILLETAUD, G., CHABOCHE, J.L. - Life time predictions in 304 stainless steel by damage approach. Conf. ASME-PVP, Orlando, 1982.

48 CHABOCHE, J.L. - Description thermodynamique et phénoménologique de la viscoplasticité cyclique avec endommagement, thèse univ. Paris VI et publication ONERA n° 1978-3.

49 MANSON, S.S., NACHTIGALL, A.J., ENSIGN, C.R., FRECHE, J.C., - Further investigation of a relation for cumulation fatigue damage in bending, Trans. of ASME, J. of Engng. for Industry, 1965.

50 CHABOCHE, J.L. - Une loi différentielle d'endommagement de fatigue avec cumulation non linéaire. Revue Française de Mécanique, n° 50-51, 1974. English translation in "Annales de l'ITBTP", HS 39, 1977.

51 SUBRAMANYAN, S., - A cumulative damage rule based on the knee point of the S.N. curve, Trans. of ASME, J. of Mat. and Technology, 1976, p. 316-321.

52   MANSON, S.S. - Some useful Concepts for the Designer in
     Treating Cumulative Damage at Elevated Temperature",
     I.C.M. 3, Cambridge, 1979, Vol. 1, pp. 13-45.

53   CHABOCHE, J.L. - Sur l'utilisation des variables d'état
     interne pour la description du comportement
     viscoplastique et de la rupture par endommagement,
     Symp. Franco-Polonais de Rhéologie et Mécanique,
     Cracovie, 1977.

54   LECKIE, F.A., HAYHURST, D.R. - Creep rupture of
     structures, Proc. Royal Soc., London, Vol. 340, 1974,
     p. 323-347.

55   SINES, G., - Behavior of metals under complex static
     and alternating stresses, Metal Fatigue, eds. G. Sines,
     J.L. Waisman (Mcgraw-Hill, New York, 1959) pp. 145-169.

56   DANG VAN, K. - Sur la résistance à la fatigue des
     métaux, Sciences et Techniques de l'Armement, 47, 3ème
     Fasc. (1973).

57   BROWN, M.W., and MILLER, K.J. - A theory for Fatigue
     Failure Under Multiaxial Stress-Strain Conditions,
     Proceedings of the Institute of Mechanical Engineers,
     Vol. 187, 1973, pp. 745-755.

58   JACQUELIN, B., HOURLIER, F., PINEAU, A. - Assessment of
     a single criterion for crack initiation under low cycle
     multiaxial fatigue loading. Application to austenitic
     stainless steel. Proceed. SMIRT Conf., Paris, Août
     1981.

59   HAYHURST, D.R. - Creep rupture under multi-axial state
     of stress, J. Mech. Phys. Solids, Vol. 20, n° 6, 1972,
     p. 381-390.

60   MURAKAMI, S., OHNO, N. - A constitutive equation of
     creep damage in polycristalline metals, IUTAM Coll.
     EUROMECH 111, "Constitutive Modelling in Inelasticity",
     Marienbad, Tchécoslovaquie, 1978.

61   MURAKAMI, S., OHNO, N. - A continuum theory of creep
     and creep damage, 3rd IUTAM Symp. on Creep in
     Structures, Leicester, 1980.

62   MURAKAMI, S. - Notion of continuum damage, mechanics
     and its application to anisotropic creep damage theory,
     ASME Pressure Vessel and Piping Int. Conf., Orlando,
     1982.

63   RIVLIN, R.S., ERICKSEN, J.L. - Stress deformation
     relations for isotropic materials, J. of Rational
     Mechanics and Analysis, Vol. 4, 1955, p. 323.

64 MURAKAMI, S., IMAIZUMI, T. - Mechanical description of creep damage state and its experimental verification, EUROMECH 147, "Damage Mechanics", Cachan, France, 1981.

65 CHABOCHE, J.L. - Le concept de contrainte effective appliqué à l'élasticité et à la viscoplasticité en présence d'un endommagement anisotrope, Coll. EUROMECH 115, Grenoble, 1979, édition CNRS, 1982.

66 DUVAUT, G. - Analyse fonctionnelle - Mécanique des milieux continus - Homogénéisation, "Theoretical and Applied Mechanics", North-Holland, 1976.

67 LITEWKA, A. - Experimental study of the effective yield surface of perforated materials, Nuclear Engineering and Design 57 (1980) 417-425.

68 CORDEBOIS, J.P., SIDOROFF, F. - Anisotropie élastique induite par endommagement, Colloque EUROMECH 115, "Comportement Mécanique des Solides Anisotropes", Grenoble, 1979.

69 LAMBA, H.S., SIDEBOTTOM, O.M. - Cyclic plasticity for nonproportional paths. Part 1 : cyclic hardening, erasure of memory, and subsequent strain hardening experiments, J. of Engng. Materials and Technology, Vol. 100, 1978, p. 98-103.

70 NOUAILHAS, D., POLICELLA, H., KACZMAREK, H. - On the description of cyclic hardening under complex loading histories, Int. Conf. on Constitutive Laws for Engineering Materials, Tucson, 1983.

71 KACHANOV, L.M. - Foundations of Fracture Mechanics, Nauka, Moscou, 1974.

72 SAVALLE, S., CAILLETAUD, G. - Microamorçage, micropropagation et endommagement, La Recherche Aérospatiale, 1982-6

Chapter 5

CREEP AND VISCOPLASTIC BRITTLE RUPTURE OF STRUCTURES BY
THE FINITE ELEMENT METHOD

O. J. A. GONCALVES F[o].
Instituto de Engenharia Nuclear - CNEN, Cidade Universitária,
Caixa Postal 2186, Rio de Janeiro, Brazil.

D. R. J. OWEN
Department of Civil Engineering, University of Wales, Swansea,
U.K.

ABSTRACT

A numerical investigation of creep and viscoplastic brittle
rupture of structures is undertaken by the finite element method.
Creep material behaviour under multiaxial stress states is
modelled by the three-dimensional generalization of the
Kachanov-Rabotnov constitutive equations due to Leckie and
Hayhurst while visco-plastic behaviour is represented by a
simplified constitutive model first introduced in this work.
Geometric nonlinear effects (large displacements and large
rotations) are accounted for by an updated Lagrangian formulation. The differential governing equations are discretised
in the space domain by isoparametric finite elements and an
implicit finite difference scheme is developed for time integration. Special attention is dedicated to the solution of the
stiff differential equations that govern damage growth.
Finally some numerical applications are presented.

238

1. INTRODUCTION

The requirement of higher pressures and temperatures in order to improve thermodynamic efficiency has led in recent years to an increased study of creep damage and rupture of heat resistant alloys used in the construction of gas turbines, turboreactors, thermal plants and thermonuclear installations. For nuclear plants in which even a small leakage could have a disastrous effect on the surroundings, this study assumes a particular importance.

Current metallurgical investigations indicate that creep damage of polycrystalline metals occurs by the nucleation and growth of grain boundary defects. The effect of their growth is to cause a progressive weakening of the material and an increase in the creep rate over a significant fraction of the structure lifetime. Rupture usually takes place by the nucleation and growth of grain boundary defects as shown by Ashby and Raj [1]. Other important works following a material science approach, whose main objective is to understand the mechanisms of nucleation, growth and rupture, are those of Dyson and McLean [2], Greenwood [3] and Cocks and Ashby [4].

The engineering approaches formerly adopted to estimate the rupture time of structures were based on determining the elastic or the steady state stress distribution and to use the corresponding maximum principal stress component together with the material uniaxial rupture data, such as those given in ASME Code Case N47-12. These approaches have proved deficient on two accounts by ignoring firstly, the influence of multiaxial stress states upon the rupture time and, secondly, the softening effect exhibited by metals over the tertiary creep range which allows stresses to be transferred from a highly to a less damaged zone and, consequently, increasing the service life of the structure.

In order to overcome these deficiencies Kachanov [5] and Rabotnov [6] introduced the concept of damage into the constitutive equations in the form of a scalar quantity referred to as the damage parameter $\underline{w}$. The Kachanov-Rabotnov equations, which in their original one dimensional form can be shown to obey the Linear Life Fraction Rule due to Robinson [7], were generalized to multiaxial stress states by Leckie and Hayhurst [8]. Their generalization assumes isotropy of damage or incorporate anisotropy of damage only with respect to the principal stress coordinates fixed in the material and in consequence it should only be applied to proportional loading conditions. For more complex stress histories due to non-proportional loadings, voids and cracks develop predominantly on the grain boundaries of certain preferred orientations corresponding to the stress state at each instant. Clearly for these cases, more elaborate theories are required introducing a damage tensor instead of a scalar.

The first characterization of damage in the form of a second-order tensor is due to Vakulenko and Kachanov [9]. The second-order model was further developed by many authors including Murakami and Ohno [10], Srinavasan et al. [11] and again Kachanov [12]. A fourth-order model was proposed by Chaboche [13]. Most of these theories, however, are restricted to formal descriptions and the relations between the tensorial variables introduced and the actual mechanism of microstructural change in the material (as well as the mechanisms through which the variables affect the process of creep and damage) have not been sufficiently clarified. It is evident furthermore that the determination of such relations will be both time-consuming and expensive. Considering these factors Trampczynski, Hayhurst and Leckie [14,15] have recently developed an experimental program to assess the validity of using the single stage variable theory for non-proportional loading conditions. The experiments were conducted on commercially pure copper manufactured to British Standards Specification BS2873-CIDI, on aluminium Cu,Fe, Ni,Mg and Si alloy manufactured to BS1472 and on Nimonic 80A. The copper and aluminium alloy were chosen because their mechanisms of failure represent extreme types of behaviour [16]. The commercially pure copper fails according to a maximum principal tensile stress criterion and the aluminium alloy by a maximum effective stress criterion. The experimental results showed that for engineering cases of non-proportional loading the deformation and rupture behaviour of the materials studied can be described by one state variable theory provided that the directional characteristics of the damage process are taken into account during the integration of the damage rate equations. For copper, the damage rate equations must be integrated separately for each set of planes perpendicular to the directions of the maximum principal tension stress. Failure occurs when the damage reaches a critical value on one set of planes. For aluminium, damage growth is in fact of a scalar nature and therefore the single state variable theory can be applied unchanged. For Nimonic 80A, life prediction under non-proportional loading is more complex as it is not sufficient to know only the multiaxial stress rupture criterion. Nevertheless, life predictions obtained using the single state variable theory tend to be on the conservative side.

Another source of nonlinearity which may have importance in rupture calculations is the change of geometry undergone by structures under loading. In high temperature design of pressure vessels, for instance, it is usual practice to assume that the geometry remains unaltered during the life of the component. In certain cases, this standard procedure leads to a considerable error in life prediction [17] which indicates that, at least for selected applications, geometric nonlinear effects should not be disregarded in rupture calculations.

In the present work the objectives are threefold:

240

(1) To develop an implicit algorithm (Runge-Kutta scheme of
     2nd order) for the numerical analysis of creep rupture
     problems by the finite element method. The multiaxial
     form of the Kachanov-Rabotnov equations proposed by Leckie
     Hayhurst is used. Geometric nonlinear effects (large dis-
     placements and large roations) are accounted for by an
     updated Lagrangian formulation. Special attention is
     dedicated to the accuracy of the integration scheme and in
     particular to the solution of the highly nonlinear differ-
     ential equations that govern damage growth.

(2) To propose a simplified model to describe multiaxial def-
     ormation and rupture of bodies undergoing elastoviscoplastic
     plastic deformation. This model is based on the results
     obtained by Lemaitre and Chaboche [18] in the study of one
     dimensional viscoplastic problems and by Leckie and
     Hayhurst [8] in the investigation of multiaxial creep
     damage in metals. The assumptions and limitations of the
     model are discussed in detail.

(3) To extend the implicit algorithm originally developed to
     study creep damage and rupture to deal with the more gene-
     ral case of elastoviscoplastic deformation. Attention is
     addressed to the problem of numerical evaluation of strain
     hardening (or softening) growth.

The layout of this work is as follows: in Section 2 the
constitutive equations and rupture criteria for the creep problem are presented. The simplified model for the elastoviscoplastic problem is also introduced and its underlying assumptions discussed. Section 3 starts with a general statement of
the initial boundary value problem investigated. After a brief
introduction of the finite element method geometric nonlinear
effects are considered using an update Lagrangian formulation.
The numerical algorithm developed is then introduced in the context of its more general elastoviscoplastic application. This
section is concluded with a discussion of the damage rate equations and in particular of a change of variable aiming to diminish the degree of nonlinearity of the original equations is
presented. In Section 4 numerical results for perforated plates
subjected to equal biaxial tension loads are presented and compared with reported experimental results. Finally in Section 5
some conclusions are drawn from this work.

2. CONSTITUTIVE EQUATIONS AND RUPTURE CRITERIA

2.1 Generalized Kachanov and Rabotnov Equations

Based on the experimental results of Johnson et al. [19]
for tension-torsion tests performed on copper and aluminium
tubes, Leckie and Hayhurst [8] proposed a generalization to
multiaxial stress states of the single state theory of Kachanov
and Rabotnov. In their work the constitutive equations

describing the creep strain rate and the damage rate can be written as, respectively,

$$\frac{\partial \varepsilon^{c}_{ij}}{\partial t} = \gamma_{f} \left(\frac{\bar{\sigma}}{1-w}\right)^{N} \frac{\partial F}{\partial \sigma_{ij}} \qquad (2.1)$$

and

$$\frac{\partial w}{\partial t} = A \frac{\sigma_{eq}^{\chi}}{(1-w)^{\phi}} \qquad (2.2)$$

where ε^{c}_{ij} is the tensor of infinitesimal strains, σ_{ij} the Cauchy stress tensor, $\bar{\sigma}$ the Von Mises, Tresca or the maximum principal stress, F a potential function, γ_{f}, N, χ, A and ϕ temperature dependent properties and σ_{eq} a scalar quantity that takes account of the rupture characteristics of the material under multiaxial stress states. In a general form

$$\sigma_{eq} = \alpha_{1}\sigma_{1} + \alpha_{2}\tau_{max} + \alpha_{3}J_{1} + \alpha_{4}\sigma_{eff} \qquad (2.3)$$

where σ_{1} is the maximum principal tensile stress, τ_{max} the maximum shear stress, J_{1} the first stress invariant, σ_{eff} the effective stress and α_{i} (i=1,4) scalar quantities such that $\alpha_{1}+\alpha_{2}+\alpha_{3}+\alpha_{4} = 1.0$. The cases ($\alpha_{1}=1$, $\alpha_{2}=\alpha_{3}=\alpha_{4}=0$) and ($\alpha_{1}=\alpha_{2}=\alpha_{3}=0$, $\alpha_{4}=1$) represent the extreme types of rupture behaviour and materials which fail under these conditions are said to obey a maximum principal tension stress rupture criterion and a maximum effective tension stress criterion, respectively. Copper ($\alpha_{1}=1$) and aluminium alloys ($\alpha_{4}=1$) are typical examples of extreme behaviours while the behaviour of other alloys, particularly creep resistant ones, fall between the two extremes. The rupture results can be most conveniently presented in terms of an isochronous surface which is the curve obtained by connecting stress states with equal rupture times. By selecting appropriate values for the α_{i} scalars the known isochronous surfaces can be represented. The general form of these curves is shown in Figure 1. To predict the variations of strain and damage for variable stress histories, Leckie and Hayhurst [20] introduced the concept of constant damage contours and suggested a testing procedure for their determination. A constant damage contour is illustrated in Figure 2 and the procedure to account for variable stress histories is shown in Figure 3. In the initial undamaged state w=0 and rupture is deemed to have occurred when w reaches some critical value, usually taken as 1.

2.2 Non-homogenuity of void growth and its influence on the creep deformation

Material damage in metals under constant uniform stress advances uniformly in the material until a certain critical stage is reached when cavities start to coalesce and to grow rapidly. It has been recently argued [21] that such local

242

features of damage growth should not affect fracture and deformation processes equally since the former is a local phenomenon
and the latter a global one. It has been argued, furthermore,
that the creep rate is less sensitive to the damage state than
the rate of damage growth.

Following Ref. [21] the non-homogenuity of damage growth
and its influence on creep deformation is taken into account in
this work by redefining eq. (2.1) as

$$\frac{\partial \varepsilon^{c}_{ij}}{\partial t} = \gamma_f \left(\frac{\bar{\sigma}}{1-cw}\right)^{N} \frac{\partial F}{\partial \sigma_{ij}} \tag{2.4}$$

where c $(0 \leq c \leq 1)$, the homogenuity factor, depends strictly
speaking, on the stress and temperature. In the absence of more
precise experimental values, however, an average value over the
stress and temperature range of interest can be adopted.

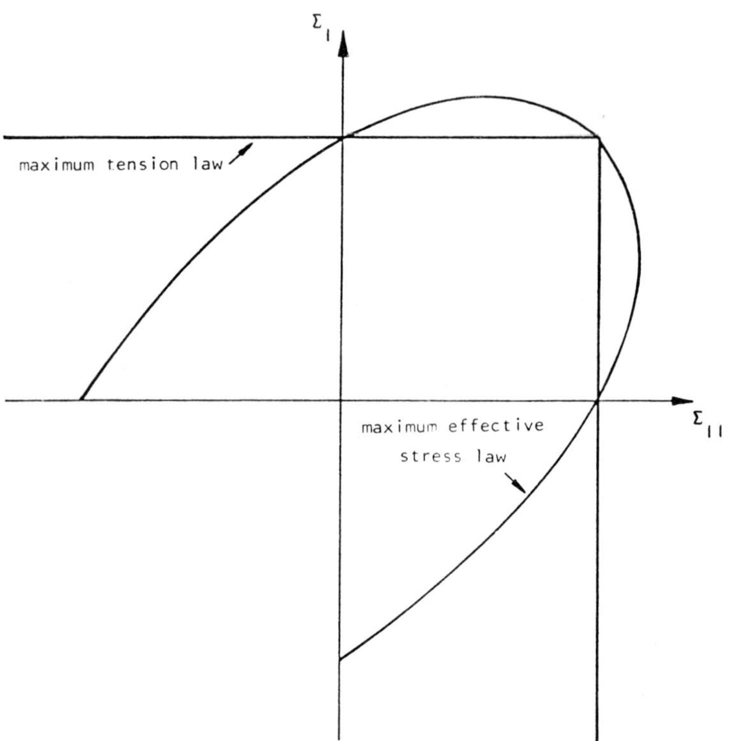

Fig. 1 _ Typical plane stress isochronous rupture locii
(Σ_I and Σ_{II} - normalized principal stress)

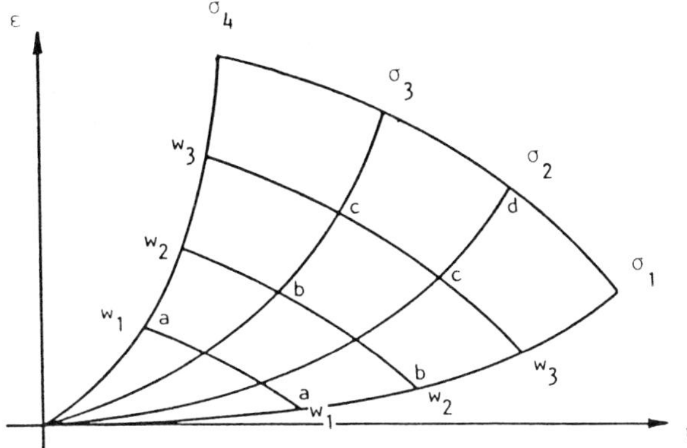

Fig. 2 _ Constant damage contours

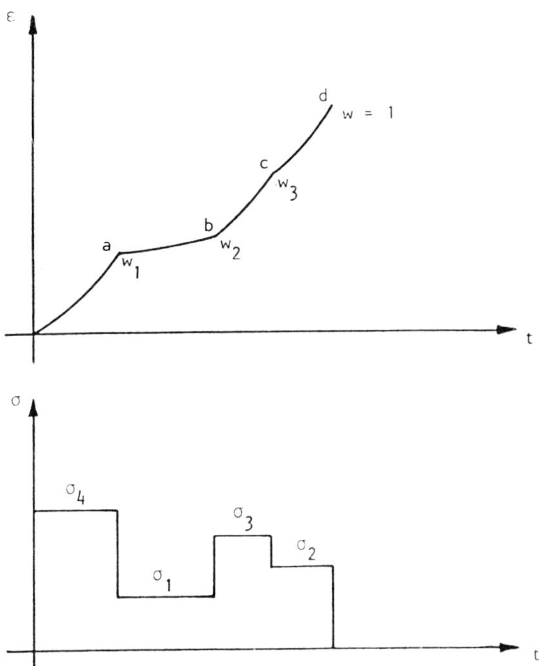

Fig. 3 _ Prediction of variable stress histories

2.3   The Elastoviscoplastic Model

Since the original propositions by Kachanov and Rabotnov
for creep rupture of metals, several studies conducted in different laboratories have shown the possibilities of their original approach. These studies initiated the development of a
new branch of Fracture Mechanics termed Continuous Damage Mechanics [22].

244

In the case of high temperature viscoplastic behaviour of
some refractory alloys, particular strain rate equations
consistent with a general thermodynamic framework, have been
developed by Lemaitre and Chaboche [18,23,24] for one-dimensional
problems. Their formulation contains many descriptive possibilities such as nonlinear hardening, Bauschinger effect,
creep and relaxation, nonlinear kinematic hardening, stabilized
cyclic behaviour, cyclic hardening or softening through isotropic hardening dependence and high temperature recovery
effects.

In the present work a simplified model is proposed to describe viscoplastic damage under multiaxial stress. It attempts
to bring together the contributions of Lemaitre and Chaboche for
the analysis of one-dimensional viscoplastic problems and of
Leckie and Hayhurst for the study of multiaxial creep rupture
problems. The basic assumptions made are:

(1)   viscoplastic strains are incompressible;

(2)   hardening is linear isotropic;

(3)   damage is isotropic;

(4)   damage and multiaxial viscoplastic rupture are overstress
      controlled processes;

(5)   elastic properties are unaffected by material damage;

(6)   creep and viscoplastic multiaxial rupture criteria are
      equivalent.

The first assumption stating that during viscoplastic rupture there is no increase in volume, follows as an extension of
the observations made by Boettner and Robertson [25] in the
course of experiments to measure volume changes during creep
rupture. The experiments showed that, until immediately before
rupture at least, the increase in volume is small and, consequently, that the assumption of zero volume-change is not seriously violated. The linear isotropic hardening rule is assumed
for simplicity. The assumption of isotropy of damage for a
body undergoing viscoplastic deformation follows the same line
of argument discussed in detail in the introductory section for
the creep phenomenon. The fourth assumption is consistent with
the basic idea that the overstress controls the viscoplastic
phenomenon. It is implicitly assumed, in consequence, that in
the elastic domain damage is neither produced or eliminated.
The fifth assumption is introduced for simplicity although the
consideration of damage effects on the elastic properties in the
sense proposed by Lemaitre and Chaboche [18], does not pose
special problems for the numerical modelling. The sixth and
final assumption is probably the most arguable one as it implies
that the same physical mechanisms control the formation, growth
and coalescence of voids in both the viscoplastic and the creep

brittle rupture processes. It is worth recalling that for most metallic alloys tested within the temperature range of 0.4 to 0.6 T_m, where T_m is the melting point of the basic metal in degrees Kelvin, the type of rupture is strongly influenced by the stress level. At high fractions of the material yield stress, failure occurs relatively quickly and is accompanied by considerable elongation and necking. At reduced stresses the time to rupture increases, the deformation diminishes and the failure tends to be of a brittle nature. In an elastoviscoplastic material, on the other hand, no viscous deformation process is activated before the threshold yield stress is reached and from then on it is only the resulting overstress (reduced stresses that are a small fraction of the total acting stresses) that controls the deformation process. Thus, provided that the total strains are small at rupture, it is conceivable that the mechanisms of brittle rupture be the same in both the viscoplastic and creep processes.

The constitutive equations proposed in this work to describe elastoviscoplastic deformations in multiaxial stress states are:

$$\frac{\partial \varepsilon_{ij}^{vp}}{\partial t} = \gamma_f \; < \; \Phi \; (f,Y,w) \; > \; \frac{\partial F}{\partial \sigma_{ij}} \qquad (2.5)$$

$$\frac{\partial w}{\partial t} = A \; < \; \frac{\hat{\sigma}_{eq}(f,Y)}{(1-w)^{\phi}} \; >^{\chi} \qquad (2.6)$$

and

$$Y = \sigma_y + H' \; \bar{\varepsilon}^{-vp} \qquad (2.7)$$

where f is the Von Mises, Tresca or maximum principal stress, Y the current value of the yield stress, F a potential function (F=f-Y), σ_y the material uniaxial yield stress, H' the hardening (H'>0) or softening (H'<0) parameter, $\bar{\varepsilon}^{-vp}$ the effective viscoplastic strain and the remaining quantities as described in Section 2.1. The notation < > implies that

$$< y(x) > = y(x), \qquad \text{for } x > 0$$
$$\qquad \qquad \qquad \qquad \qquad \qquad (2.8)$$
$$< y(x) > = 0, \qquad \qquad x \leq 0$$

The function $\Phi(f,Y,w)$ is given by the following normalized expression

$$\Phi(f,Y,w) = \left(\frac{f-Y}{(1-w)Y} \right)^{N} \qquad (2.9)$$

where N is a material constant. The $\hat{\sigma}_{eq}$ function is written, according to the 5th and 6th assumptions as

$$\hat{\sigma}_{eq} = \alpha_1 \hat{\sigma}_1 + \alpha_2 \hat{\tau}_{max} + \alpha_3 \hat{J}_1 + \alpha_4 \hat{\sigma}_{eff} \qquad (2.10)$$

where $\hat{\sigma}_q$, $\hat{\tau}_{max}$, $\hat{J}_1$ and $\hat{\sigma}_{eff}$ and the α_i (i=1,4) have the same meaning as in Section 2.1 except that they refer to the over-stress only, which is indicated by the superposed hat. The tensor of overstress acting in any particular point subject to a stress state σ_{ij} is given by

$$\hat{\sigma}_{ij} = \langle 1 - \frac{Y}{f} \rangle \; \sigma_{ij} \tag{2.11}$$

One result which proves very useful when programming equations (2.5) to (2.11) is that σ_{eq} is a linear function of σ_{ij}, i.e.,

$$\sigma_{eq}(\lambda\sigma_{ij}) = \lambda\sigma_{eq}(\sigma_{ij}), \; \forall\lambda \in \mathbb{R} \tag{2.12}$$

This property is demonstrated below by direct substitution: (in the indicial notation employed, δ_{ij} is the Kronecker delta and Einstein's summation convention is invoked).

Let σ'_{ij} be the deviatoric component of the stress tensor σ_{ij} and J'_2 and J'_3 its second and third invariants, respectively, i.e.

$$\sigma'_{ij} = \sigma_{ij} - \frac{1}{3} \delta_{ij} \; \sigma_{kk} \tag{2.13}$$

$$J'_2 = \frac{1}{2} \sigma'_{ij} \; \sigma'_{ij} \tag{2.14}$$

$$J'_3 = \frac{1}{3} \sigma'_{ij} \; \sigma'_{jk} \; \sigma'_{ki} \tag{2.15}$$

Following a formulation due to Nayak [26] the maximum principal stress and the maximum shear stress are written as

$$\sigma_1 = \frac{2}{\sqrt{3}} \sqrt{J'_2} \; \sin\left(\theta + \frac{2\pi}{3}\right) + \frac{J_1}{3} \tag{2.16}$$

$$\tau_{max} = \frac{2}{\sqrt{3}} \sqrt{J'_2} \left[\sin\left(\theta + \frac{2\pi}{3}\right) - \sin\left(\theta + \frac{4\pi}{3}\right)\right] \tag{2.17}$$

where J_1 is the first stress invariant and θ is given by

$$\theta = \sin^{-1}\left(- \frac{3\sqrt{3}}{2} \; \frac{J'_3}{J'_2{}^{3/2}}\right) \tag{2.18}$$

With the help of expressions (2.13) to (2.18) each term of $\sigma_{eq}(\lambda\sigma_{ij})$ in eq. (2.10) can now be evaluated. First it is noted that

$$\sigma'_{ij}(\lambda\sigma_{ij}) = \lambda\sigma_{ij} - \frac{1}{3} \delta_{ij} \lambda\sigma_{kk} = \lambda\sigma'_{ij}(\sigma_{ij}) \tag{2.19}$$

$$J_1(\lambda\sigma_{ij}) = \lambda\sigma_{ii} = \lambda J_1(\sigma_{ij}) \tag{2.20}$$

$$J'_2(\lambda\sigma_{ij}) = \frac{1}{2}(\lambda\sigma'_{ij})(\lambda\sigma'_{ij}) = \lambda^2 \, J'_2(\sigma_{ij}) \tag{2.21}$$

$$J'_3(\lambda\sigma_{ij}) = \frac{1}{3}(\lambda\sigma'_{ij})(\lambda\sigma'_{jk})(\lambda\sigma'_{ki}) = \lambda^3 \, J'_3(\sigma_{ij}) \tag{2.22}$$

and

$$\begin{aligned}
\theta(\lambda\sigma_{ij}) &= \sin^{-1}\left[-\frac{3\sqrt{3}}{2}\ \frac{J'_3(\lambda\sigma_{ij})}{J'_2{}^{3/2}(\lambda\sigma_{ij})}\right] = \\[2mm]
&= \sin^{-1}\left[-\frac{3\sqrt{3}}{2}\ \frac{\lambda^3 \, J'_3(\sigma_{ij})}{\lambda^3 \, J'_2{}^{3/2}(\sigma_{ij})}\right] = \\[2mm]
&= \theta(\sigma_{ij})
\end{aligned} \tag{2.23}$$

therefore

$$\begin{aligned}
\sigma_1(\lambda\sigma_{ij}) &= \frac{2}{3}\sqrt{J'_2(\lambda\sigma_{ij})}\ \sin\left(\theta(\lambda\sigma_{ij}) + \frac{2\pi}{3}\right) + \frac{J_1(\lambda\sigma_{ij})}{3} = \\[2mm]
&= \frac{2}{3}\ \lambda\sqrt{J'_2(\sigma_{ij})}\ \sin\left(\theta(\sigma_{ij}) + \frac{2\pi}{3}\right) + \lambda\,\frac{J_1(\sigma_{ij})}{3} = \\[2mm]
&= \lambda\ \sigma_1(\sigma_{ij})
\end{aligned} \tag{2.24}$$

$$\begin{aligned}
\tau_{max}(\lambda\sigma_{ij}) &= \frac{2}{\sqrt{3}}\sqrt{J'_2(\lambda\sigma_{ij})}\left[\sin\left(\theta(\lambda\sigma_{ij}) + \frac{2\pi}{3}\right) - \right.\\[2mm]
&\quad \left. - \sin\left(\theta(\lambda\sigma_{ij}) + \frac{4\pi}{3}\right)\right] = \\[2mm]
&= \frac{2}{3}\ \lambda\sqrt{J'_2(\sigma_{ij})}\left[\sin\left(\theta + \frac{2\pi}{3}\right) - \sin\left(\theta + \frac{4\pi}{3}\right)\right] = \\[2mm]
&= \lambda\ \tau_{max}(\sigma_{ij})
\end{aligned} \tag{2.25}$$

$$J_1(\lambda\sigma_{ij}) = \lambda\sigma_{ii} = \lambda J_1(\sigma_{ij}) \tag{2.26}$$

$$\begin{aligned}
\sigma_{eff}(\lambda\sigma_{ij}) &= \sqrt{3\,J'_2(\lambda\sigma_{ij})} = \sqrt{3\,\lambda^2\,J'_2(\sigma_{ij})} = \\[2mm]
&= \lambda\ \sigma_{eff}(\sigma_{ij})
\end{aligned} \tag{2.27}$$

It is now evident from expressions (2.24) to (2.27) that

$$\sigma_{eq}(\lambda\sigma_{ij}) = \lambda\sigma_{eq}(\sigma_{ij}) \tag{2.28}$$

248

Finally in concluding this section it is worth noting that the constitutive equations (2.1) to (2.3) proposed by Leckie and Hayhurst for the analysis of creep damage and rupture in multiaxial stress states are readily obtained as a particular case of the more general elasto-viscoplastic model suggested in this work (eqs. (2.5) to (2.8)) by setting the threshold yield stress σ_y and the hardening (softening) parameter H' to zero.

3. PROBLEM STATEMENT AND FINITE ELEMENT APPROXIMATION

3.1 Problem statement

Let Ω be a bounded region in R^n (n=2,3) with a piecewise smooth boundary Γ and the ordered pair (x,t) denote the position of a general point x in R^n at time t. The quasi-static initial boundary value problem of interest is governed by the following equations where, vector and tensor fields on Ω are written in standard cartesian notation, a comma indicates differentiation with respect to the cartesian coordinates, a superposed dot time differentiation and the summation convention for repeated indices is employed:

- Equilibrium Equations:

$$\sigma_{ij,j} + b_i = 0 \tag{3.1}$$

where σ_{ij} is the Cauchy stress tensor and b_i the body forces.

- Constitutive Equations:

$$\dot{\sigma}_{ij} = C_{ijkl} \, (\dot{\varepsilon}_{kl} - \dot{\varepsilon}_{kl}^{vp}) \tag{3.2}$$

$$\dot{\varepsilon}_{kl}^{vp} = \dot{\varepsilon}_{kl}^{vp} \, (\sigma_{kl}, \varepsilon_{kl}^{vp}, w) \tag{3.3}$$

$$\dot{w} = \dot{w} \, (\sigma_{kl}, \varepsilon_{kl}^{vp}, w) \tag{3.4}$$

where C_{ijkl} is the isotropic tensor of elastic constants and ε_{kl} the tensor of infinitesimal strains defined by

$$\varepsilon_{kl} = \frac{1}{2} (\mu_{k,1} + \mu_{1,k}) \tag{3.5}$$

where μ_k is the kth component of the displacement vector, ε_{kl}^{vp} the viscoplastic component of the infinitesimal strain tensor and w the damage state variable. Specific forms for the viscoplastic strain rate and the damage rate equations are given by eqs. (2.5) to (2.10).

- Boundary Conditions

$$\mu_i(x,t) = \bar{\mu}_i \, (x,t), \; \forall x \epsilon \Gamma_1, \; t \epsilon [0,T] \tag{3.6}$$

$$n_j(x)\,\sigma_{ij}(x,t) = \bar{p}_i(x,t),\ \forall x \epsilon \Gamma_2,\ t \epsilon [0,T] \tag{3.7}$$

where $\bar{\mu}_i(x,t)$ are prescribed displacements on Γ_1, $\bar{p}_i(x,t)$ prescribed surface tractions on Γ_2, n_j the unit outward normal vector to Γ_2, Γ_1 and Γ_2 subregions of Γ such that $\Gamma_1 \cup \Gamma_2 = \Gamma$ and $\Gamma_1 \cap \Gamma_2 = \emptyset$.

- Initial Conditions

$$\mu_i(x,0) = \mu_{oi}(x) \qquad , \qquad \forall x \epsilon \Omega \tag{3.8}$$

$$w(x,0) = w_o(x) \qquad , \qquad \forall x \epsilon \Omega \tag{3.9}$$

$$\sigma_{ij}(x,0) = \sigma_{oij}(x) \qquad , \qquad \forall x \epsilon \Omega \tag{3.10}$$

where $\mu_{oi}(x)$, $w_o(x)$ and $\sigma_{oij}(x)$ are the initial given data which are compatible with the boundary data and satisfy the equilibrium condition.

The solution to the quasi-static initial boundary value problem described consists in finding a scalar field $w(x,t)$, a vector field $\mu_i(x,t)$ and a tensor field $\sigma_{ij}(x,t)$ which satisfy eqs. (3.1) to (3.10) for all $x\epsilon\Omega$ and $t\epsilon[0,T]$, T>0.

3.2 Finite Element Approximation

In this section the finite element approximation to the solution of the initial boundary value problem described in the previous section is considered. Vectorial notation is also introduced.

3.2.1 Basic Concepts

The fundamental concept of the finite element method is the discretization of the actual domain Ω into a set of simple elements of finite size. Each element is connected to adjacent ones at a finite number of points termed the nodal points of the element. The basic step in the analysis is the unique description of the unknown function $\underline{u}$ (the displacement field in this work) in terms of m parameters μ_i, associated generally with the values of the unknown function at the nodal points, in the form

$$\underline{u} = \sum_{i=1}^{m} N_i \mu_i = \underline{N}\,\underline{U} \tag{3.11}$$

where N_i are the global shape functions dependent on the spatial coordinates and m is equal to the total number of nodal points in Ω. With the displacements known the strains at any point can

250

be obtained by introducing eq. (3.11) into eq. (3.5) yielding
a relationship of the form

$$\underline{\varepsilon} = \underline{B}\ \underline{U} \tag{3.12}$$

where $\underline{B}$ is a matrix composed generally of the derivatives of
the shape functions with respect to the cartesian coordinates.

The finite element equations corresponding to the initial
boundary value problem described can be derived by using energy
or variational principles, or by applying weighted residual
methods directly to the governing differential equations [27].
By considering the principle of virtual work, the overall equilibrium condition at any instant t can be written as [28]

$$\int_{\Omega} \underline{B}^T\ \underline{\sigma}\ d = \int_{\Omega} \underline{N}^T\ \underline{b}\ d\Omega + \int_{\Gamma_2} \underline{N}^T\ \underline{\bar{p}}\ d\Gamma \tag{3.13}$$

where $\underline{b}$ and $\underline{\bar{p}}$ are the vector counterparts of b_i and $\bar{p}_i$ in eqs.
(3.1) and (3.7), respectively. The right-hand side of eq.
(3.13) is known as the vector of global loads.

If the shape functions N_i are chosen satisfying the condition that the resulting displacement field be continuous on Ω
(C(o) continuity condition) the integrals in eq. (3.13) can be
calculated as the sum of the contributions of each individual
element, i.e.

$$\sum_{i=1}^{NE} \int_{\Omega_e} \underline{B}^T_e\ \underline{\sigma}\ d\Omega_e = \sum_{i=1}^{NE} \left(\int_{\Omega_e} \underline{N}^T_e\ \underline{b}\ d\Omega_e + \int_{\Gamma_{2_e}} \underline{N}^T_e\ \underline{\bar{p}}\ d\Gamma \right)$$
$$\tag{3.14}$$

where NE is the total number of elements into which the domain
Ω is discretized and the subscript e identifies an elemental
quantity. It is evident that from many elements there is no
contribution to the second term on the right-hand side of eq.
(3.14).

In this study eight and twenty noded isoparameteric elements satisfying the C(o) condition are used for the spatial
discretization. In the isoparametric approach [27] the geometry of the element is also interpolated from its nodal values
by the same shape functions used to approximate the displacement field. The isoparametric elements have been chosen because
they have been proved efficient in a wide variety of nonlinear
problems. With the continuity condition satisfied the standard
procedures for assembling the structure matrices [27] are
invoked and attention is restricted in the following sections
to the derivation of the matrices corresponding to a single
element. For simplicity of notation the subscript e is abandoned.

3.2.2 Geometric Nonlinear Effects

As discussed earlier (Section 1) geometric nonlinear effects, mainly large displacements, may assume in some selected applications an important role in the design of metallic components against brittle rupture. In Sections 3.1 and 3.2.1, for simplicity of exposition, it was implicitly assumed that the displacements and strains were infinitesimal so that it was possible to consider the domain configuration to be practically unaltered throughout the solution. When large displacements take place it is evident that this assumption can no longer be sustained and that special formulations are required.

First, it is noted that the governing differential equations (3.1) to (3.10) must always refer to the current configuration $\Omega_t(\Gamma_t)$. As this configuration is not known a priori, the solution is obtained by using an incremental procedure. In principle any of all the previous configurations could be used as a reference configuration but in practice the choice lies essentially between two approaches which have been termed total Lagrangian (T.L.) and updated Lagrangian (U.L.) formulations [29]. In the T.L. formulation all variables are referred to the initial undeformed configuration while in the U.L. description all variables are referred to the last calculated configuration. Clearly, if appropriate transformations are performed both formulations yield the same numerical results. Therefore, the choice of a formulation depends basically on its numerical efficiency for the solution of the problem under consideration.

In the present study the U.L. formulation has been adopted. This formulation leads to simpler strain-displacement expressions and is probably the most natural to use. As is common in most available computer programs, the material constants are not transformed when updating the reference configuration. These transformations increase the computer time drastically and moreover, as they must be performed at each integrating point, they induce an artificial anisotropy into the material. The omission of the material constant transformations is acceptable provided that the strains remain small throughout the response.

The initial boundary value problem of eqs. (3.1) to (3.10) is now rewritten in an incremental form. First let the solution be known at all discrete time stations 0, t_1, t_2, ... t_n. Then by considering the principle of virtual work the overall equilibrium condition at time t_{n+1} is expressed as

$$\int_{\Omega_{t_{n+1}}} (\sigma_{ij})_{n+1} \, (\delta\varepsilon_{ij})_{n+1} \; d\Omega = (\delta W_e)_{n+1} \qquad (3.15)$$

where σ_{ij} is the Cauchy stress tensor, ε_{ij} the tensor of infinitesimal strains (eq.(3.5)), W_e the work done by the external forces and the symbol δ denotes a virtual variation. The

252

external virtual work term δW_e is given by

$$(\delta W_e)_{n+1} = \int_{\Omega_{t_{n+1}}} (b_i)_{n+1} (\delta\mu_i)_{n+1} \, d\Omega + \int_{\Gamma_{2_{t_{n+1}}}} (\bar{p}_i)_{n+1}$$

$$(\delta\mu_i)_{n+1} \, d\Gamma \qquad (3.16)$$

where b_i are the body forces, $\bar{p}_i$ the prescribed surface tractions on Γ_2 and $\delta\mu_i$ the virtual displacements.

In eqs. (3.15) and (3.16) all quantities are referred to the domain configuration at time t_{n+1}. By using the notion of stress-strain energy conjugate [30,31], eq. (3.15) is transformed to

$$\int_{\Omega_{t_n}} (S_{ij})_{n+1/n} (\delta e_{ij})_{n+1/n} = (\delta W_e)_{n+1/n} \qquad (3.17)$$

where S_{ij} is the 2nd Piola-Kirchhoff stress tensor, and e_{ij} the Green-Lagrange strain tensor. The subscript n+1/n indicates that a quantity of time t_{n+1} is calculated with reference to the last known configuration Ω_{t_n}, while the subscript n (or n+1) indicates that a quantity is evaluated at time t_n (or t_{n+1}) with reference to the current configuration Ω_{t_n} (or $\Omega_{t_{n+1}}$).

In the U.L. formulation the Green-Lagrange strains are given by

$$e_{ij} = \varepsilon_{ij} + \eta_{ij} = \frac{1}{2}(\mu_{i,j} + \mu_{j,i}) + \frac{1}{2}(\mu_{k,i}\,\mu_{k,j}) \qquad (3.18)$$

and the external virtual work by

$$(\delta W_e)_{n+1/n} = \int_{\Omega_{t_n}} (b_i)_{n+1/n} (\delta\mu_i)_{n+1/n} \, d\Omega$$

$$+ \int_{\Gamma_{2_{t_n}}} (\bar{p}_i)_{n+1/n} (\delta\mu_i)_{n+1/n} \, d\Gamma \qquad (3.19)$$

The second Piola-Kirchhoff stresses $(S_{ij})_{n+1/n}$ are written in incremental form as

$$(S_{ij})_{n+1/n} = (\sigma_{ij})_n + (\Delta S_{ij})_{n+1/n} \qquad (3.20)$$

where $(\sigma_{ij})_n$ are the Cauchy stresses at time t_n determined from

$$(\sigma_{ij})_n = \frac{\rho_n}{\rho_{n-1}} (x_{i,k})_{n/n-1} (S_{kl})_{n/n-1} (x_{j,l})_{n/n-1} \qquad (3.21)$$

where ρ_n and ρ_{n-1} are the specific mass of the body at times t_n and t_{n-1}, respectively, and $(x_{i,k})_{n/n-1}$ and $(x_{j,l})_{n/n-1}$ are the derivatives of the cartesian coordinates x_i and x_j at time t_n with respect to the coordinates x_k and x_l at time t_{n-1}.

The constitutive relation between the stress and strain increments is obtained from eq. (3.2) as

$$(\Delta S_{ij})_{n+1/n} = C_{ijkl} (\Delta \varepsilon^E_{kl})_{n+1/n} \qquad (3.22)$$

where C_{ijkl} is the elasticity tensor calculated, as discussed before, at the initial undeformed configuration and $(\Delta \varepsilon^E_{kl})_{n+1/n}$ the elastic component of the linear part of the Green-Lagrange strain increment $(\Delta e_{kl})_{n+1/n}$.

Substituting eqs. (3.20) and (3.22) into eq. (3.17) yields

$$\int_{\Omega_{t_n}} C_{ijkl} (\Delta \varepsilon^E_{kl})_{n+1/n} (\delta e_{ij})_{n+1/n} \, d\Omega +$$

$$+ \int_{\Omega_{t_n}} (\sigma_{ij})_n (\delta \eta_{ij})_{n+1/n} \, d\Omega = (\delta W_e)_{n+1/n} -$$

$$- \int_{\Omega_{t_n}} (\sigma_{ij})_n (\delta \varepsilon_{ij})_{n+1/n} \, d\Omega \qquad (3.23)$$

Equation (3.23) can not be solved directly since it is non-linear in the displacement increments. An approximate solution is then obtained by linearising eq. (3.23) under the assumption that $(\delta e_{ij})_{n+1/n} = (\delta \varepsilon_{ij})_{n+1/n}$. Thus,

$$\int_{\Omega_{t_n}} C_{ijkl} (\Delta \varepsilon^E_{kl})_{n+1/n} (\delta \varepsilon_{ij})_{n+1/n} \, d\Omega +$$

$$+ \int_{\Omega_{t_n}} (\sigma_{ij})_n (\delta \eta_{ij})_{n+1/n} \, d\Omega = (\delta W_e)_{n+1/n} -$$

$$- \int_{\Omega_{t_n}} (\sigma_{ij})_n (\delta \varepsilon_{ij})_{n+1/n} \, d\Omega \qquad (3.24)$$

which is, finally, the incremental form sought for the original initial boundary value problem.

3.2.3 Finite Element Incremental Equations

Using the isoparametric concept, briefly introduced in Section 3.2.1, to evaluate the displacement derivatives in the integrals, the finite element approximation to eq. (3.24)

254

becomes,

$$\int_{\Omega_{t_n}} (\underline{B}_L)_n^T \ \underline{D} \ \Delta\underline{\varepsilon}^E_{n+1/n} \ d\Omega + (\underline{K}_{NL})_n \ \Delta\underline{U}_{n+1/n} = \Delta\underline{R}_{n+1/n} \tag{3.25}$$

where $\underline{B}_L$ is the linear strain-displacement transformation matrix, $\underline{D}$ the elasticity matrix, $\Delta\underline{\varepsilon}^E$ the elastic component of the Green-Lagrange strain increment, $\underline{K}_{NL}$ the nonlinear strain incremental stiffness matrix, $\Delta\underline{U}$ the vector of incremental nodal displacements and $\Delta\underline{R}$ the vector of incremental nodal forces. Considering eqs. (3.24) and (3.25) each term in eq. (3.24) is approximated in eq. (3.25) as:

$$\int_{\Omega_{t_n}} C_{ijkl} \ (\Delta\varepsilon^E_{kl})_{n+1/n} \ (\delta\varepsilon_{ij})_{n+1/n} \ d\Omega \rightarrow \int_{\Omega_{t_n}} (\underline{B}_L)_n^T$$

$$\underline{D}(\Delta\varepsilon^E)_{n+1/n} \ d\Omega \tag{3.27}$$

$$\int_{\Omega_{t_n}} (\sigma_{ij})_n \ (\delta\eta_{ij})_{n+1/n} \ d\Omega \rightarrow (\underline{K}_{NL})_n \ \Delta\underline{U}_{n+1/n} =$$

$$= \left[\int_{\Omega_{t_n}} (\underline{B}_{NL})_n^T \ \underline{\sigma}^* \ (\underline{B}_{NL})_n \ d\Omega \right] \Delta\underline{U}_{n+1/n} \tag{3.28}$$

$$(\delta W_e)_{n+1/n} - \int_{\Omega_{t_n}} (\sigma_{ij})_n \ (\delta\varepsilon_{ij})_{n+1/n} \rightarrow \Delta\underline{R}_{n+1/n} =$$

$$= \int_{\Omega_{t_n}} \underline{N}^T \ \underline{b}_{n+1/n} \ d\Omega + \int_{\Gamma_{2_{t_n}}} \underline{N}^T \ \underline{\bar{p}}_{n+1/n} \ d\Gamma -$$

$$- \int_{\Omega_{t_n}} (\underline{B}_{NL})_n^T \ \underline{\sigma}_n \ d\Omega \tag{3.29}$$

where $\underline{B}_{NL}$ is the nonlinear strain-displacement transformation matrix, $\underline{\sigma}$ the vector of Cauchy stresses, $\underline{\sigma}^*$ a matrix formed by the components of the Cauchy stress tensor and the remaining quantities have been defined previously.

It is interesting to note that eq. (3.29) can be recast as

$$\Delta\underline{R}_{n+1/n} = \left[\int_{\Omega_{t_n}} \underline{N}^T \ \underline{b}_n \ d\Omega + \int_{\Gamma_{e_{t_n}}} \underline{N}^T \ \underline{\bar{p}} \ d\Gamma - \int_{\Omega_{t_n}} (\underline{B}_L)_n^T \right.$$

$$\left. \underline{\sigma}_n \ d\Omega \right] + \left[\int_{\Omega_{t_n}} \underline{N}^T \ \Delta\underline{b}_{n+1/n} \ d\Omega + \right.$$

$$+ \int_{\Gamma_{2_{t_n}}} \underline{N}^T \, \Delta\bar{\underline{p}}_{n+1/n} \; d\Gamma \Biggr) = \underline{\psi}_n + \Delta\underline{f}_{n+1/n} \tag{3.30}$$

where $\underline{\psi}_n$ are the out-of-balance (residual) forces at time t_n and $\Delta\underline{f}_{n+1/n}$ the variation of the external forces during the time interval Δt_n ($\Delta t_n = t_{n+1} - t_n$).

3.3 Time Integration Algorithm

In this Section an implicit algorithm is proposed for the time inegration of the algebraic system represented by eq. (3.25).

The first step in the development of the solution consists of the evaluation of the elastic component of the linear term of the Green-Lagrange strain increment. From the basic assumption that the strain increments can be decomposed into elastic and viscoplastic parts it follows that

$$(\Delta\underline{\varepsilon}^E)_{n+1/n} = (\Delta\underline{\varepsilon})_{n+1/n} - (\Delta\underline{\varepsilon}^{vp})_{n+1/n} =$$

$$= (\underline{B}_L)_n \, (\Delta\underline{U})_{n+1/n} - (\Delta\underline{\varepsilon}^{vp})_{n+1/n} \tag{3.31}$$

The viscoplastic strain increment $\Delta\underline{\varepsilon}^{vp}$ is calculated using an implicit scheme as

$$(\Delta\underline{\varepsilon}^{vp})_{n+1/n} = \left[(1-\theta) \, \dot{\underline{\varepsilon}}_n^{vp} + \theta(\dot{\underline{\varepsilon}}^{vp})_{n+1/n} \right] \tag{3.32}$$

where θ, $0 \leq \theta \leq 1$, defines several scheme strategies such as, for instance, Euler forward or fully explicit ($\theta=0$), Implicit-Trapezoidal ($\theta=1/2$), Galerkin ($\theta=2/3$) and Euler backward or fully implicit ($\theta=1$).

In order to evaluate the viscoplastic strain rate terms in eq. (3.32) it is convenient first to rewrite the constitutive eq. (2.5) as

$$\dot{\underline{\varepsilon}}^{vp} = \gamma(\gamma_f, \, w) < \Phi \, (f,Y) > \frac{\partial F}{\partial \sigma_{ij}} \tag{3.33}$$

$$\gamma(\gamma_f, \, w) = \gamma_f / (1 - cw)^N \tag{3.34}$$

$$\Phi(f,Y) = \left(\frac{f-Y}{Y} \right)^N \tag{3.35}$$

The viscoplastic strain rate $(\dot{\underline{\varepsilon}}^{vp})_{n+1/n}$ is now obtained by using a truncated Taylor series expansion as (note that by definition, eq. (3.21), $\underline{\sigma}_n \equiv \underline{S}_n$)

256

$$\left(\underline{\dot{\varepsilon}}^{vp}\right)_{n+1/n} \simeq \left(\underline{\dot{\varepsilon}}^{vp}\right)_n + \left(\frac{\partial \underline{\dot{\varepsilon}}}{\partial \underline{S}}\right)_n \Delta \underline{S}_{n+1/n} + \left(\frac{\partial \underline{\dot{\varepsilon}}}{\partial \gamma}\right)_n \Delta \gamma_{n+1/n} +$$

$$\left(\frac{\partial \underline{\dot{\varepsilon}}}{\partial Y}\right)_n \Delta Y_{n+1/n} \tag{3.36}$$

where each partial derivative of $\underline{\dot{\varepsilon}}$ is given, considering eqs. (3.33) to (3.35), by

$$\left(\frac{\partial \underline{\dot{\varepsilon}}}{\partial \underline{S}}\right)_n = \gamma_n \left(\Phi_n \left(\frac{\partial}{\partial \underline{S}} \left(\frac{\partial F}{\partial \underline{S}}\right)_n\right)_n + \left(\frac{\partial \Phi}{\partial \underline{S}}\right)_n \left(\frac{\partial F}{\partial \underline{S}}\right)_n^{\mathbf{T}}\right) = \underline{H}_{1n} \tag{3.37}$$

$$\left(\frac{\partial \underline{\dot{\varepsilon}}}{\partial \gamma}\right)_n = \Phi_n \left(\frac{\partial F}{\partial \underline{S}}\right)_n = \underline{H}_{2n} \tag{3.38}$$

$$\left(\frac{\partial \underline{\dot{\varepsilon}}}{\partial Y}\right)_n = \gamma_n \left(\frac{\partial \Phi}{\partial Y}\right)_n \left(\frac{\partial F}{\partial \underline{S}}\right)_n = \underline{H}_{3n} \tag{3.39}$$

Substituting eqs. (3.36) to (3.39) into eq. (3.32) yields, after some algebraic manipulation,

$$\left(\Delta \underline{\varepsilon}^{vp}\right)_{n+1/n} = \underline{\dot{\varepsilon}}_n^{vp} \Delta t_n + \underline{C}_n \Delta \underline{S}_{n+1/n} + \underline{E}_n \Delta \gamma_{n+1/n} +$$

$$+ \underline{X}_n \Delta Y_{n+1/n} \tag{3.40}$$

where

$$\underline{C}_n = \underline{C}_n(\underline{S}, \gamma) = \theta \Delta t_n \underline{H}_{1n} \tag{3.41}$$

$$\underline{E}_n = \underline{E}_n(\gamma) = \theta \Delta t_n \underline{H}_{2n} \tag{3.42}$$

$$\underline{X}_n = \underline{X}_n(\underline{S}, \gamma, Y) = \theta \Delta t_n \underline{H}_{3n} \tag{3.43}$$

Introducing eq. (3.40) into eq. (3.31) leads to the following expression for the elastic strain increment

$$\left(\Delta \underline{\varepsilon}^E\right)_{n+1/n} = (\underline{B}_L)_n \Delta \underline{U}_{n+1/n} - \left[\underline{\dot{\varepsilon}}_n^{vp} \Delta t_n + \underline{C}_n \Delta \underline{S}_{n+1/n} + \right.$$

$$\left. \underline{E}_n \Delta \gamma_{n+1/n} + \underline{X}_n \Delta Y_{n+1/n}\right] \tag{3.44}$$

The 2nd Piola Kirchhoff stress increment is obtained from eq. (3.22) as

$$\Delta \underline{S}_{n+1/n} = \underline{D} \left(\Delta \underline{\varepsilon}^E\right)_{n+1/n} \tag{3.45}$$

Substituting eq. (3.45) into eq. (3.44) yields

$$(\Delta \underline{\varepsilon}^E)_{n+1/n} = (\underline{I} + \underline{C}_n D)^{-1} \left((\underline{B}_L)_n \, \Delta \underline{U}_{n+1/n} - \underline{\dot{\varepsilon}}_n^{vp} \, \Delta t_n - \right.$$
$$\left. - \underline{E}_n \, \Delta \gamma_{n+1/n} - \underline{X}_n \, \Delta Y_{n+1/n} \right) \qquad (3.46)$$

Introducing eq. (3.46) into the incremental equilibrium equation (3.25) and solving for the displacement increment results in the following equation system

$$\Delta \underline{U}_{n+1/n} = (\underline{K}_T)_n^{-1} \, \Delta \underline{V}_{n+1/n} \qquad (3.47)$$

where $\underline{K}_T$, termed the tangential stiffness, is given by

$$(\underline{K}_T)_n = \int_{\Omega_{t_n}} (\underline{B}_L)_n^T \, \hat{\underline{D}}_n \, (\underline{B}_L)_n \, d\Omega + (\underline{K}_{NL})_n \qquad (3.48)$$

where $\hat{\underline{D}} = (\underline{D}^{-1} + \underline{C}_n)^{-1}$

and $\Delta \underline{V}_{n+1/n}$, termed the incremental pseudo-load vector, is given by

$$\Delta \underline{V}_{n+1/n} = \Delta \underline{R}_{n+1/n} + \int_{\Omega_{t_n}} (\underline{B}_L)_n^T \, \hat{\underline{D}}_n \left(\underline{\dot{\varepsilon}}_n^{vp} \, \Delta t_n + \underline{E}_n \, \Delta \gamma_{n+1/n} + \right.$$
$$\left. + \underline{X}_n \, \Delta Y_{n+1/n} \right) d\Omega \qquad (3.49)$$

Equations (3.47) can not be solved directly because the increments of the γ and Y functions are yet unknown. Estimates of these increments are now discussed.

Considering eq. (3.34) $\Delta \gamma_{n+1/n}$ is readily obtained as

$$\Delta \gamma_{n+1/n} = c N \gamma_f \, \Delta w_{n+1/n} / (1 - c w_n)^{N+1} \qquad (3.50)$$

where the increment of damage $\Delta w_{n+1/n}$ is predicted using constitutive eq. (2.6) and an Euler forward scheme as

$$\Delta w_{n+1/n} = A \frac{(\hat{\sigma}_{eq})_n^X}{(1 - w_n)^\phi} \, \Delta t_n \qquad (3.51)$$

The increment of the Y function, on the other hand, is obtained from eq. (2.7) as

$$\Delta Y_{n+1/n} = \beta_n \, H' \, \Delta \bar{\varepsilon}_{n+1/n}^{vp} \qquad (3.52)$$

where the increment of the effective viscoplastic strain component $\Delta \bar{\varepsilon}_{n+1/n}^{vp}$ is predicted from constitutive eq. (2.5) and an Euler forward scheme as

$$\Delta \bar{\varepsilon}_{n+1/n}^{vp} = \dot{\bar{\varepsilon}}_n \, \Delta t_n \qquad (3.53)$$

Also, the weighting factor β_n is given by

$$\beta_n = 1\Big/\left[1 - \Theta\Delta t_n\ \gamma_n\ H'\left(\frac{\partial\Phi}{\partial Y}\right)_n\right] \tag{3.54}$$

The weighting factor β_n is a simple extension to multiaxial stress states of a result derived in the context of the equivalent one-dimensional elastoviscoplastic damage problem. This study is briefly summarized below:

First let

$$\Delta\varepsilon^{vp}_{n+1/n} = \left[(1-\Theta)\dot{\varepsilon}^{vp}_n + \Theta\dot{\varepsilon}^{vp}_{n+1/n}\right]\Delta t_n \tag{3.55}$$

$$\Delta Y_{n+1/n} = H'\ \Delta\dot{\varepsilon}^{vp}_{n+1/n} \tag{3.56}$$

$$\dot{\varepsilon}^{vp}_{n+1/n} = \dot{\varepsilon}^{vp}_n + \left(\frac{\partial\dot{\varepsilon}^{vp}}{\partial\sigma}\right)_n \Delta\sigma_{n+1/n} + \left(\frac{\partial\dot{\varepsilon}^{vp}}{\partial\gamma}\right)_n \Delta\gamma_{n+1/n} +$$

$$+ \left(\frac{\partial\dot{\varepsilon}^{vp}}{\partial Y}\right)_n \Delta Y_{n+1/n} \tag{3.57}$$

Substituting eqs. (3.56) and (3.57) into eq. (3.55) yields

$$\Delta\varepsilon^{vp}_{n+1/n} = \frac{\Delta t_n}{(1-\alpha_n)}\left[\dot{\varepsilon}^{vp}_n + \Theta\gamma_n\left(\frac{\partial\Phi}{\partial\sigma}\right)_n \Delta\sigma_{n+1/n} + \Theta\Delta t_n\ \Delta\gamma_{n+1/n}\right] \tag{3.58}$$

where $\alpha_n = \Theta\Delta t_n\ \gamma_n\ H'\left(\frac{\partial\Phi}{\partial Y}\right)_n$.

Now let the increment of the Y function be approximated as

$$\Delta Y_{n+1/n} = \beta_n\ H'\ \dot{\varepsilon}^{vp}_n\ \Delta t_n \tag{3.59}$$

which is the one-dimensional expression equivalent to the three-dimensional eqs. (3.52) and (3.53).

Introducing eqs. (3.57) and (3.59) into eq. (3.55) yields

$$\Delta\varepsilon^{vp}_{n+1/n} = \Delta t_n\left[(1+\alpha_n\beta_n)\dot{\varepsilon}^{vp}_n + \Theta\gamma_n\left(\frac{\partial\Phi}{\partial\sigma}\right)_n \Delta\sigma_{n+1/n} + \right.$$

$$\left. + \Theta\Delta t_n\ \Delta\gamma_{n+1/n}\right] \tag{3.60}$$

By comparing eqs. (3.58) and (3.60) it is clear that for a unique value of $\Delta\varepsilon^{vp}_{n+1/n}$ to be obtained it is necessary that firstly

$$(1+\alpha_n\beta_n) = 1/(1-\alpha_n) \rightarrow \beta_n = 1/(1-\alpha_n) \tag{3.61}$$

and secondly, that the 2nd and 3rd terms in eq. (3.60) be both multiplied by $(1/(1-\alpha_n))$. The first condition leads to the β_n expression given by eq. (3.54) while the second one implies that the E_n vector and the C_n matrix in the three-dimensional formulation (eqs. 3.41 and 3.42) must also be weighted by the β_n factor, i.e.

$$E_n = E_n(\gamma) = \theta \Delta t_n \beta_n H_{1n} \tag{3.62}$$

$$C_n = C_n(\underline{\sigma},\gamma) = \theta \Delta t_n \beta_n H_{3n} \tag{3.63}$$

The inclusion of the weighting factor β_n has the advantage that, in principle, there is no subsequent need of verifying and, eventually, abandoning a particular timestep solution as a consequence of a bad prediction for the Y function increment. The evaluation of β_n, on the other hand, although having to be performed at each integrating point, is computationally inexpensive because it only requires information already available at time t_n and simple arithmetic operations.

With the evaluation of the pseudo forces completed, the displacement increments are finally found from eq. (3.47) and, subsequently, the strain increments from eqs. (3.40) and (3.31) and the 2nd Piola-Kirchhoff stress increments from eq. (3.22). With these increments known, the displacements, strains and stresses (2nd Piola-Kirchhoff and Cauchy) are updated to their final values at time t_{n+1} as previously indicated. It remains only to discuss the updating of the damage state variable. This is done by using the predictor-corrector scheme described below where the superscripts, p, c and a denote predicted, corrected and accepted values, respectively

$$w^p_{n+1/n} = w_n + \frac{A(\hat{\sigma}_{eq})_n^\chi}{(1-w_n)^\phi} \Delta t_n \tag{3.64}$$

$$w^c_{n+1} = w_n + \frac{A(\hat{\sigma}_{eq})_{n+1}^\chi}{(1-w^p_{n+1/n})^\phi} \Delta t_n \tag{3.65}$$

$$w^a_{n+1} = \frac{1}{2}(w^p_{n+1/n} + w^c_{n+1}) \tag{3.66}$$

The updated values at time t_{n+1} are accepted and the solution advanced to the next time station provided that two accuracy controls to be discussed in Section 3.3.1 are satisfied. Rupture of a particular point in the structure is deemed to have occurred when the damage state variable reached the critical value 1 (in practice, for numerical convenience, the

260

critical value is taken as 0.99). The residual stress existing at this point at the instant it is declared ruptured is automatically redistributed to the remaining non-ruptured elements under the assumption that the real redistribution process is both instantaneous and elastic. The analysis proceeds until a sufficiently large damaged zone develops characterizing the global failure of the structure.

3.3.1 Time Step Length: Stability and Accuracy Considerations

A proper stability and accuracy analysis of the implicit algorithm proposed is mathematically rather complex, due to the inclusion of two state variables (viscoplastic deformation and damage) in the constitutive equations, and has not been attempted. Instead, empirical rules were established to select the time step length and to provide, at least, an indirect control over the stability and accuracy of the solution.

The first rule to select the magnitude of the time step consists in limiting the maximum amount of incremental viscoplastic strain to a fraction of the total accumulated strain according to

$$\Delta t_{n+1} = \min. \ (\tau \bar{\varepsilon}_n / \dot{\bar{\varepsilon}}_n^{vp}) \qquad (3.67)$$

where τ is a pre-specified value, $\bar{\varepsilon}$ and $\dot{\bar{\varepsilon}}^{vp}$ the effective values of the total strain and of the viscoplastic strain rate, respectively, and the minimum value of Δt_{n+1} is taken with respect to all integrating points in the domain. The second rule limits the change in the time step length between any two intervals, i.e.

$$\Delta t_{n+1} \leq k \ \Delta t_n \qquad (3.68)$$

where k is a given constant. Numerical experience to date indicates that suitable values for τ and k are in the range $0.01 \leq \tau \leq 0.15$ and $1.5 \leq k \leq 3.0$ respectively, although there are no fixed criteria for their specification.

Stability and accuracy of the solution are controlled by requiring that at any time step the Cauchy stress variation and the error in the damage increment prediction at each integrating point should not exceed pre-specified tolerances, i.e.,

$$(1) \quad \left| \frac{\| \underline{\sigma}_{n+1} \| - \| \underline{\sigma}_n \|}{\| \underline{\sigma}_n \|} \right| \leq toler_1 \qquad (3.69)$$

$$(2) \quad \left| \frac{\Delta w_{n+1/n}^a - \Delta w_{n+1/n}^p}{\Delta w_{n+1/n}^p} \right| \leq toler_2, \ w_n \geq 0.1 \qquad (3.70)$$

where $\| \ \|$ denotes the usual Euclidean norm, $| \ |$ the absolute value, toler_1 and toler_2 the accuracy tolerance and $\Delta w^p_{n+1/n}$ and $\Delta w^a_{n+1/n}$, the predicted and accepted damage increment values, respectively. The damage increment terms are readily obtainable from eqs. (3.64) and (3.66) as

$$\Delta w^p_{n+1/n} = w^p_{n+1/n} - w_n = A \frac{(\hat{\sigma}_{eq})^{\chi}_{n+1}}{(1-w_n)^{\phi}} \qquad (3.71)$$

$$\Delta w^a_{n+1/n} = w^a_{n+1} - w_n = A \left[\frac{(\hat{\sigma}_{eq})^{\chi}_n}{(1-w_n)^{\phi}} + \frac{(\hat{\sigma}_{eq})^{\chi}_{n+1}}{(1-w_{n+1/n})^{\phi}} \right] \frac{\Delta t_n}{2} \qquad (3.72)$$

If either condition (1) or condition (2) is violated at any integrating point the current time step length is reduced to 1/10th of its value and the time step solution repeated. A maximum of two reanalyses are allowed for any time step after which, if the accuracy controls are still not satisfied (usually a rare situation if τ and k are carefully chosen) the solution is halted and the last calculated results stored on file. This procedure gives the analyst an opportunity to examine the deformation and damage histories and to decide whether to restart the analysis from the current solution at time t_n adopting a time step length smaller than the one automatically calculated or, if restart facilities are normally provided, from any of the previous time station solutions stored.

3.4 Damage Rate Equations

The differential equation governing the damage growth (eq. (2.6)) is highly nonlinear with respect to both the overstress and damage variables. These characteristics lead to a type of equation that is referred to as stiff in mathematical terms. Some of the difficulties involved in the integration of this type of equation are discussed by Krieg [32] and several strategies for its numerical integration are compared by Kumar et al. [33], Hayhurst and Krzeczkowski [34] and Savalle and Culié [35] in the context of Hart's unified theory of inelastic deformation, creep deformation and rupture in elastic solids, and viscoplastic cyclic flow with damage, respectively.

In the present work the damage rate equation is solved, as discussed before, by a predictor-corrector scheme with an automatic time step length control. Prior to the start of the solution process, however, a change of variable is performed with the objective of diminishing the degree of nonlinearity of the original equation. In this section this change of variable is discussed and the recast form of the constitutive equations presented.

The new variable is defined by

262

$$z = (1+w)^{(1+\phi)} \tag{3.73}$$

and its introduction into eq. (2.6) yields

$$\dot{z} = -A(1+\phi)\hat{\sigma}^{\chi}_{eq} \tag{3.74}$$

Equation (3.74) shows that the rate of the new variable z is a function of the overstress only, whereas the rate of the initial variable w is a function of both the overstress and damage state. This fact underlines the main advantage of the z variable approach which is best illustrated in the particular case of constant stress histories. For such situations, as indicated in Figure 4, z becomes a linear function for its whole domain of definition $(1 \geq z \geq 10^{-2(1+\phi)})$ and its increment $\Delta z_{n+1/n}$ during any time step can be obtained in an exact manner by the simplest Euler scheme. The w curve, on the other hand, still exhibits a strong nonlinearity (due to the damage rate dependence on the damage state) and therefore a precise evaluation of its increment $\Delta w_{n+1/n}$ will require more elaborate time integration schemes, smaller time step lengths or rather cumbersome analytical expressions. For variable stress histories, the use of the z variable in conjunction with the predictor-corrector scheme given by eqs. (3.64) to (3.66) can be justified following a similar line of argument. To this end it is only necessary to note that during the predictor and corrector phases the overstresses $\hat{\sigma}_n$ and $\hat{\sigma}_{n+1}$ are taken as constants during the time interval Δt_n.

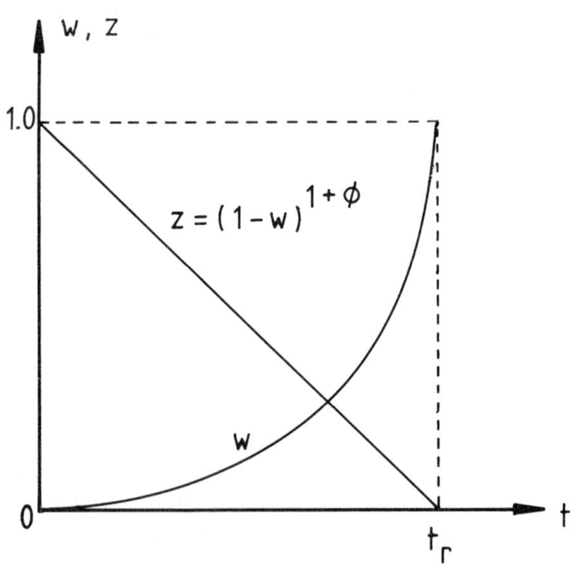

Fig. 4 _ W and Z curves for constant stress

Considering the new variable z defined by eq. (3.73) and also eqs. (3.33) to (3.35), the constitutive equations (3.2) to (3.4) are rewritten as,

$$\dot{\underline{\sigma}} = \underline{D} \, (\underline{\varepsilon} - \underline{\dot{\varepsilon}}^{vp}) \tag{3.75}$$

$$\underline{\dot{\varepsilon}}^{vp} = \frac{\gamma_f}{1 - c + cz^{1/(1+\phi)_)N}} \left\langle \left(\frac{f-Y}{Y}\right)^N \right\rangle \frac{\partial F}{\partial \underline{\sigma}} \tag{3.76}$$

$$\dot{z} = - A(1+\phi)\hat{\sigma}_{eq}^{\chi} \tag{3.77}$$

Having introduced the new variable and recast the constitutive equations accordingly, this section is concluded by discussing the derivation of the transformation equation (3.73). Let

$$z^r = (1-w)^\phi \tag{3.78}$$

where r is a parameter to be determined.

Differentiating eq. (3.78) with respect to time and substituting the w term by the value given by eq. (2.6) results in

$$rz^{r-1} \, \dot{z} = - \, \phi A \, \frac{(\hat{\sigma}_{eq})^{\chi}}{(1-w)} \tag{3.79}$$

Introducing eq. (3.78) into eq. (3.79) yields

$$\dot{z} = - \, \frac{\phi A (\hat{\sigma}_{eq})^{\chi}}{rz^{(r/\phi)+r-1}} \tag{3.80}$$

The r parameter is now determined by setting the exponent of z in eq. (3.80) to zero. Thus,

$$r = \phi/(1+\phi) \tag{3.81}$$

Substituting eq. (3.81) into eq. (3.78) yields, finally the transformation equation (3.73).

4. NUMERICAL APPLICATIONS

In order to illustrate some of the items discussed in the previous sections, numerical solutions are obtained for the creep rupture behaviour of perforated plates subjected to equal biaxial tensile load and compared with the results of rupture experiments carried out by Hayhurst, Morrison and Leckie [36].

The plates considered are square in form, contain a central circular hole (Figure 5) and are made of aluminium alloy (BS1472)

and commercially pure copper (BS2873-CIDI). The geometric data
is listed in Table 1 together with the average value of the
normal stress σ_N applied over the minimum cross-section (σ_N =
qb/(h(b-a)). For both materials sets of uniaxial creep curves
and creep rupture data for steady uniaxial and biaxial loads
are presented in Refs. [37] and [38], respectively. All relevant uniaxial data for the aluminium and copper alloys studied
are compiled in Table 2 and correspond to temperatures of 210±
1°C and 250±1°C, respectively.

TABLE 1 - Geometric dimensions (mm.) and applied stress on the
minimum cross-section (N/mm²)

	Aluminium	Copper
Circular hole		
diameter (a)	6.27	6.35
width (b)	30.23	30.23
thickness (h)	4.60	5.99
applied stress (σ_N)	57.96	40.02

TABLE 2 - Material properties

Property	Aluminium	Copper
E	60030.0 N/mm²	66240.0 N/mm²
ν	0.3	0.3
γ_f	1.15×10^{-17}/hr	3.21×10^{-12}/hr
N	6.9	5.0
A	$1.79 \times 10^{-14} (N/mm^2)^{-6.9}$/hr	$1.89 \times 10^{-7} (N/mm^2)^{-5.0}$/hr
χ	6.48	3.19
ϕ	9.50	6.0
c	1.0	1.0
M	- 0.2	- 0.43

The material constant M, first appearing in Table 2,
refers to a polynomial term in time, t^M, which has been included
in the creep strain rate and damage rate equations in order to
account for primary creep effects [37]. This polynomial term
can be most conveniently dealt with by introducing a new time
scale t*, defined by $t* = t^{(M+1)}$. When recast in terms of t*
the creep strain rate and the damage rate equations become

independent of t* but the A and γ_f constants now appear both multiplied by the factor $1/(M+1)$.

   As mentioned before, the rupture behaviour of the aluminium and copper alloys when subjected to biaxial stress states are closely approximated by the maximum effective stress and the maximum principal stress rupture criteria, respectively. For the particular alloys studied, the values of the α_i parameters (eq. (2.3)) are listed in Table 3 [16].

TABLE 3 - Rupture behaviour. α_i parameters

Parameter	Aluminium	Copper
α_1	0.0	0.848
α_2	0.0	0.0
α_3	0.090	0.064
α_4	0.910	0.088

   Due to the symmetry of the geometry and loading only one single quadrant of the plates need be considered. A finite element mesh consisting of 100 eight-noded isoparametric elements was used to discretise one quadrant of each plate and is illustrated in Figure 5. The distribution and size of the elements were chosen so that the expected regions of higher stress concentrations were split up by a greater number of smaller elements.

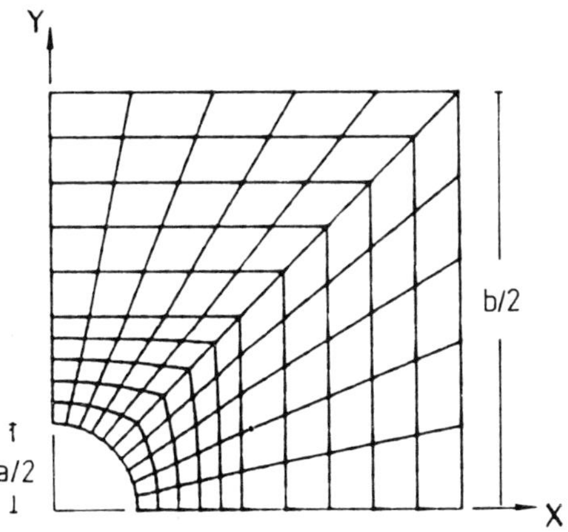

Fig. 5 _ Plate's geometry and finite element mesh

For time integration the Implicit-Trapezoidal scheme ($\theta=\frac{1}{2}$) was used and the timestep length controlling parameters, τ and k were taken as τ = 0.05 and k = 2.50. The accuracy tolerances $toler_1$ and $toler_2$ were imposed as 0.10 and 0.05, respectively.

The computed and experimentally determined values of the rupture times (t_R) and of the normalized representative rupture stresses (σ_R) are compared in Table 4. The representative rupture stress for an arbitrary structure is defined as the stress that causes an uniaxial creep specimen of the same material to have the same lifetime as the structure.

TABLE 4 - Rupture times t_R and normalized representative stresses σ_R/σ_N

Material	t_R (hours)		σ_R/σ_N (N/mm^2)	
	experimental	computed	experimental	computed
Aluminium	416	$\sim$ 700	1.05	0.98
Copper	356	$\sim$ 190	0.94	1.05

As indicated in Table 4 the normalized representative stress computed is approximately 7% lower for the aluminium plate and 9% higher for the copper plate. Since basically the same results have been obtained for each problem using two different finite element meshes (one with 60 elements and the other with 100 elements) and also that stringent accuracy tolerances have been imposed for the solution with refined mesh, it is not expected that any further refinements in these directions would reduce the observed differences. The damage and constitutive equations, on the other hand, are expected to give good lifetime predictions since for the aluminium alloy damage is of a scalar nature (Table 3) and for the copper alloy a close examination of the stress history throughout the solution clearly indicates that the plate is proportionally loaded. Hence the difference between the computed and experimental values of the representative rupture stress (and consequently of the rupture times) is probably due to the scatter in the material data.

Figures 6 and 7 shows the variation of the normal stress σ_{yy} at the minimum cross section y=0 at different time stations for the aluminium and copper plates, respectively. At time t=0 the stresses and creep deformation rates are highest at the edge of the hole. As time increases the stresses are redistributed due to the primary and tertiary creep deformations. It is seen in these figures that the distribution of stress across the minimum section of the plate becomes almost constant within a short period of time. This observation justifies, at least for high

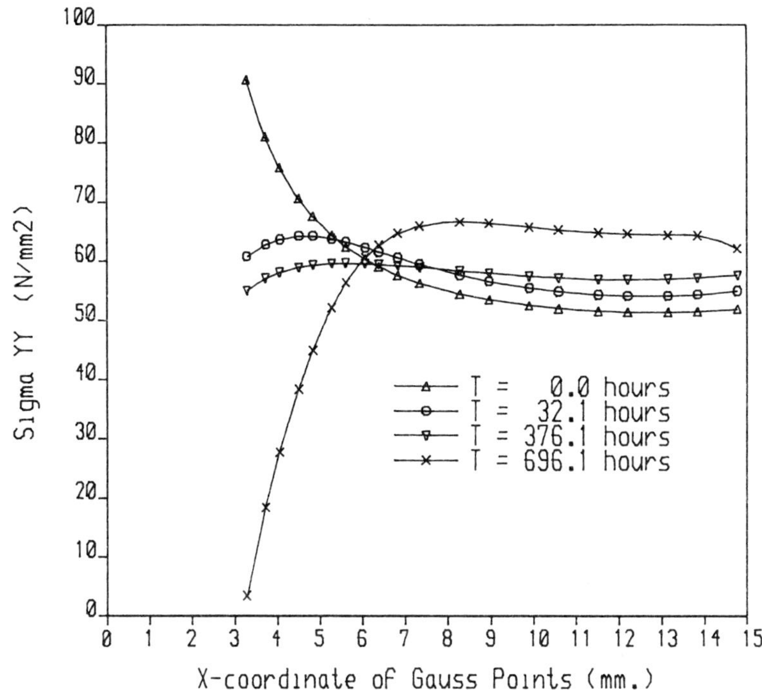

Fig. 6 _ Variation of σ_{yy} with time at minimum cross-section Y=0.0 (Aluminium plate)

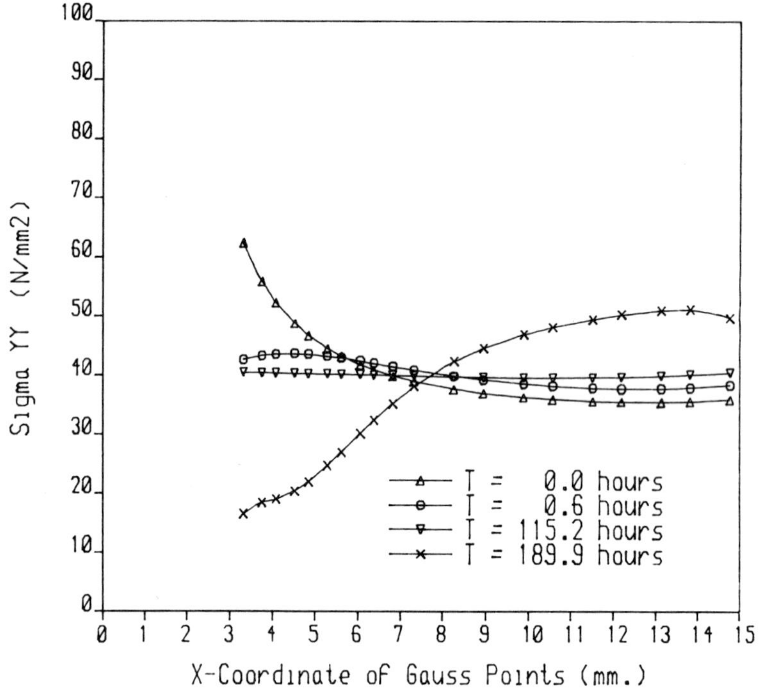

Fig. 7 _ Variation of σ_{yy} with time at minimum cross-section Y = 0.0 (Copper plate)

values of N, the assumption frequently used in approximate solutions [8,38] that for a given structural geometry the energy
dissipation rate throughout the deforming part of the structure
rapidly achieves a constant value. At the onset of tertiary
creep a strong stress redistribution occurs with the crack
front moving from the edge of the hole outwards.

Figures 8 and 9 show damage contours at large fractions
of the plate's lifetime. For both plates, the damaged zones
are symmetrical about the center of the hole. Rupture first
takes place at the edge of the hole with the crack front propagating outward along the diagonal directions. These observations are in accordance with the results of metallographic
examinations [36].

8. CONCLUDING REMARKS

In the course of this work on the numerical analysis of
creep and viscoplastic brittle rupture of structures several
observations were made which now merit further attention:-

(1) - Material damage in structures subject to complex loading
      histories is of a tensorial nature [39,40]. Most of the
      tensorial theories which have emerged to date [9-13,41],
      however, are restricted to formal descriptions and the
      relationships between the tensorial variables introduced
      and the actual mechanisms of microstructural change in
      the material have not been sufficiently clarified. The
      multiaxial generalization of the scalar theory of Kachanov
      and Rabotnov [5,6] proposed by Leckie and Hayhurst [8],
      on the other hand, does describe the behaviour of materials
      subjected to proportional loading and, for engineering
      cases of non-proportional loading, recent experimental
      results [14,15] have shown that deformation and rupture
      can be also described by a one state variable theory
      provided that some modifications be introduced to account
      for the different damage behaviour.

(2) - The constitutive model proposed in the present work to
      represent elastoviscoplastic damage behaviour under multi-
      axial stress states which, basically, attempts to bring
      together the contributions of Lemaitre and Chaboche [18]
      for the analysis of one-dimensional viscoplastic damage
      and of Leckie and Hayhurst [8] for the study of multi-
      axial creep deformation and rupture problems, must yet be
      validated experimentally. To this end it seems that the
      same type of tests used for studying multiaxial creep
      damage in elastic solids [16,42] could be employed.

(3) - Geometric nonlinear effects, mainly large displacements
      and large rotations may play, in selected applications,
      an important role in structural design against brittle
      failure [17] and should not be disregarded in rupture

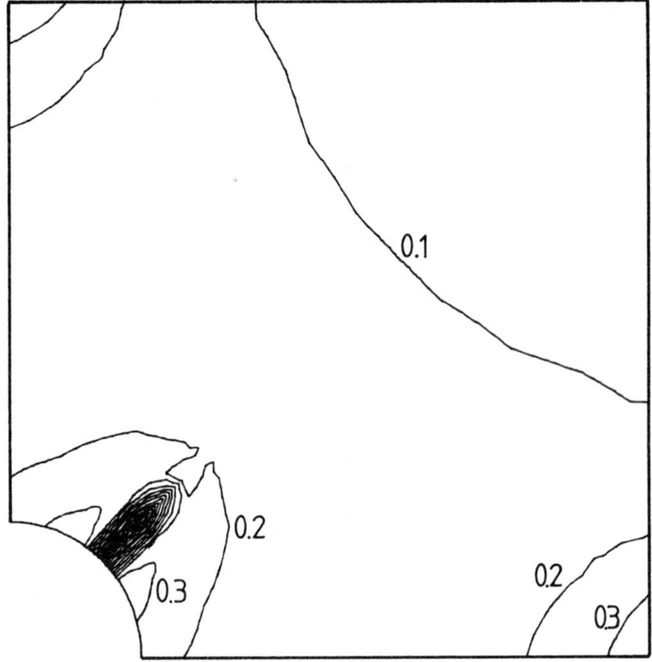

Fig. 8 _ Damage contours at time 698.8 hours (Aluminium plate)

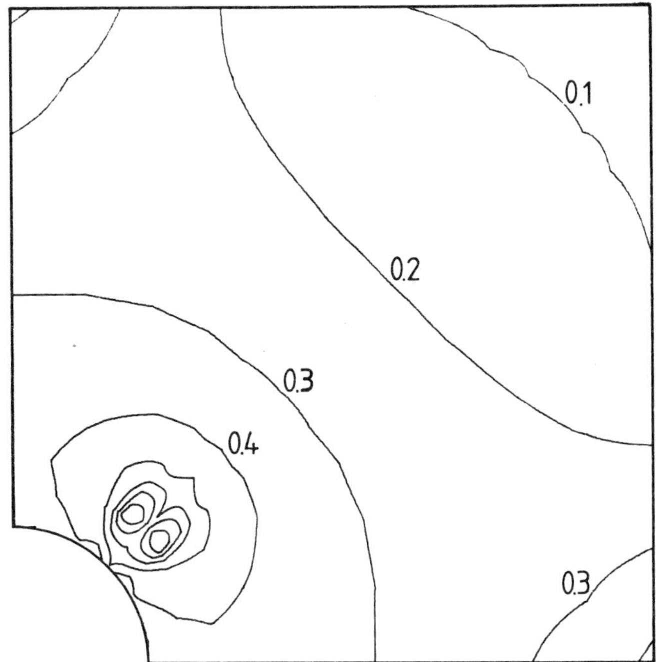

Fig. 9 _ Damage contours at time 188.0 hours   (Copper plate)

calculations. An updated Lagrangian formulation has been adopted in this work since it leads to simpler strain-displacement relationships and is probably the most natural one to use. The tensor of material constants is not transformed when updating the reference configuration since these transformations increase the computer time drastically and, moreover, as they must be performed at each integrating point, they induce an artificial anisotropy into the material. This omission is acceptable provided that the strains remain small throughout the solution.

(4) - The fundamental decomposition of motion into elastic and viscoplastic components converts the inelastic rate process directly into an equivalent driving force process which is entirely controlled by the viscoplastic deformation and damage growth law on the constitutive level. The implicit algorithm developed for time integration is, mathematically speaking, a single-step method which can be regarded as a special type of the Runge-Kutta second-order scheme [43]. In the formulation, at each time step, the driving force which represents the structural behaviour is unknown beforehand and required the prediction and eventual correction of the increment of the two state variablesincluded in the analysis. The viscoplastic deformation (or Y function) growth estimation is made using a weighted Euler scheme which, in principle, eliminates the need of a subsequent correction. The weighting factor has been derived in the context of the equivalent one dimensional elastoviscoplastic damage problem and its inclusion has been shown to be computationally inexpensive. The damage (or γ function) growth is calculated using a first order predictor-corrector scheme with automatic time step length control. The predictor phase is performed using the simplest Euler scheme. Prior to solving the damage rate equation a change of variable has been introduced and shown to possess certain advantages.

(5) - Since the inelastic rate process investigated is path dependent, the drifting error (i.e. the accumulation of numerical integration errors) can only be contained within an acceptable tolerance by reanalysing with a reduced time increment length for all time steps where specified accuracy criteria have been violated [44]. A proper stability and accuracy analysis of the implicit algorithm developed is mathematically rather complex, due to the inclusion of two state variables in the constitutive equations, and has not been attempted. Instead, empirical relationships were used to select the time step length and limits suggested for the corresponding time step length controlling parameters. Additionally a restart facility has been incorporated in the computer program written, so that the analyst can examine the deformation history at inter-

mediate time stations and alter some of the pre-selected time integration parameters before proceeding further with the solution process.

(6) - Numerical experience to date indicates, as exemplified by the applications considered in this work and also else-where [45], that qualitative results, both in terms of strain and damage distributions can be obtained and that reasonable estimates of time to rupture (considering the highly nonlinear nature of the physical problem and the scatter normally associated with the material data) are also possible. In common with most types of nonlinear analyses, the demands on computer resources are high, especially when dealing with large and complex structures. In certain applications where viscoplastic deformation is expected to be confined to localized regions in the struc-ture important computational savings can be achieved by using substructuring techniques [46,47]. One such tech-nique has been developed in conjunction with the implicit algorithm described and is documented in detail in Ref. [48].

9.   REFERENCES

1.   ASHBY, M. F. and RAJ, M. - Proc. Conf. Mechanics and Physics, The Metal Society, Institute of Physics, Cambridge 1975.

2.   DYSON,   B. F. and McLEAN, D. - Metal Science, Vol.2, pp. 37, 1977.

3.   GREENWOOD, G. W. - Microstructure and the design of alloys, Int. Conf. on Metals, Cambridge, Vol.2, pp. 91, 1973.

4.   COCKS, A. C. F. and ASHBY, M. F. - IUTAM Symposium on Creep of structures, ed. A. R. S. Ponter and D. R. Hayhurst, Leicester, 1980.

5.   KACHANOV, L. M. - Time of the fracture process under creep conditions.   IZV. Akad. Nauk. USSR, Otdgel. Tekh. Nauk. No.8, pp. 26-31, 1958.

6.   RABOTNOV, Y. N. - Creep rupture, Proc. XII Int. Cong. Appl. Mech., Stanford, 1968.

7.   ROBINSON, E. L. - Effects of temperature variation on the long-term rupture strength of steels, ASME, J. of Pressure Vessel Technology, Vol.47, pp. 777-784, 1952.

8.   LECKIE, F. A. and HAYHURST, D. R. - Creep rupture of struc-tures, Proc. R. Soc. London A340, pp. 323-347, 1974.

9. VAKULENKO, A. A. and KACHANOV, L. M. - Continuum theory
    of cracked media, Mekh. Tverdogo Tela, Vol.6, pp. 159-166,
    1971.

10. MURAKAMI, S. and OHNO, N. - A constitutive equation of
    creep damage in polycrystalline metals, IUTAM Colloquium
    Euromech 111, Marienbad, 1978.

11. SRINAVASAN, M. G. et al. - The distributed damage theory
    and its application in dynamically loaded structures, 16th
    Annual Meeting S.E.S., Northwestern University, pp. 24-25,
    1979.

12. KACHANOV, L. M. - Continuum model of medium with cracks,
    J. Engng. Mech. Division, Proc. ASCE, Vol. 106, pp. 1039-
    1051, 1980.

13. CHABOCHE, J. L. - Le concepte de constrainte effective
    applique a l'elasticite et a viscoplasticite en presence
    d'end$_o$mmagement anisotrope, Colloquium Euromech 115,
    Grenoble, 1979.

14. TRAMPCZYNSKI, W. A., HAYHURST, D. R. and LECKIE, F. A. -
    Creep rupture of copper and aluminium under non-
    proportional loading, J. Mech. Phys. Solids, Vol. 29, No.
    5/6, pp. 353-374, 1981.

15 HAYHURST, D. R., TRAMPCZYNSKI, W. A. and LECKIE, F. A., -
    Creep rupture under non-proportional loading, Acta
    Metallurgica, Vol.28, pp. 1171-1183, 1980.

16. HAYHURST, D. R. - Creep rupture under multiaxial states
    of stress. J. Mech. Phys. Solids, Vol.20, pp. 381-390,
    1972.

17. LECKIE, F. A. - The role of geometry change in high
    temperature design, Int. J. Mech. Sci. Vol.24, No.4,
    pp. 245-250, 1982.

18. LEMAITRE, J. and CHABOCHE, J. L. - Aspect Phénoménologique
    de la rupture par endommagement. Journal de Mécanique
    Appliquée, Vol.2, No.3, pp. 317-365, 1978.

19. JOHNSON, A. E., HENDERSON, J. and KHAN, B. - Complex
    stress creep, relaxation and fracture of metallic alloys,
    Ch.4, H.M.S.O., London, 1962.

20. LECKIE, F. A. and HAYHURST, D. R. - Constitutive equations
    for creep rupture, Acta Metallurgica, Vol. 25, pp. 1059 -
    1070, 1977.

21. MURAKAMI, S. and OHNO, N. - A continuum theory of creep
    and creep damage, IUTAM Symp. Creep in Structures, ed.
    A. R. S. Ponter and D. R. Hayhurst, Leicester, 1980.

22. JANSON, J. and HULT, J. - J. Méca. Appliquée, 1977.

23. LEMAITRE, J. - Damage modelling for prediction of plastic or creep fatigue failure in structures. Paper L5/1, Proc. 5th SMIRT, Berlin, 1979.

24. CHABOCHE, J. L. - Sur l'utilisation des variables d'etat interne pour la description du comportement viscoplastique et de la rupture par endommagement. Symp. Franco-Polonais de Rhéologie et Mécanique, Cracovic, 1977.

25. BOETTNER, R. C. and ROBERTSON, W. D. - Trans. metall. Soc. A.I.M.E. 221, p. 613, 1961.

26. NAYAK, G. C. and ZIENKIEWICZ, O. C. - Convenient form of stress invariants for plasticity. J. of the Struct. Div. Proc. of A.S.C.E. pp. 949-953, 1972.

27. GALLAGHER, R. H. - Finite element analysis - fundamentals. Prentice-Hall, 1975.

28. OWEN, D. R. J. and HINTON, E. - Finite elements in plasticity: theory and practice. Pineridge Press, Swansea, U.K., 1980.

29. BATHE, K. J., RAMM, E. and WILSON, E. L. - Finite element formulations for large deformation dynamic analysis, Int. J. Num. Meth. Engng. Vol.9, pp. 353-386, 1975.

30. FUNG, Y. C. - Foundations of solid mechanics. Prentice-Hall, Englewood Cliffs, New Jersey, 1965.

31. MALVERN, L. E. - Introduction to the mechanics of a continuum medium. Prentice-Hall, Englewood Cliffs, New Jersey, 1969.

32. KRIEG, R. D. - Numerical integration of some new unified plasticity-creep formulation, Paper M6/4, Proc. 4th SMIRT San Francisco, 1977.

33. KUMAR, V., MORJARIA, M. and MUKHERJEE, S. - Numerical integration of some stiff constitutive models of inelastic deformation, ASME, J. of Engng. Materials and Technology, Vol. 102, pp. 92-96, Jan. 1980.

34. HAYHURST, D. R. and KRZECZKOWSKI, A. J. - Numerical solution of creep problems, Comp. Meth. Appl. Mech. Engng. Vol. 20, pp. 151-171, 1979.

35. SAVALLE, S. and CULIÉ, J. P. - Méthodes de calcul associés aux lois de comportement cyclique et d'endommagement, Rech. Aérosp., No.5, pp. 263-278, 1978.

36. HAYHURST, D. R., MORRISON, C. J. and LECKIE, F. A. - The effect of stress concentrations on the creep rupture of tension panels, ASME, J. of Appl. Mech., Vol. 42, pp. 613-618, 1975.

37. HAYHURST, D. R., DIMMER, P. R. and CHERNUKA, M. W. - Estimates of the creep rupture life of structures using the finite element method, J. Mech. Phys. Solids, Vol.23, pp. 335-355, 1975.

38. RABOTNOV, Y. N. - Advances in creep design: the A. E. Johnson Memorial Volume, ed. A. I. Smith and A. M. Nicholson, pp. 3-19, Applied Science, London, 1971.

39. LECKIE, F. A. and ONAT, E. T. - Tensorial nature of damage measuring internal variables, IUTAM Symposium, Selis, France, pp. 140-155, 1981.

40. ONAT, E. T. - Representation of inelastic behaviour in the presence of anisotropy and of finite deformations, Ch.5, in Recent Advances in Creep and Fracture of Engineering Materials and Structures, ed. B. Wilshire and D. R. J. Owen, Pineridge Press, Swansea,1982.

41. KRAJCINOVIC, D. and FONSEKA, G. U. - The continuous damage theory of brittle materials, ASME, J. of Appl. Mech. Vol.48, pp. 809-824, 1981.

42. HAYHURST, D. R., LECKIE, F. A. and HENDERSON, J. T. - Design of notched bars for creep-rupture testing under tri-axial stresses, Int. J. Mech. Sci., Vol.19, pp. 147-159, 1977.

43. GERALD, C. F. - Applied Numerical Analysis, 2nd edition, Addison-Wesley, 1978.

44. ARGYRIS, J. H., VAZ, L. E. and WILLAM, K. J. - Improved solution methods for inelastic rate problems, Comp. Meth. App. Mech. Engng., Vol.16, pp. 231-277, 1978.

45. OWEN, D. R. J. and GONCALVES F°., O. J. A. - A numerical investigation of creep brittle rupture under proportional loading using a state variable theory, Proc. 1[er] Simposium Nacional sobre Aplicaciones del Metodo de los Elementos Finitos en Ingenieria, eds. E. Onate, E. Alonso and M. Casteleiro, Barcelona, 1982.

46. DODDS, Jr., R. H. and LOPEZ, L. A. - Substructuring in linear and nonlinear analysis, Int. J. Num. Meth. Engng. Vol.15, pp. 583-597, 1980.

47. ROW, D. G. and POWELL, G. H. - A substructure technique
for nonlinear static and dynamic analysis, Rep. UCB/EER/
C78-15, University of California, Berkeley, 1978.

48. OWEN, D. R. J. and GONCALVES F°., O. J. A. - Substructuring
techniques in material nonlinear analysis, Comp. and
Struct., Vol.15, No.3, pp. 201-213, 1982.

Chapter 6

BREE DIAGRAMS FOR ALTERNATIVE LOADING SEQUENCES

H. W. Ng* and D. N. Moreton**

* National Nuclear Corporation Ltd., Knutsford, Cheshire.
   Formerly the Dept. of Mechanical Engineering, University of
   Liverpool.
** Dept. of Mechanical Engineering, University of Liverpool,
   Liverpool L69 3BX.

SUMMARY

The logic of Bree [1], used to identify the various boundaries of strain behaviour of a thin walled cylinder subjected
to continuous pressure loading and cyclic thermal loading, is
presented. An alternative approach to this problem is given
and shown to yield the same result. When the loading sequence
differs from that prescribed by Bree this method of solution
is found to be simpler to apply and has been used to find the
boundaries of strain behaviour for (a) Continuous thermal
loading with cyclic pressure loading, (b) Cyclic pressure and
thermal loading "out of phase" and (c) "in phase". All of the
original assumptions and approximations adopted by Bree have
been used throughout this work.

1. INTRODUCTION

An almost inevitable consequence of high temperature
design is the presence of thermal gradients within structures.
Such gradients will be complex in form and lead to difficulties in assessing the resulting stresses, particularly when
yielding of the structural material occurs. Furthermore, it
is unlikely that these thermal stress will exist in isolation
and the presence of "mechanical loads" will further complicate
the analysis. It is therefore not surprising that the work of
Bree [1] has become so widely adopted*. Briefly Bree considered the case of a thin walled tube subjected to a continuous
internal pressure together with a cyclic through-wall thermal
gradient. The cylinder material was taken to be elastic/
perfectly-plastic and the thermal stress taken to be linearly

---

* A review of this work and the literature following from it
  is given by Ng and Moreton [2].

distributed through the cylinder wall. By 'neglecting' the axial stress in the cylinder, Bree was able to formulate sufficiently accurate equilibrium equations to permit the analysis to proceed and identify the strain behaviour of the cylinder i.e. elastic, shakedown, plasticity or ratchetting. This work resulted in what has become known as the 'Bree Diagram' i.e. a plot of the fictitious elastic thermal stress against the pressure stress showing boundaries of the various modes of strain behaviour.

Because of its simplicity, the diagram has found wide-spread use, particularly by designers, although it has perhaps been used incorrectly on occasions. For example, if both thermal and pressure loading are cyclic and are applied and removed in-phase the boundary of ratchetting given by the Bree diagram is significantly in error for this sequence of loading. If the Bree diagram were used for this 'in phase' case at $\sigma_t/\sigma_y = 2$ the maximum permissible pressure stress to avoid ratchetting would be found to be $\sigma_p/\sigma_y = 0.5$. The corresponding figure found using the present analysis for in-phase loading is $\sigma_p/\sigma_y = 0.875$.

Since there will be many applications for which Bree type diagrams may be required, the authors have pursued the ideas of Bree but altered the loading sequence. Initially, considerable problems were encountered in determining the strain behaviour because of the complexity of the through thickness stress distributions. However, a solution method was developed which consisted of formulating the various equations of equilibrium and imposing boundary conditions to identify the regimes of strain behaviour. Although this method invariably led to more lengthy algebra than Bree encountered it was found to be reliable and conceptually less difficult.

2. SOLUTION OF BOUNDS FOR THE BREE [1] LOADING SEQUENCE

Although the bounds of the various regimes of strain behaviour have been given by Bree, this solution is included to demonstrate the technique used for the alternative loading sequences presented later. In this analysis the theorems of Frederick and Armstrong [3] are invoked. These theorems state that, regardless of the initial state of stress which may be present in a structure, a unique steady cyclic stress state will be achieved after a number of cycles. Implied here are the conditions that yielding occurs during the first loading cycle and that it may be necessary to apply several cycles of load before the stresses converge to a unique steady state. Thus, the analysis of the structure for a small number of cycles is considered sufficient to determine the steady cyclic stress state.

In this analysis, it is only necessary to study the stress changes over the first full cycle of loading (i.e. over the first two half cycles). All of the simplifying assumptions used by Bree are used here and of particular importance are the assumptions of elastic/perfectly-plastic and time independent material properties.

Starting with a given temperature variation through the thickness of the tube, ΔT, the maximum fictitious elastic thermal stress may be written as

$$\sigma_t = \frac{E\ \alpha\ \Delta T}{2(1-\nu)}$$

where ν is Poisson's ratio and the term $(1-\nu)$ is used to account for the axial stress component in the cylinder wall.

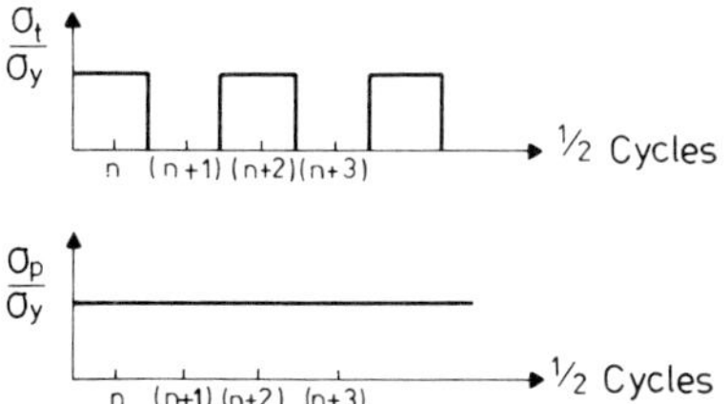

FIG. 1.   Bree (1) Loading Sequence.

The loading sequence of Bree is shown in Fig. 1. Two possible through thickness stress distributions for Bree loading are shown in Figs. 2a and 2b. The first, Fig. 2a, shows the condition where yielding has occurred on the outer surface and progressed past the mid wall. The second, Fig. 2b, shows additional yielding occurring on the inner surface of the cylinder. These two types of stress distribution will be referred to as Case 1 and Case 2 referring to Figs. 2a and 2b respectively.

It may be noted that these stress distributions are drawn using different thermal stress gradients. Both gradients are however given the slope K which may be defined in two ways without conflict.

$$K = \frac{2\sigma_t}{t} \quad \text{or} \quad K = \frac{2\sigma_y}{b-a}$$

Both Figs. 2a and 2b show stress distributions for cycles n and (n+1) of the loading sequence, it being assumed that a steady state exists. We may now establish equilibrium equations for the two cases as follows.

282

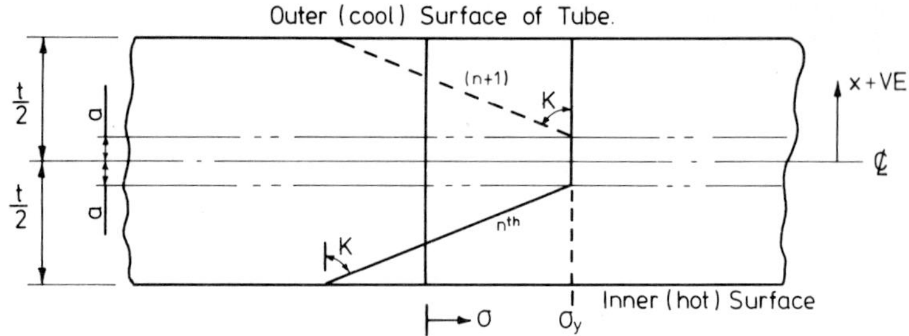

FIG. 2a   Stress Distribution Under R₁ Ratchetting.

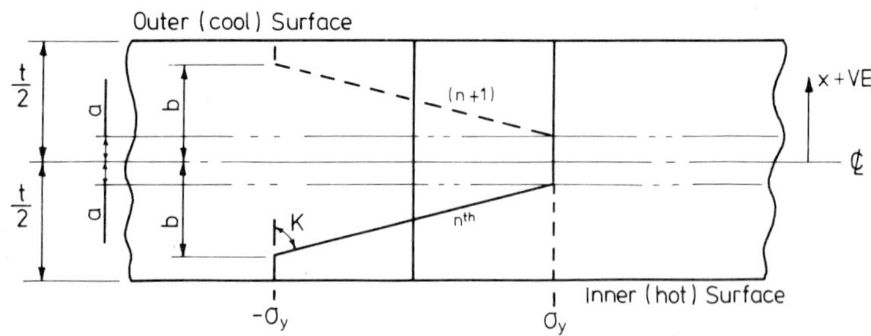

FIG. 2b   Stress Distribution Under R₂ Ratchetting

<u>Case 1</u>: For the nth and $(n+1)^{th}$ cycles equilibrium requires

$$\sigma_p t = \int_{-t/2}^{+t/2} \sigma(x)\, dx \tag{1}$$

where $\sigma(x)$ represents the state of stress at any position, x, through the thickness of the cylinder wall. For Case 1 we may define $\sigma(x)$ as:

$$\sigma(x) = \sigma_y - K(a - x) \quad \text{for } \frac{-t}{2} < x < -a$$

$$\text{and } \sigma(x) = \sigma_y \quad \text{for } -a < x < \frac{t}{2}$$

Substituting these stress conditions into equation (1) gives

$$\sigma_p t = \int_{-t/2}^{-a} (\sigma_y - K(a-x))dx + \int_{-a}^{t/2} \sigma_y\, dx$$

which may be evaluated as

$$\sigma_p = \frac{-K}{2t}\left(a + \frac{t}{2}\right)^2 + \sigma_y \tag{2}$$

and this will be used as the equilibrium equation for Case 1 type stress distributions as depicted in Fig. 2a.

Case 2: For Case 2 type stress distribution the material at the inner surface of the cylinder has yielded and the stress conditions become

$$\sigma(x) = -\sigma_y \qquad \text{for } -\frac{t}{2} < x < -b$$

$$\sigma(x) = -\sigma_y + K(b+x) \quad \text{for } -b < x < -a$$

$$\sigma(x) = \sigma_y \qquad \text{for } -a < x < \frac{t}{2}$$

Substituting into equation (1) gives

$$\sigma_p t = \int_{-\frac{t}{2}}^{-b} -\sigma_y \, dx + \int_{-b}^{-a} (-\sigma_y + K(b+x))dx + \int_{-a}^{t/2} \sigma_y \, dx$$

Integrating and simplifying yields

$$\sigma_p t = \sigma_y 2a + \frac{K}{2}(b-a)^2$$

and multiplying by 2K throughout we have

$$4\sigma_p \sigma_t = 4\sigma_y Ka + K^2(b-a)^2 \qquad (3)$$

which will be used as the equilibrium equation for Case 2 type stress distributions as depicted by Fig. 2b.

We now seek to define boundaries on a $\sigma_t/\sigma_y - \sigma_p/\sigma_y$ plot which separates modes of strain behaviour. Firstly, however, it is useful to make some observations concerning ratchetting.

Referring to Fig. 3 the contributions to the total strain (i.e. elastic + plastic + thermal) are shown for the nth and $(n+1)^{th}$ cycles. The centre column shows the change of strain which occurs when the thermal gradient is removed. From this illustration it is a simple matter to deduce the magnitude of the steady state ratchet strain for Case 1 ratchetting

$$\text{i.e. } \delta = 2aK = 2a \frac{2\sigma_t}{t}$$

from equation (2) we can evaluate a as

$$a = -\frac{t}{2}\left[1 - 2\sqrt{(\sigma_y - \sigma_p)/\sigma_t}\right]$$

and hence the strain accumulated on each cycle is

284

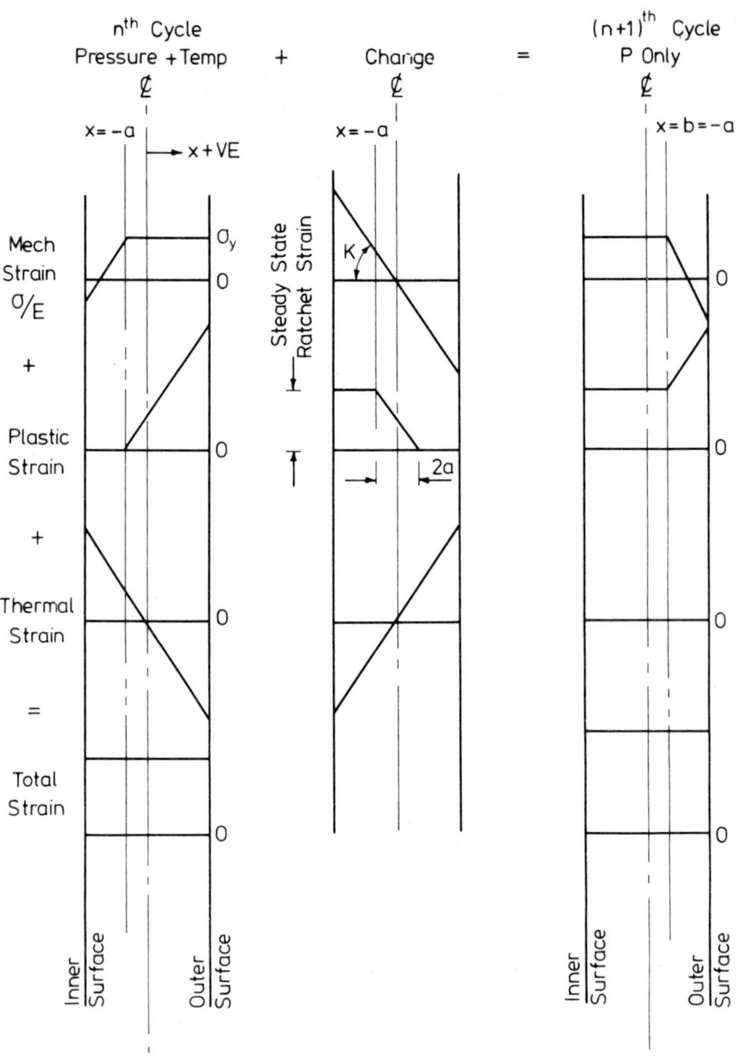

FIG.3 Ratchetting R₁ for Bree Loading Sequence.

$$\delta = \frac{Kt}{E} \left(1 - 2 \frac{\sqrt{(\sigma_y - \sigma_p)}}{\sigma_t} \right) \qquad (4)$$

If yielding occurs on both surfaces of the cylinder simultaneously (i.e. Case 2 behaviour) a similar procedure yields

$$\delta = \frac{(2\sigma_t)(\sigma_p - \sigma_y)}{E} \frac{}{\sigma_y} \frac{}{\sigma_t} \tag{5}$$

Clearly, the boundary of Case 1 ratchetting (R1) may be found by letting $\delta = 0$ in equation (4). Then

$$\frac{2\sqrt{\sigma_y - \sigma_p}}{\sigma_t} = 1$$

or

$$\frac{\sigma_t}{4\sigma_y} + \frac{\sigma_p}{\sigma_y} = 1 \tag{6}$$

which then represents the boundary of R1 and non-ratchetting behaviour. This was the approach used by Bree in his original analysis. The same result would be obtained by letting $a = 0$ in equation (2).

$$\text{i.e.} \quad \sigma_p = \frac{-K}{2t} \cdot \frac{t^2}{4} + \sigma_y$$

and by substituting $K = \dfrac{2\sigma_t}{t}$ we obtain

$$\frac{\sigma_t}{4\sigma_y} + \frac{\sigma_p}{\sigma_y} = 1$$

This is a useful observation since it now permits us to find boundaries of ratchetting more simply. Qualitatively, it means that ratchetting occurs when material at the centre of the cylinder wall yields repeatedly during both the nth and the $(n+1)^{th}$ cycles. Fig. 3 shows this aspect clearly.

To obtain the boundary of ratchetting for Case 2 behaviour we may use the equilibrium equation (3) and substitute $a = 0$, thus

$$4\sigma_p\sigma_t = K^2 b^2$$

and defining

$$K = \frac{2\sigma_y}{b-a} = \frac{2\sigma_y}{t}$$

then

$$\sigma_p\sigma_t = \sigma_y^2 \tag{7}$$

286

which represents a boundary of R2 ratchetting.

The boundary separating Case 1 and Case 2 behaviour can be identified since at the onset of Case 2 behaviour the inner surface of the cylinder will be on the verge of compressive yielding (on the nth cycle). The thermal stress gradient may be defined two ways, i.e.

$$K = \frac{2\sigma_y}{(b-a)} = \frac{2\sigma_t}{t} \text{ hence } (b-a) = \frac{t\sigma_y}{\sigma_t}$$

The onset of compressive yielding on the inner surface may be written as:

$$b = t/2$$

and using the equilibrium condition for Case 2, i.e.

$$\sigma_p t = \sigma_y 2a + \frac{K(b-a)^2}{2}$$

with the conditions above for K and b, we have after simplifying -

$$\sigma_t(\sigma_y - \sigma_p) = \sigma_y^2 \tag{8}$$

which is thus the boundary separating R1 and R2 and indeed S1 and S2. It would also separate P1 and P2 if P1 existed.

The boundary of purely elastic behaviour may be found using the equilibrium equation of Case 1 behaviour and letting a = t/2 which can be seen from Fig. 2a as an extension of the yield interface to the outer surface. Thus

$$\sigma_p = \frac{-K}{2t}\left(a + \frac{t}{2}\right)^2 + \sigma_y \quad \text{with } a = t/2$$

$$\text{and } K = \frac{2\sigma_t}{t}$$

which simplifies to

$$\sigma_p + \sigma_t = \sigma_y \tag{9}$$

which is thus the boundary of purely elastic behaviour.

We have so far not considered the case of alternating plasticity. During this mode of inadaptation the mid-surface stress has not reached yield but the stresses at both inner and outer surfaces have exceeded $2\sigma_y$ (fictitious elastic). Thus, the material beneath the surfaces alternately yields in

opposite senses at every half cycle. The loop width is given
by Bree as

$$z \;=\; (|x| - c).\frac{2\sigma_t}{Et}$$

where $c = \dfrac{\sigma_y}{\sigma_t}.t$ and indicates the depth of penetration of the
material (see [1]).

To find the bound of alternating plasticity we need simply
put $z = 0$ and $c = t/2$. Thus

$$|x| \;=\; t/2$$

and

$$\frac{\sigma_t}{\sigma_y} \;=\; 2 \tag{10}$$

This represents the bound of alternating plasticity but
clearly at slightly lower σ_t/σ_y values the outer surfaces will
yield on the first cycle. On subsequent cycles shakedown will
occur and thus equation (10) represents the boundary between
P2 and S2.

Alternatively, we can write down the equilibrium equation,
as before at the $(n+1)^{th}$ cycle. The stress distribution for
the nth and $(n+1)^{th}$ cycles are illustrated in Fig. 4.

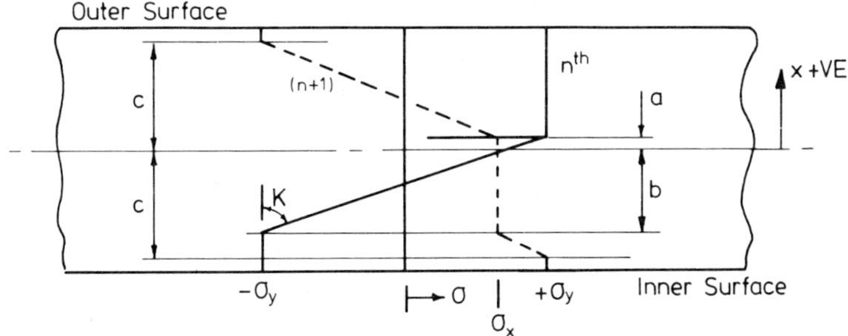

FIG. 4   Stress Distribution Under P2 Alternating Plasticity.

288

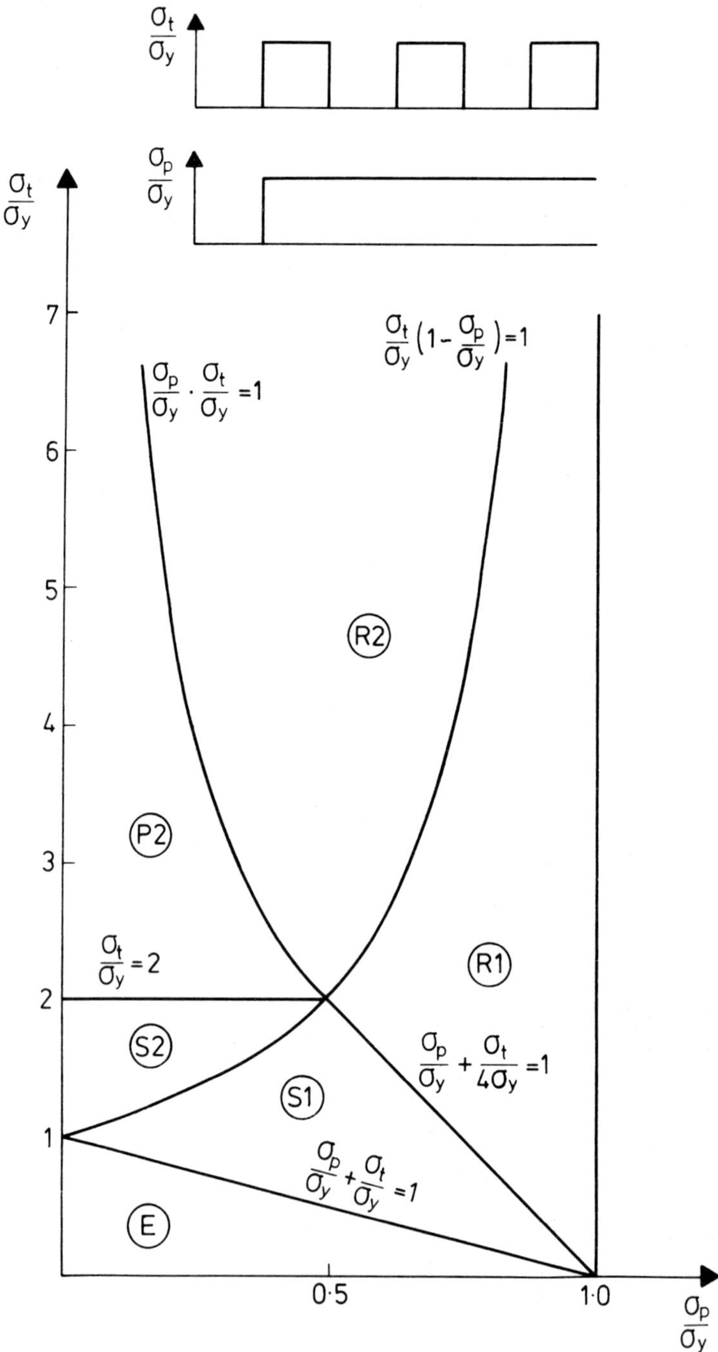

FIG. 5 "The Bree Diagram".

For the $(n+1)^{th}$ cycle we have:

$$\sigma_p t = \int_c^{t/2} -\sigma_y \, dx + \int_a^c (-\sigma_y + K(c - x)) dx + \int_{-b}^a \sigma_x dx$$

$$+ \int_{-c}^{-b} (\sigma_y - K(b + x)) dx + \int_{-t/2}^{-c} \sigma_y dx$$

where σ_x is used to represent the unknown stress over the range $-b < x < a$. After integration and considerable algebraic manipulation this equation reduces to

$$\sigma_p t = \sigma_y c + \frac{\sigma_x^2 - \sigma_y^2}{K} \qquad (11)$$

Substituting $K = 2\sigma_y/c$ and noting that at the boundary of plasticity $c = t/2$ we have

$$\sigma_x = \sqrt{4\sigma_p \sigma_y - \sigma_y^2}$$

Substituting back into (11) with $K = 2.\sigma_t/t$ we find

$$\sigma_p \left(1 - \frac{2\sigma_y}{\sigma_t}\right) = \frac{\sigma_y}{2} \left(1 - \frac{2\sigma_y}{\sigma_t}\right)$$

from which $\left(1 - \frac{2\sigma_y}{\sigma_t}\right)\left(\sigma_p - \frac{\sigma_y}{2}\right) = 0$

so that

$$\frac{\sigma_t}{\sigma_y} = 2 \qquad (12)$$

or

$$\frac{\sigma_p}{\sigma_y} = \frac{1}{2} \qquad (13)$$

We see that equation (12) is Bree's bound to the P2 regime and because of the previously obtained boundaries for R1/R2, S1/S2 and S1/R1 equation (13) is reduced to being a point solution only at $\sigma_t/\sigma_y = 2$, $\sigma_p/\sigma_y = 1/2$.

Thus we have now derived, through the various equilibrium equations, all the bounds which Bree identified and for completeness these bounds are shown in Fig. 5. This method of solution will now be used for obtaining similar boundaries for alternative loading sequences.

3. CONTINUOUS THERMAL LOADING WITH ALTERNATING PRESSURE LOAD

Following from the Bree loading sequence we now consider

290

the case of a thin-walled cylinder subjected to a continuous
thermal gradient through the cylinder wall together with an
alternating internal pressure load. This loading sequence is
shown in Fig. 6.

All of the assumptions adopted in the previous section are
invoked for this analysis.

We firstly note that the pressure loading must be such
that

$$0 < \frac{\sigma_p}{\sigma_y} < 1$$

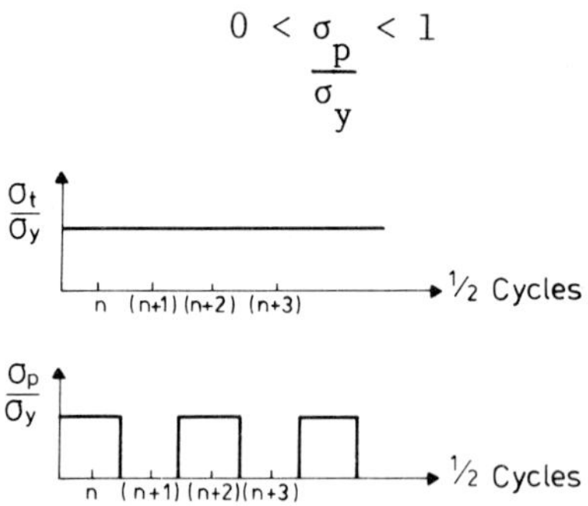

FIG. 6   Continuous Thermal Loading with Alternating Pressure Loading.

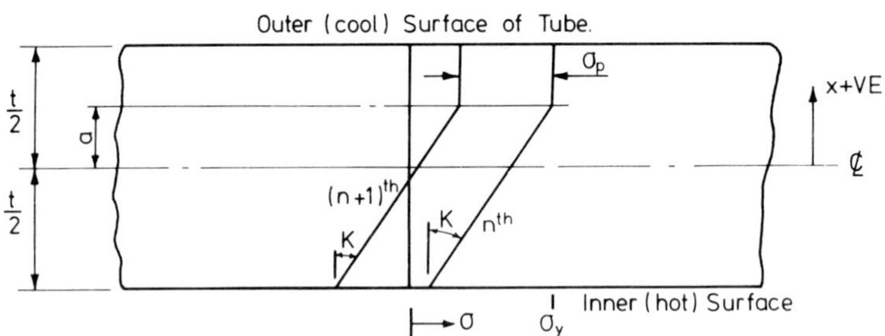

FIG. 7a   Stress Distribution Under S₁ Shakedown.

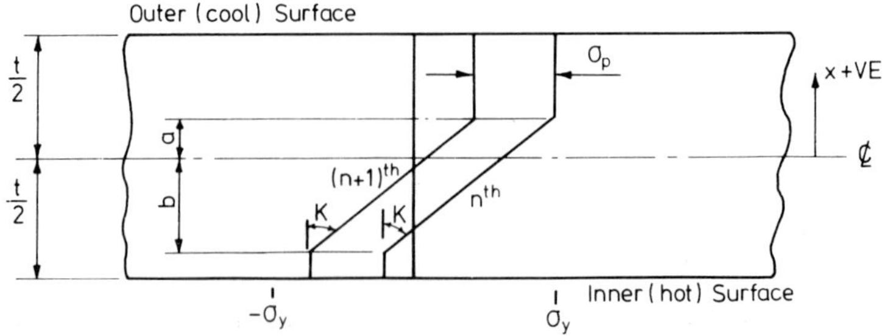

FIG. 7b   Stress Distribution Under S₂ Shakedown.

since if this inequality is not satisfied we have either external pressure or a collapse condition.

We stipulate that the thermal stress is never removed and when the pressure is not acting the thermal stress must be self-equilibriating. This stress may therefore be thought of as a residual stress and we may now take advantage of the Bleich-Melan [4,5] shakedown theorem. This theorem, in essence, states that, "If any self equilibriating set of residual stresses can be found, such that the addition of elastic stresses, results in stresses which do not violate the yield criterion, then the structure will eventually shakedown".

It is clear therefore that for this sequence of loading the behaviour of the cylinder must be either elastic or shakedown. The cylinder may yield on one or both surfaces during the first few cycles but during the steady state the application and removal of the pressure load will simply move the stress distribution bodily from left to right by an amount σ_p as indicated by Figs. 7a and 7b, these showing types S1 and S2 shakedown.

We will however proceed as before and formulate equilibrium equations for Case 1 and Case 2 behaviour.

Case 1:   The general equilibrium equations for the nth and $(n+1)^{th}$ half cycles are obtained through Fig. 7a, and at the nth half cycle:

Equilibrium requires that

$$\sigma_p t = \int_{-t/2}^{t/2} \sigma(x)\, dx$$

for which the stresses and limits are

$$\sigma(x) = \sigma_y - K(a-x) \quad \text{for} \quad \frac{-t}{2} < x < +a$$

$$\sigma(x) = \sigma_y \quad \text{for} \quad a < x < t/2$$

Hence $\sigma_p t = \int_{-t/2}^{a} ((\sigma_y - K(a-x)))\, dx + \int_{a}^{t/2} \sigma_y\, dx$

Integrating and simplifying gives

$$\sigma_p t = \sigma_y t - \frac{K}{2}\left(a + \frac{t}{2}\right)^2 \tag{14}$$

At the $(n+1)^{th}$ half cycle, when the pressure loading is removed the equilibrium equation is found to be exactly the same.

<u>Case 2</u>: Referring now to Fig. 7b which depicts the stress distributions for the nth and $(n+1)^{th}$ half cycles for which the material on the inner surface has yielded, then equilibrium requires that for the nth cycle:

$$\sigma_p t = \int_{-t/2}^{t/2} \sigma(x)\, dx$$

for which the stresses and limits are now

$$\sigma(x) = \sigma_p - \sigma_y \qquad \text{for } \frac{-t}{2} < x < -b$$

$$\sigma(x) = \sigma_p - \sigma_y + K(b-x) \quad \text{for } -b < x < a$$

$$\sigma(x) = \sigma_y \qquad \text{for } a < x < \frac{t}{2}$$

which after substituting, integrating and simplifying yields

$$\sigma_p t = \frac{K}{2}(a+b)^2 + \sigma_p \left(a + \frac{t}{2}\right) - 2a\, \sigma_y \qquad (15)$$

For the $(n+1)^{th}$ cycle a similar procedure gives the same result.

Equations 14 and 15 are then the equilibrium equations for Cases 1 and 2 for the nth and $(n+1)^{th}$ cycles of loading. We now proceed to find the various boundaries of strain behaviour.

Firstly, for purely elastic behaviour we require that $a = +t/2$. Using the equilibrium equation for Case 1

$$\sigma_p t = \sigma_y t - \frac{K(a + \frac{t}{2})^2}{2}$$

substituting $a = t/2$ and $K = 2\sigma_t/t$ we have

$$\frac{\sigma_p}{\sigma_y} + \frac{\sigma_t}{\sigma_y} = 1 \qquad (16)$$

which is the boundary for purely elastic behaviour.

We have previously noted that during an operating cycle, outside the elastic domain, the stress distribution for either Case 1 or Case 2 behaviour simply moves bodily by an amount σ_p. Thus no material in the cylinder wall yields repeatedly from one cycle to the next. The possibility of ratchetting or plasticity is therefore excluded and we need only investigate shakedown conditions.

Two types of shakedown are possible, being Case 1 or Case 2. These may be identified from the equilibrium equations since at the onset of Case 2 behaviour b = t/2.

Thus using equation (15) with b = t/2 and noting that

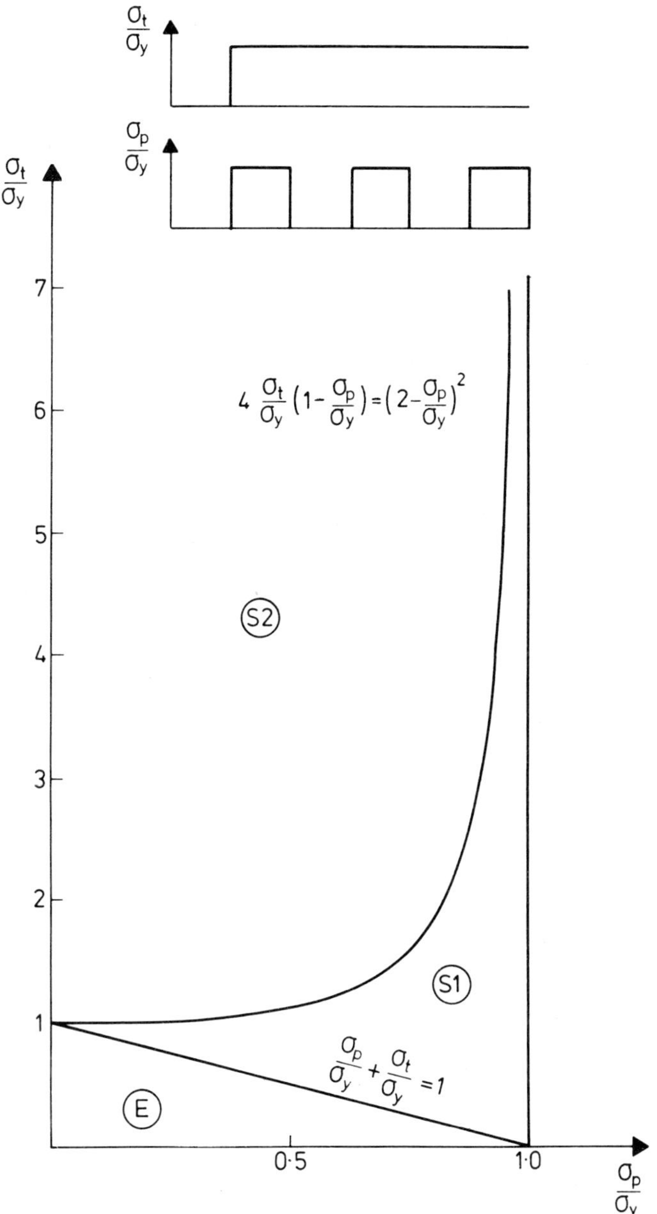

FIG. 8   The Bree Diagram for Continuous Thermal Load with Alternating Pressure Load.

294

$$K(a+b) = 2\sigma_y - \sigma_p \quad \text{and} \quad K = 2\sigma_t/t$$

we find after substitution and simplification that

$$\frac{\sigma_t}{\sigma_y} = (2 - \frac{\sigma_p}{\sigma_y})^2 / 4(1 - \frac{\sigma_p}{\sigma_y}) \qquad (17)$$

which is the boundary between S1 and S2 shakedown.

The 'Bree Diagram' for this sequence of loading is given as Fig. 8. Only two boundaries exist - that bounding elastic behaviour and that separating S1 and S2 behaviour.

4. CYCLIC THERMAL AND PRESSURE LOADING OUT OF PHASE

In this section we consider the strain behaviour of a thin walled cylinder subjected to alternating thermal and pressure loadings such that these loads are out of phase. Such a loading sequence is illustrated in Fig. 9 and we retain all of the simplifications and assumptions adopted by Bree. In addition, it is assumed that a load (either thermal or pressure) is completely removed before the application of the second load. Since no time dependence is including in the material properties, a finite time may elapse before the application of the second load. As before, we will develop the equilibrium equations, and from these obtain the bounds separating the various modes of strain behaviour.

Figs. 10a and 10b show the stress distributions for the nth and $(n+1)^{th}$ cycles for alternating plasticity regimes P1 and P2. As before we will deal with Case 1 and Case 2 behaviour.

$\underline{\text{Case 1}}$: For the thermal load acting alone, i.e. the nth cycle, equilibrium requires that

$$0 = \int_{-t/2}^{t/2} \sigma(x) \, dx$$

where the stresses and limits are:

$$\sigma(x) = -\sigma_y \qquad \text{for } -t/2 < x < -a$$

$$\sigma(x) = -\sigma_y + K(a + x) \quad \text{for } -a < x < +a$$

$$\sigma(x) = \sigma_y \qquad \text{for } a < x < t/2$$

Hence

$$0 = \int_{-t/2}^{-a} -\sigma_y \, dx + \int_{-a}^{+a}(-\sigma_y + K(a+x)) \, dx + \int_{a}^{t/2} \sigma_y \, dx$$

Integrating, putting in the limits and simplifying gives

$$Ka = \sigma_y \qquad (18)$$

For the pressure loading, acting alone i.e. the $(n+1)^{th}$ cycle

$$\sigma_p t = \int_{-t/2}^{t/2} \sigma(x)\ dx$$

for which $\sigma(x)$ and its limits are now

$$\sigma(x) = \sigma_y \qquad for\ -t/2 < x < -b$$

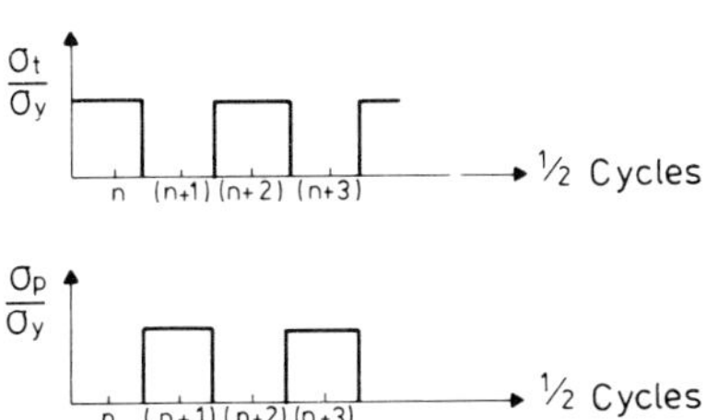

FIG.9   Cyclic Thermal and Pressure Loading Out of Phase.

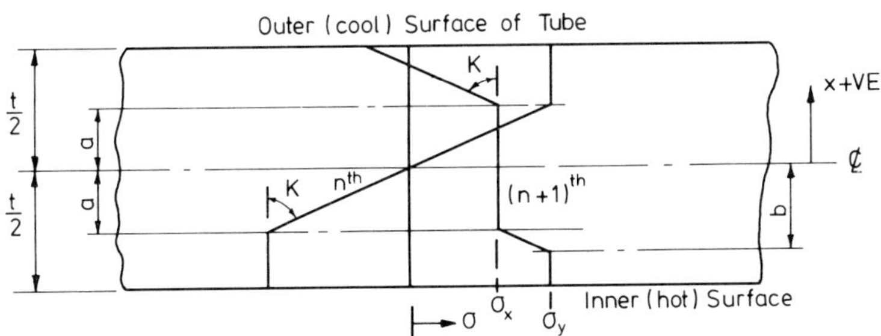

FIG.10a   Stress Distribution Under P₁ Plasticity

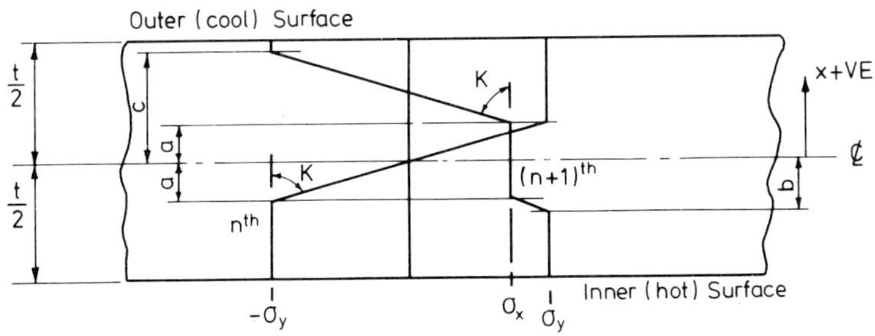

FIG.10b   Stress Distribution Under P₂ Plasticity.

$$\sigma(x) = \sigma_y - K(b + x) \text{ for } -b < x < -a$$

$$\sigma(x) = \sigma_x \qquad\qquad \text{ for } -a < x < a$$

$$\sigma(x) = \sigma_x - K(x - a) \text{ for } a < x < t/2$$

Substituting, integrating and simplifying gives

$$\sigma_p t = \sigma_y(t - a) - \frac{K(a^2 + (b-a)^2}{2} + \frac{t}{2}(\frac{t}{2} - 2a)) + \sigma_x(\frac{t}{2} + a) \qquad (19)$$

Equations (18) and (19) are thus the equilibrium equations for Case 1 behaviour.

<u>Case 2</u>: The equilibrium equations are obtained with reference to Fig. 10b.

For the nth cycle, i.e. thermal loading alone

$$0 = \int_{-t/2}^{t/2} \sigma(x)\, dx$$

for which the stresses and limits are

$$\sigma(x) = -\sigma_y \qquad\qquad \text{ for } -t/2 < x < -a$$

$$\sigma(x) = -\sigma_y + K(a + x) \text{ for } -a < x < +a$$

$$\sigma(x) = \sigma_y \qquad\qquad \text{ for } a < x < t/2$$

Integrating and simplifying gives as before

$$Ka = \sigma_y$$

For the pressure load acting alone, i.e. the $(n+1)^{th}$ cycle

$$\sigma_p t = \int_{-t/2}^{t/2} \sigma(x)\, dx$$

for which the stresses and limits are

$$\sigma(x) = \sigma_y \qquad\qquad \text{ for } -t/2 < x < -b$$

$$\sigma(x) = \sigma_y - K(b + x) \text{ for } -b < x < -a$$

$$\sigma(x) = \sigma_x \qquad\qquad \text{ for } -a < x < +a$$

$$\sigma(x) = \sigma_x - K(x - b) \text{ for } a < x < c$$

$$\sigma(x) = -\sigma y \qquad\qquad \text{ for } c < x < t/2$$

Integrating and simplifying gives

$$\sigma_p t = \sigma_y(c - a) - \frac{K}{2}(a^2 + (b-a)^2 + c(c - 2a)) + \sigma_x(c + a) \quad (20)$$

Equations (18)-(20) are thus the equilibrium equations for Case 1 and Case 2 behaviour.

We seek now to obtain the bounds for the various regimes of strain behaviour. For completely elastic behaviour we simply let $a = t/2$ so that no yielding may occur during the first cycle of thermal loading. Hence, from equation (18)

$$\frac{Kt}{2} = \sigma_y$$

and writing $K = 2\sigma_t/t$ we have

$$\frac{\sigma_t}{\sigma_y} = 1 \quad (21)$$

This thus represents the boundary of purely elastic behaviour. Clearly, if yielding occurs as a result of pressure loading the entire cross section will become plastic and this is thus a collapse condition, represented by

$$\frac{\sigma_p}{\sigma_y} = 1$$

which is the right hand boundary of the diagram.

For the cylinder to shakedown to elastic conditions we require that yielding should occur during the nth cycle but that no yielding should occur on the $(n+1)^{th}$ cycle. We have therefore the neccesary conditions that $a < t/2$ and $b = t/2$ for the boundary of shakedown behaviour. Considering only the $(n+1)^{th}$ cycle this latter condition then requires that the stress distribution is fully elastic at every material point. The regime will clearly be of S2 type since yielding occurred on both surfaces during the nth cycle.

From Fig. 10a we note that

$$K(b - a) = \sigma_y - \sigma_x$$

and from equation (18)

$$Ka = \sigma_y$$

then $\sigma_x = 2\sigma_y - \sigma_t$ since $K = 2\sigma_t/t$

Substituting for σ_x in equation (20) gives after simplification

$$\frac{\sigma_t}{\sigma_y} = 2 - \frac{\sigma_p}{\sigma_y} \tag{22}$$

which is the boundary of S2 behaviour - there being no S1 regime.

Next, the boundary of ratchetting will be considered. For a combination of σ_p/σ_y and σ_t/σ_y which gives rise to type 1 ratchetting, the equilibrium equations (18) and (19) hold although the parameter 'a' in the nth and $(n+1)^{th}$ cycles can no longer be the same. If a_n and a_{n+1} now refer to the positions of 'a' during the nth and $(n+1)^{th}$ half cycles respectively then the overlap width 'd' where tensile yielding occurs on both successive half cycles will be

$$d = a_{n+1} - a_n$$

with the prior provision that $\sigma_x = \sigma_y$ then if d is to be less than zero

$$i.e. \quad a_{(n+1)} < a_n$$

ratchetting will not occur.

The condition $d = 0$ is sufficient to identify the boundary of ratchetting/non-ratchetting behaviour.

From equation (18) putting $a = a_n$

$$Ka_n = \sigma_y$$

From equation (19), putting $a = a_{n+1}$, $\sigma_x = \sigma_y$ and $Kb = Ka_n = Ka_{n+1}$ (this latter condition arising from the condition of $\sigma_x = \sigma_y$) then

$$\sigma_p t = \sigma_y t - \frac{K}{2}(a_{(n+1)} - \frac{t}{2})^2$$

$$\text{or} \quad a_{n+1} = \frac{2}{K}\sqrt{\sigma_t(\sigma_y - \sigma_p)} + \frac{t}{2} \tag{23}$$

As for the case of shakedown it is necessary that

$$a_{n+1} < \frac{t}{2}$$

and we must therefore take the -ve root in equation (23).

If we now substitute $\frac{\sigma_y}{K} = a_n = a_{(n+1)}$ in equation (23) we obtain

$$4\sigma_t(\sigma_y - \sigma_p) = (\sigma_y - \sigma_t)^2$$

and dividing throughout by σ_y^2 we have finally

$$\frac{4\sigma_t}{\sigma_y}(1 - \frac{\sigma_p}{\sigma_y}) = (1 - \frac{\sigma_t}{\sigma_y})^2 \qquad (24)$$

which is thus the boundary of type 1 ratchetting/non-ratchetting behaviour.

The boundary between type 1 and type 2 behaviour may be obtained by observing that when the outer surface of the cylinder is on the point of compressive yielding, then $c = t/2$.

From Fig. 10b we see that

$$K(b - a) = \sigma_y - \sigma_x$$

$$\qquad (25)$$

$$\text{and} \quad K(c - a) = \sigma_y + \sigma_x$$

Since $c = \frac{t}{2}$ and $\frac{Kt}{2} = \sigma_t$

then $Kc = \sigma_t$

and similarly $Ka = \sigma_y$

We may solve equations (25) for σ_x and b giving

$$\sigma_x = \sigma_t - 2\sigma_y$$

$$Kb = 4\sigma_y - \sigma_t$$

Returning to the equilibrium equation for Case 2 behaviour at the $(n+1)^{th}$ half cycle - i.e. equation (20) and substituting the above we eventually obtain

$$\frac{\sigma_t}{\sigma_y} = \frac{4}{(2 - \frac{\sigma_p}{\sigma_y})} \qquad (26)$$

which represents the boundary separating type 1 and type 2 behaviour. This boundary is seen to separate R1 and R2 and also P1 and P2.

We have identified the boundary between type 1 ratchetting and plasticity (equation (24)) and now seek the boundary of type 2 ratchetting and plasticity. In this case there will be compressive yielding on the outer surface. The equation

$$d = a_{n+1} - a_n$$

still applies here and σ_x must be equal to σ_y. From equation (18) which is also applicable for case 2 behaviour

$$a_n = \frac{\sigma_y}{K}$$

and from equation (25)

$$K(b-a) = 0$$

The parameter 'c' is obtained from equation (25) such that

$$Kc = 2\sigma_y + Ka_{(n+1)}$$

Upon substituting the above in equation (20) and simplifying we obtain

$$a_{(n+1)} = \left(\frac{\sigma_p \sigma_t}{\sigma_y} - \sigma_y \right)/K \qquad (27)$$

For the boundary of ratchetting behaviour, $d = 0$ and hence

$$a_n = a_{(n+1)}$$

therefore

$$\left(\frac{\sigma_p \sigma_t}{\sigma_y} - \sigma_y \right)/K = \sigma_y/K$$

or

$$\sigma_p \sigma_t = 2\sigma_y^2$$

or

$$\frac{\sigma_p \sigma_t}{\sigma_y^2} = 2 \qquad (28)$$

which is thus the boundary between R2 and P2 behaviour.

The five boundaries of strain behaviour for the 'out of phase' loading sequence are illustrated in Fig. 11.

Finally, for this sequence of loading, we see from Fig. 11

that between the P2 and S2 regimes a Pl regime exists. Unlike
the plasticity regime of the Bree diagram, this form of alternating plasticity is one in which repeated yielding occurs on
the inside surface of the cylinder. Thus if fatigue failure
occurs, it will be at the inside surface of the cylinder that

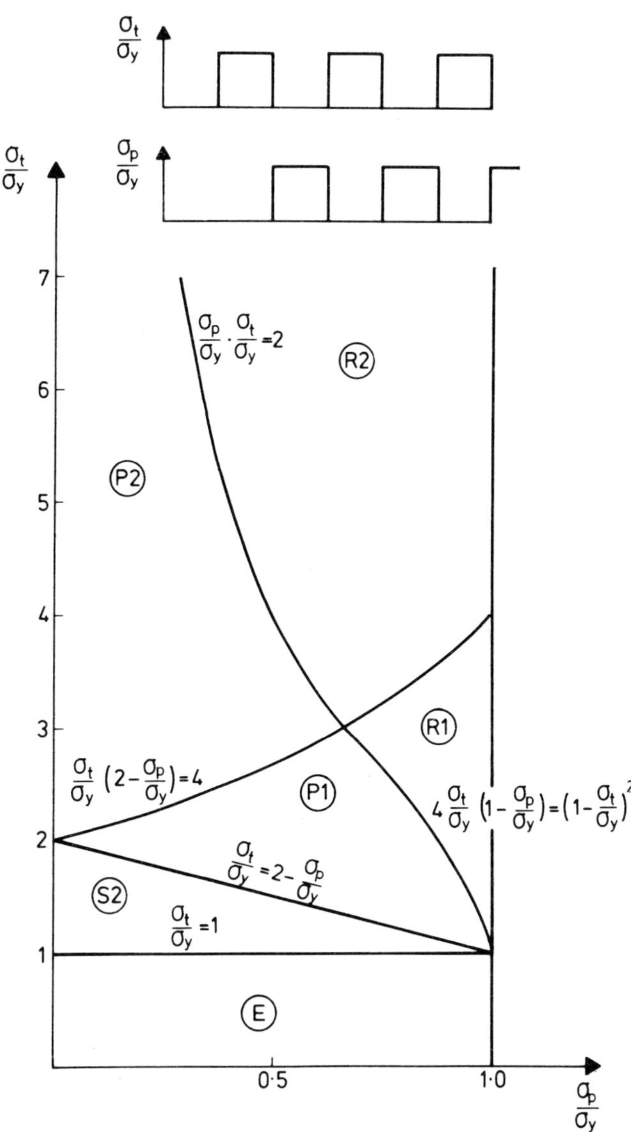

FIG. 11 The Bree Diagram for Cyclic Thermal and Pressure
Loading – Out of Phase.

302

cracks will first appear. Since this surface may not be available to inspection techniques designs which utilise this regime should perhaps be avoided.

5. CYCLIC THERMAL AND PRESSURE LOADING IN-PHASE

Of the three alternative loading sequences this is perhaps the sequence having most practical significance. The loading sequence is shown in Fig. 12. During the nth half cycle, during which both loads are imposed, the equilibrium equations are similar to those for the Bree sequence at the same half cycle. Equations (2) and (3), the equilibrium equations for

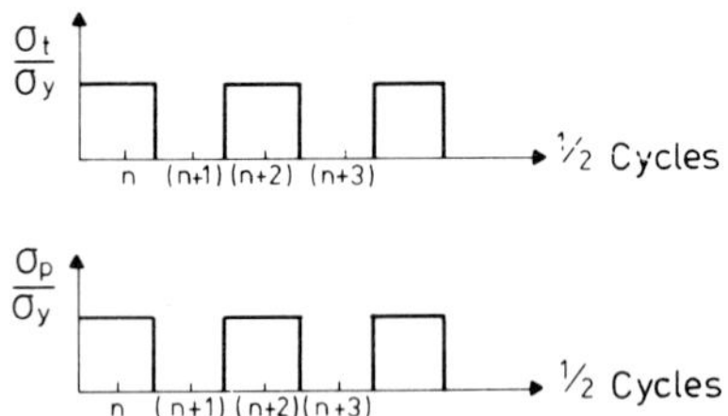

FIG. 12   Cyclic Thermal and Pressure Loading In Phase.

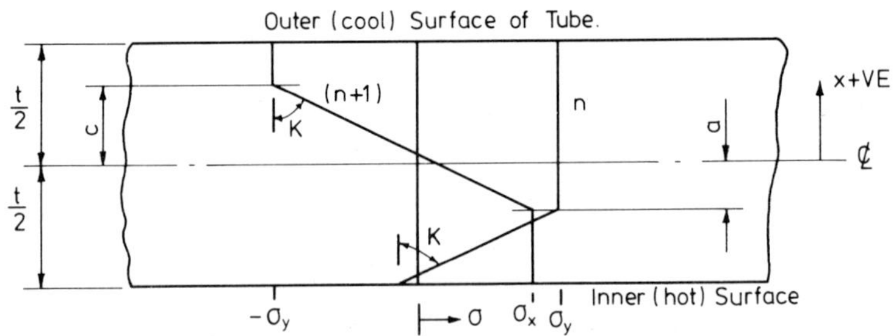

FIG. 13a   Stress Distribution Under P₁ Plasticity.

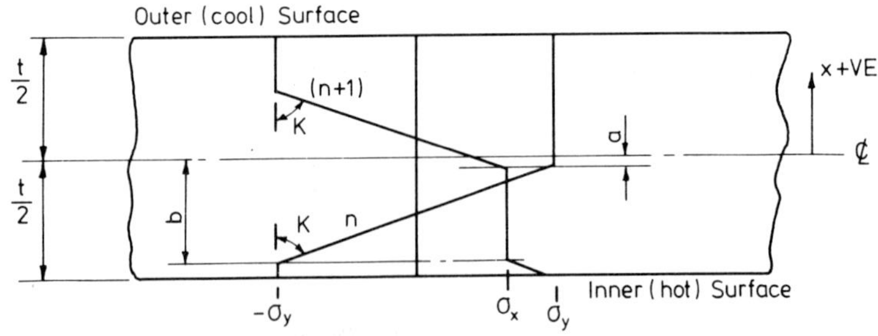

FIG. 13b   Stress Distribution Under P₂ Plasticity.

Cases 1 and 2 at the nth half cycle are therefore used. The stress distribution for the P1 and P2 regimes are shown in Figs. 13a and 13b for the nth and $(n+1)^{th}$ half-cycles.

As before, the yielding of the cylinder on the inside surfaces is used to identify Case 1 and Case 2 behaviour. The term σ_x is used to represent the unknown elastic stress required for equilibrium during the unloaded half-cycle.

<u>Case 1</u>: Equation (2) is repeated here,

$$\sigma_p = \frac{-K}{2t}(a + \frac{t}{2})^2 + \sigma_y \qquad (29)$$

as the equilibrium for the nth half-cycle.

At the $(n+1)^{th}$ half-cycle, equilibrium requires that

$$0 = \int_{-t/2}^{t/2} \sigma(x)dx$$

for which the stresses and limits are

$$\sigma(x) = \sigma_x \qquad \text{for } \frac{-t}{2} < x < -a$$

$$\sigma(x) = \sigma_x - K(x+a) \quad \text{for } -a < x < c$$

$$\sigma(x) = -\sigma_y \qquad \text{for } c < x < t/2$$

so that

$$0 = \int_{-t/2}^{-a} \sigma_x dx + \int_{-a}^{c} (\sigma_x - K(x+a))dx + \int_{c}^{t/2} - \sigma_y dx$$

which after integration and simplification gives

$$0 = 2\sigma_x K(\frac{t}{2} + c) - K^2(a+c)^2 - 2\sigma_y K(\frac{t}{2} - c) \qquad (30)$$

which is thus the equilibrium equation for the $(n+1)^{th}$ cycle.

<u>Case 2</u>: The equilibrium equation for the nth half cycle has already been obtained for the Bree sequence of loading. We need only rewrite equation (3) as

$$4\sigma_p \sigma_t = 4\sigma_y Ka + K^2(b-a)^2 \qquad (31)$$

to be the equilibrium equation for the nth half cycle.

At the $(n+1)^{th}$ half cycle, the summation of forces gives

$$0 = \int_{-t/2}^{t/2} \sigma(x)dx$$

for which the stresses and limits are

$$\sigma(x) = \sigma_x - K(x+b) \text{ for } -b < x < -t/2$$

$$\sigma(x) = \sigma_x \qquad \text{for } -a < x < -b$$

$$\sigma(x) = \sigma_x - K(x+a) \text{ for } c < x < -a$$

$$\sigma(x) = -\sigma_y \qquad \text{for } \frac{t}{2} < x < c$$

After substitution, integration and simplification the above is evaluated as

$$0 = 2\sigma_x K(\frac{t}{2} + c) - K^2((a + c)^2 - (b - \frac{t}{2})^2) - 2\sigma_y K(\frac{t}{2} - c) \quad (32)$$

which is the equilibrium equation for the $(n+1)^{th}$ half cycle for Case 2 behaviour.

Before proceeding to find the various boundaries of strain behaviour it is useful to firstly obtain some simple expressions for a and b.

For all cases considered

$$K(b - a) = 2\sigma_y \qquad (33)$$

Solving equations (31) and (33) for a and b gives

$$Ka = (\sigma_p \sigma_t - \sigma_y^2)/\sigma_y \qquad (34)$$

$$\text{and} \quad Kb = (\sigma_p \sigma_t + \sigma_y^2)/\sigma_y \qquad (35)$$

For fully elastic behaviour, no yielding may occur during the first half-cycle. Substituting a = t/2 in equation (29) gives

$$\sigma_p = -Kt/2 + \sigma_y$$

or

$$\sigma_p + \sigma_t = \sigma_y \qquad (36)$$

which is therefore the boundary of purely elastic behaviour.

For the boundary of ratchetting of type 1 behaviour, the overlap width d must again be equal to zero.

Hence $a_{(n+1)} - a_n = 0$

from equation (29)

$$a_n = (2\sqrt{(\sigma_y - \sigma_p)/\sigma_t} + 1)\sigma_t/K$$

and from equation (30)

$$0 = 4\sigma_y(2\sigma_y - Ka_{(n+1)}) - 4\sigma_y^2$$

where

$$K(a_{(n+1)} + c) = 2\sigma_y$$

therefore

$$a_{(n+1)} = \sigma_y/K$$

Hence

$$\sigma_y - (2\sqrt{(\sigma_y - \sigma_p)\sigma_t} + 1)\sigma_t = 0$$

or

$$4\frac{\sigma_t}{\sigma_y}(1 - \frac{\sigma_p}{\sigma_y}) = (1 - \frac{\sigma_t}{\sigma_y})^2 \qquad (37)$$

which is the same as equation (24) - the boundary of Rl for the 'out of phase' case.

For Case 2 behaviour, the boundary of ratchetting is obtained in a similar manner to that used for the out of phase case. The conditions that $\sigma_x = \sigma_y$ and $a_n = a_{n+1}$ are applied.

Firstly, considering the nth half cycle, equation (34) gives

$$a_n = (\frac{\sigma_p \sigma_t}{\sigma_y^2} - 1)\sigma_y/K$$

and at the $(n+1)^{th}$ half cycle

$$K(a_{(n+1)} + c) = \sigma_y + \sigma_x = 2\sigma_y$$

$$Kc = 2\sigma_y - Ka_{(n+1)}$$

and

$$K(\frac{t}{2} - b) = \sigma_y - \sigma_x = 0$$

therefore

$$Kb = \sigma_t$$

Substituting the above into equation (30) and simplifying gives

$$Ka_{(n+1)} = \sigma_y$$

and for the ratchetting boundary

$$a_n = a_{n+1}$$

therefore

$$\left(\frac{\sigma_p \sigma_t}{\sigma_y^2} - 1\right)\sigma_y = \sigma_y$$

or

$$\frac{\sigma_p \sigma_t}{\sigma_y^2} = 2 \tag{38}$$

which is the boundary for Case 2 ratchetting, and is again the same as that for the out of phase case.

Turning now to the boundaries of shakedown, we observe that for Case 1 behaviour for all material points to behave elastically, after initial yielding the parameter 'c' must equal $t/2$ and

$$Ka(a + c) = \sigma_y + \sigma_x$$

Also, from equation (29)

$$Ka = 2\sqrt{\sigma_t(\sigma_y - \sigma_p)} + \sigma_t$$

and solving for σ_x gives

$$\sigma_x = -\sigma_y + \frac{Kt}{2} + 2\sqrt{\sigma_t(\sigma_y - \sigma_p)} + \sigma_t$$

and since $Kt = 2\sigma_t$

$$\sigma_x = -\sigma_y + 2\sqrt{\sigma_t(\sigma_y - \sigma_p)} + 2\sigma_t$$

Substituting the above into equation (30) and simplifying gives

$$\frac{\sigma_t}{\sigma_y} = 2 - \frac{\sigma_p}{\sigma_y} \tag{39}$$

which is the bound of the S1 shakedown regime.

For the boundary of the S2 regime we again take $c = t/2$ in order that all material points behave elastically after initial yielding.

Since $K(a + c) = \sigma_y + \sigma_x$

then

$$\sigma_x = Ka + \frac{Kt}{2} - \sigma_y = \left(\frac{\sigma_p \sigma_t}{\sigma_y^2} - 2\right)\sigma_y + \sigma_t$$

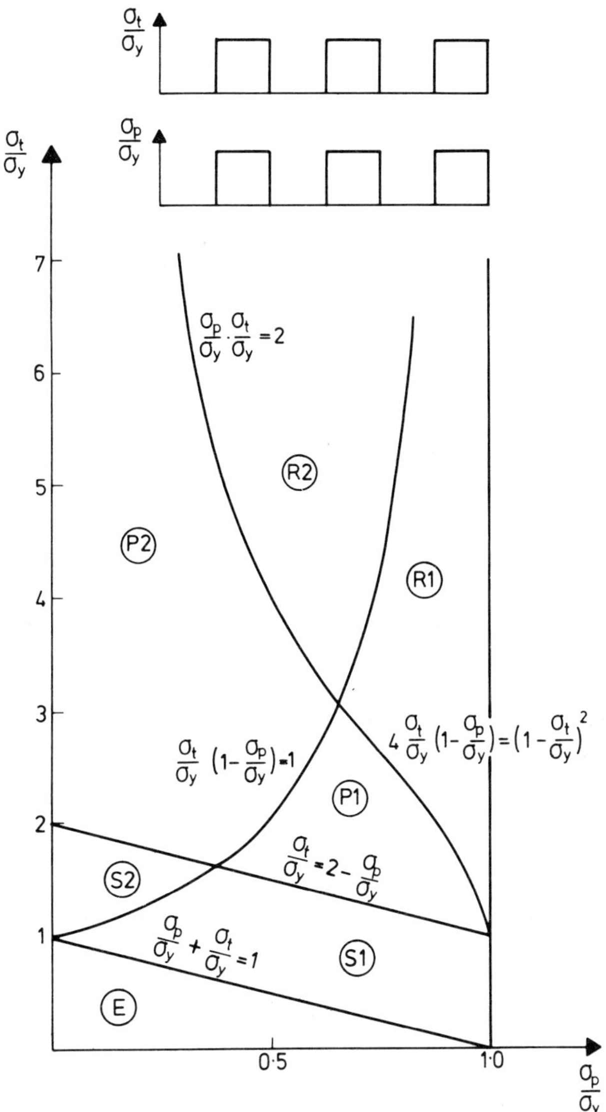

FIG. 14 The Bree Diagram for Cyclic Thermal and Pressure Loading — In Phase.

substituting the above and equations (34) and (35) into equation (32) and simplifying gives

$$\frac{\sigma_t}{\sigma_y} = 2 - \frac{\sigma_p}{\sigma_y}$$

which is the same linear boundary as has been obtained above for the S1 regime.

Finally, the boundary separating Case 1 and Case 2 behaviour is determined by the onset of compressive yielding on the inside surface of the cylinder. That is, at the boundary b = t/2.

Using equation (35)

$$\frac{Kt}{2} = (\frac{\sigma_p \sigma_t}{\sigma_y^2} + 1)\sigma_y$$

or

$$\frac{\sigma_t}{\sigma_y} = (\frac{\sigma_p \sigma_t}{\sigma_y^2} + 1)$$

or

$$\frac{\sigma_t}{\sigma_y} (1 - \frac{\sigma_p}{\sigma_y}) = 1 \tag{40}$$

which is the boundary separating types 1 and 2 behaviour - i.e. separating P1/P2, S1/S2 and R1/R2. A plot of these boundaries is given as Fig. 14.

6. DISCUSSION

In this discussion the four loading sequences will be referred to as 'Bree', C.T. (Continuous Thermal), I.P. (In Phase) and O.P. (Out of Phase).

Table 1 shows the various criteria used in obtaining the three alternative diagrams.

The classification of the seven stress regimes (E, S1, S2, P1, P2, R1 and R2) has been made on the basis of initial yielding behaviour and subsequent cyclic behaviour. Table 1 shows the criteria used and clearly there are some differences in the criteria for the different loading sequences. For example, in the alternating plasticity regime P1 of the O.P. and I.P. sequences, cyclic plasticity occurs on the inner surface of the cylinder during the O.P. sequence but on the outer surface in the I.P. sequence. Analytically, this is of no consequence but in practice, any low cycle fatigue damage will be difficult to identify in a cylinder which has been

Loading Sequence	Stress Distribution	Interface Conditions	Equation of Bound	Stress Regimes Differentiated
CT	Case 1	$a = {}^t\!/2$	$\dfrac{\sigma_t}{\sigma_y} + \dfrac{\sigma_p}{\sigma_y} = 1$	E/S1
"	"	$a = 0$	$\dfrac{\sigma_p}{\sigma_y} = 1$	S1/Collapse
"	Case 2	$b = {}^t\!/2$ $a + b = \dfrac{(2\sigma_y - \sigma_p)}{K}$	$\dfrac{\sigma_t}{\sigma_y} = \dfrac{(2 - \dfrac{\sigma_p}{\sigma_y})}{4(1 - \dfrac{\sigma_p}{\sigma_y})}$	S1/S2
OP	Case 1	$b = {}^t\!/2$ $K(b-a) = \sigma_y - \sigma_x$ $Ka = \sigma_y$	$\dfrac{\sigma_t}{\sigma_y} = 2 - \dfrac{\sigma_p}{\sigma_y}$	P1/S2
"	"	$a = {}^t\!/2$	$\dfrac{\sigma_t}{\sigma_y} = 1$	S2/E
"	"	$a_{(n+1)} = a_n$ $\sigma_x = \sigma_y$	$4 \cdot \dfrac{\sigma_t}{\sigma_y}(1 - \dfrac{\sigma_p}{\sigma_y}) = (1 - \dfrac{\sigma_t}{\sigma_y})^2$	R1/P1
"	Case 2	$c = {}^t\!/2$ $K(b-a) = \sigma_y - \sigma_x$ $K(c-a) = \sigma_y + \sigma_x$ $Ka = \sigma_y$	$\dfrac{\sigma_t}{\sigma_y} = \dfrac{4}{(2 - \dfrac{\sigma_p}{\sigma_y})}$	P1/P2 R1/R2
"	"	$a_{n+1} = a_n$ $\sigma_x = \sigma_y$	$\dfrac{\sigma_p}{\sigma_y} \cdot \dfrac{\sigma_t}{\sigma_y} = 2$	P2/R2
IP	Case 1	$a = {}^t\!/2$	$\dfrac{\sigma_p}{\sigma_y} + \dfrac{\sigma_t}{\sigma_y} = 1$	E/S1
"	"	$K(a+c) = \sigma_y + \sigma_x$ $c = {}^t\!/2$	$\dfrac{\sigma_t}{\sigma_y} = 2 - \dfrac{\sigma_p}{\sigma_y}$	S1/P1

Loading Sequence	Stress Distribution	Interface Conditions	Equation of Bound	Stress Regimes Differentiated
IP	Case 1	$a_{n+1} = a_n$ $\sigma_x = \sigma_y$	$\dfrac{4\sigma_t}{\sigma_y}\left(1 - \dfrac{\sigma_p}{\sigma_y}\right)$ $= \left(1 - \dfrac{\sigma_t}{\sigma_y}\right)^2$	P1/R1
"	Case 2	$c = t/2$ $K(a+c) = \sigma_y + \sigma_x$	$\dfrac{\sigma_t}{\sigma_y} = 2 - \dfrac{\sigma_p}{\sigma_y}$	S2/P2
"	"	$b = t/2$	$\dfrac{\sigma_t}{\sigma_y}\left(1 - \dfrac{\sigma_p}{\sigma_y}\right) = 1$	S1/S2 P1/P2 R1/R2
"	"	$a_{n+1} = a_n$ $\sigma_x = \sigma_y$	$\dfrac{\sigma_t}{\sigma_y} \cdot \dfrac{\sigma_p}{\sigma_y} = 2$	P2/R2

Table 1

subjected to an O.P. loading sequence. The I.P. loading sequence seems to be inherently safer since any fatigue damage should be visible from the outside of the cylinder. This differentiation was not noted by Bree since for that loading sequence equal damage occurs on both surfaces. However, in the P2 regime of the IP and OP sequences this is not the case. Cyclic plasticity is present on both surfaces but the loop widths are not equal.

The ratchetting behaviour in I.P., O.P. and the Bree sequences follow a similar trend. However, for the regime R1 of the Bree sequence, tensile yielding occurs at every alternate half-cycle with no associated cyclic plasticity. In the R2 regime, ratchetting is accompanied by plasticity. For the I.P. and O.P. sequences in R1, plasticity occurs on both inner and outer surfaces. In R2 both surfaces suffer plasticity in addition to ratchetting.

The analysis of the C.T. sequence results in a particularly interesting diagram - i.e. one in which only elastic and shakedown regimes exist. If this analysis is conducted over the first few cycles of load (rather than during the steady state) it is found that ratchet strains are initially accumulated. However, the magnitude of these strains is small and it appears that for this sequence of loading, whatever the magnitude of the thermal stress, the cylinder will suffer no damage. Clearly, this conclusion is reached following assumptions that the yield stress is unaffected by temperature and that the material is elastic/perfectly-plastic. In practice, the material will strain-harden and the diagram will not be bounded at $\sigma_p/\sigma_y = 1$ (see e.g. [2]), and the yield stress will be reduced by the continuous thermal gradient.

In this work, we have made no mention of ratchetting rates or quantified the damage due to alternating plasticity. This work is in hand by both analytical and by numerical techniques.

7. CONCLUSIONS

Boundaries defining the strain behaviour of a thin walled tube subjected to a thermal gradient and internal pressure have been obtained. The sequences of loading considered have been continuous pressure with cyclic thermal gradient (the Bree sequence), continuous thermal gradient with cyclic pressure load, cyclic pressure and thermal gradient "in phase" and "out of phase".

ACKNOWLEDGEMENTS

The work reported here was conducted with financial support from the National Nuclear Corporation Limited. Their permission to publish this work is gratefully acknowledged.

REFERENCES

1.   BREE, J. - 'Elastic-Plastic Behaviour of Thin Tubes Subjected to Internal Pressure and Intermittent High-Heat Fluxes with Applications to Fast-Nuclear Reactor Fuel Elements', J. of Strain Analysis, 1967, Vol. 2, 3, pp. 226-238.

2.   NG, H. W. and MORETON, D. N. 'The Bree Diagram - Origins and Literature: Some Recent Advances Concerning Experimental Verification and Strain-Hardening Materials', Recent Advances in Creep and Fracture of Engineering Materials and Structures, (Eds. Wilshire and Owen), Pineridge Press, Swansea, 1982, pp. 185-230.

3.   FREDERICK, C.O. and ARMSTRONG, P. J. - 'Convergent Internal Stresses and Steady Cyclic States of Stress', J. of Strain Analysis, Vol. 1, 2, 1966, p. 154.

4, BLEICH, H. - 'Uber die Bern'essling Statisch un Bestimmter
    Strahltragwerkerkeunter Beruckischtigung dis Elastisch-
    Plastischen Verhaltens des Bausstoffes', Der Bauingenieur,
    Vols. 19/20, 1932, p. 261.

5. MELAN, E. - 'Theorie Statisch Urbestmimter Systeme aus
    Ideal-Plastischen Baustoff', Sitzbar. Akad. Wiss, Wien.,
    IIa, 147, 1938, pp. 73-87.

Chapter 7

THE REFERENCE STRESS METHOD AND ITS ROLE IN HIGH TEMPERATURE
DESIGN

J T BOYLE

Dept of Mechanics of Materials, University of Strathclyde,
Glasgow, G1 1XJ

SUMMARY

The main objective of this chapter is to give a detailed
state of the art review of the reference stress method and to
debate its possible usefulness in high temperature design in
the light of some recent studies by the author.   The application of the reference stress concept to structures under
steady and combined loading, thermal loading, cyclic loading,
and relaxation conditions will be described together with its
extension to creep rupture and buckling.   Some implications
for the use of reference stress concepts in pressure vessel
design will be considered.

1.   INTRODUCTION

To the stress analyst, the prediction of the behaviour of
structural components subject to material creep should be the
most frustrating experience.   There are drawbacks at almost
every stage of the analysis.   Initially the analysis must be
founded upon some suitable mathematical model of the material
behaviour.   At present, there is little general agreement as
to the most suitable model - at best this would be largely
phenomenological and reflect only some aspects of the observed
behaviour.   Moreover, there is the associated need to establish numerical values of any parameters required for the
chosen model for a specific material.   Since creep test data
is notoriously prone to scatter, only "best" estimates
according to some criterion can be obtained and the fitted
curves are thus in final form only mathematically idealised
approximations to the real behaviour.   Sadly, this basic fact
is often forgotten.   Once some model of the material behaviour has been adopted, the method of stress analysis is
selected.   Here solutions based on an application of finite
element techniques can be expensive, and consequently given
the uncertainty in the basic material idealisation, of
arguable value.   In addition, the lack of repeatable experi-

314

mental investigations on real components (due to the difficulty
and timescale of creep testing) has not allowed currently
available finite element packages to be benchmarked against
anything but the most elementary component under simple loading
histories. We are merely given the assurance that it should
work in principle. There is little evidence that nonlinear,
time-dependent finite element models of complex structures
which are so successful in linear elasticity, can have a similar success if material inelasticity is assumed, and there is
no apparent simple means of 'sizing' the result for accuracy.
Finally, it is also worth noting that initial states of
residual stress imposed on a structure during manufacture are
unknown in practice and difficult to take into account in the
analysis. At this point the reader may be tempted to retire
gracefully to work at something with more chance of success.

The deficiencies in the above procedure are obvious and
there has been a concurrent search for simple, approximate
methods of assessing component creep behaviour - at least
sufficient for design purposes. One such method - the Reference Stress Method - has gained some prominence since it
addresses many of the deficiencies outlined above. It aims at
minimising the effect of uncertainties in the material model
in the first instance and, in addition, suggests a simple
approach to the estimation of component behaviour. In the
writer's experience, it is one of the few results in creep
mechanics which contains an element of surprise - if all else
fails, it should be treasured for this alone.

In a recent book by the writer [1], the fundamental underlying concepts concerning the Reference Stress Method as it is
applied to steady creep are introduced, but the idea is taken
no further owing to the conjectural nature of the subject.
Here it is our intention to examine the method in more detail
and in particular its application to a wider class of creep
behaviour and consequential application to design.

In Section 2, the Reference Stress Method in steady creep
is developed with reference to a simple example. The reader
is advised to study this in some detail since the reference
stress concept is quite subtle. Following this in Section 3,
these concepts are generalised to other aspects of creep
behaviour and in Section 4 the success of this is discussed by
comparison with available experimental evidence. Finally, in
section 5 the use of the method in design is outlined.

2   THE REFERENCE STRESS METHOD IN STEADY CREEP

The Reference Stress Method was originally developed to deal
with the uncertainties which arise in predicting the steady
creep of a structure or component. Nevertheless, the concepts
which are derived for steady creep have been applied to other
aspects of material creep behaviour, as we will see.

Initially we will examine the concepts behind the Reference
Stress Method for steady creep.

2.1 A convenient constitutive relation for steady creep

In the steady creep of a structure the stresses and
deformation rates are constant in time and the material is
presumed to be in its secondary phase. If we measure the
strain rate in secondary creep from several creep tests at
different stress and temperature levels, and plot the result
in log/log form, then we typically obtain a graph as shown in
Fig 1.

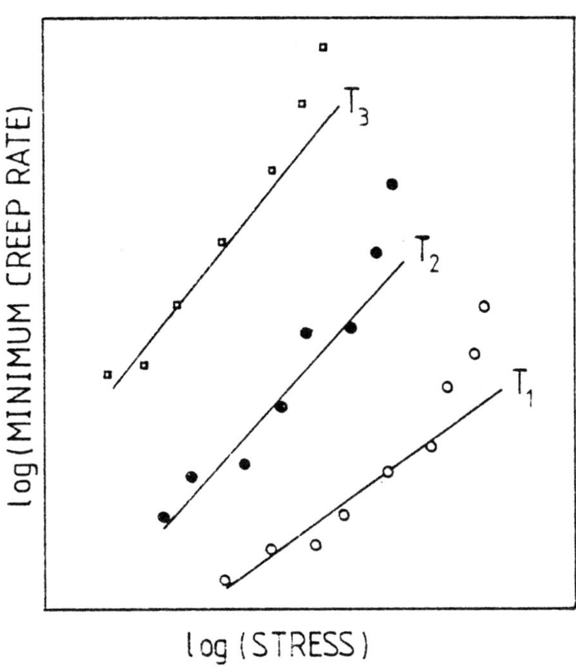

Fig 1 : Typical uniaxial creep data

Clearly, smooth lines are not obtained and there is a consider
-able scatter.   Here we will adopt a phenomenological description of material creep behaviour.   That is, we will relate
the observables - in this case stress, strain rate and temperature - through an equation of state.   Thus we assume an
equation of state for secondary creep in the separable form

$$\dot{\varepsilon} = f_1(T) f_2(\sigma) \tag{1}$$

316

where $\dot{\varepsilon}$ is the strain rate, σ is stress and T is temperature. From observation and physical considerations it is found that creep strain rate increases very nearly in an exponential manner with temperature according to

$$f_1(T) = \exp(-\Delta H/RT) \qquad (2)$$

where ΔH is a measured physical constant called the activation energy, R is Boltzmann's constant and T the absolute temperature.

The stress dependence is more difficult to determine because of scatter (Fig 1). However if we look at the plot of log strain rate versus log stress, we see that except for high stresses, the "mean" line for a given temperature is sensibly linear. Thus we may write approximately

$$f_2(\sigma) = A\sigma^n \qquad (3)$$

excluding high stress, where A and n are material constants which are found by simple curve fitting to the measured data. This relationship is known as Norton's Law. It is good enough for most engineering purposes, but, as we have seen, there is some deviation from linearity on the log/log plot at high stress. To provide a more generally applicable relation -ship it has been suggested that Norton's Law be modified as

$$f_2(\sigma) = B(\sinh(\gamma\sigma))^n$$

or

$$f_2(\sigma) = C(\sigma - \sigma_f)^n$$

where B, γ, C and σ_f are new material constants. These relations are known as Garofalo's Law and Wilshire's (friction stress) Law respectively. Other possibilities are variously

$$f_2(\sigma) = \sinh(\alpha\sigma) \qquad \text{(Prandtl's Law)}$$

$$f_2(\sigma) = \exp(\beta\sigma) \qquad \text{(Dorn's Law for high stress)}$$

Although most of these expressions have some basis in materials science, in practice the various material constants must be determined for a specific material, either from available data or from new testing. Because of scatter there is an element of uncertainty here. Let us look at this more closely:

Suppose we have chosen Norton's Law to represent the secondary creep of some material - specifically that shown in Fig 2. The required parameters B and n for the isothermal law

$$\dot{\varepsilon} = B\sigma^n$$

can be initially determined using one of three methods:

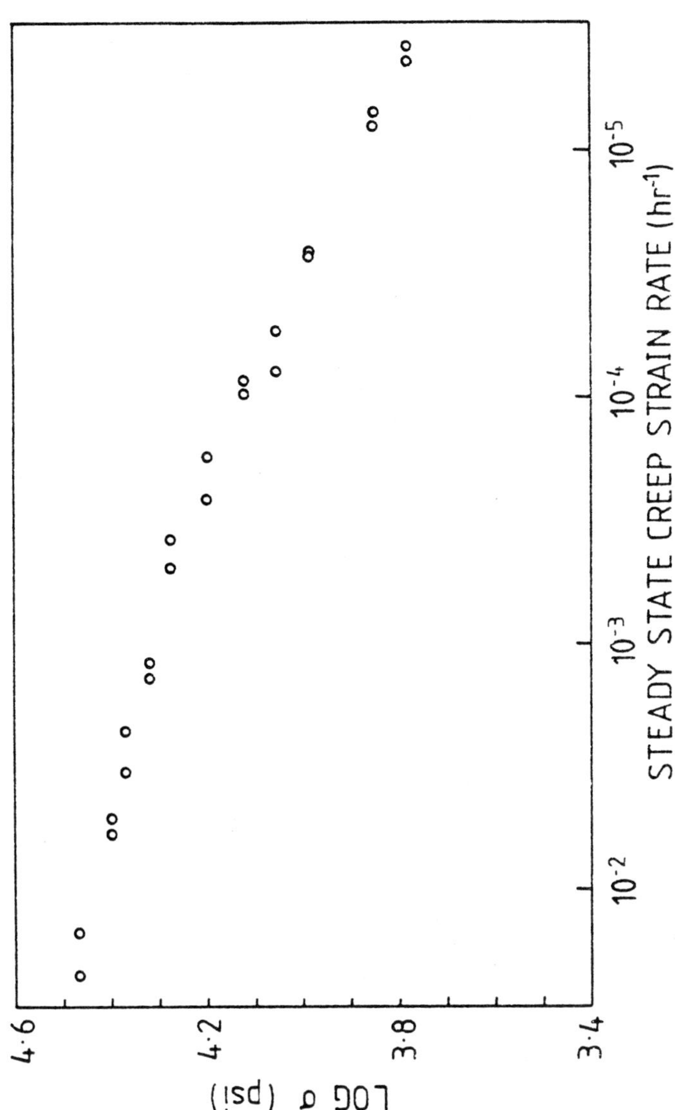

Fig 2 : Steady creep data for 316SS at 1300°F
         Ref: F. Garofalo - Paper 30 Proc I Mech E Conf
         on "Thermal Loading and Creep", London, 1963.

318

(1) By simply fitting a straight line in a least squares
sense to the log-transformed data.

(2) This is not strictly correct since we have log-transformed
the data and, with the assumption that the statistical
variance in each observation is the same, we should weight each
term in the least squares sum (Ref [1]).

(3) Alternatively, we can carry out a nonlinear least squares
fit to the untransformed data.

Using each of these methods the following estimates of the
parameters are obtained for the date of Fig 2,

Method	B	n
1	4.78×10^{-25}	5.06
2	1.61×10^{-38}	8.07
3	1.21×10^{-37}	7.87

We have three noticeably different estimates.   Others could
be derived if we adopted different criteria of "best fit" and,
indeed, if we started from a different set of creep data for
the material.   In practice, the designer would in all likeli-
hood use the result of the simplest calculation - fitting a
straight line to the log-transformed data.

The effect this uncertainty has on the calculated
stresses and deformations in a typical structure is crucial.
We examine here a simple example which can be verified by the
reader.

2.2 Steady creep of a three-bar truss

Consider a simple plane truss composed of pin-jointed
bars of uniform cross-sectional area A subject to a vertical
load Q which lies in the plane of the truss as shown in Fig 3.

If σ_1, σ_2 and σ_3 are the stresses in the bar, then
the equilibrium equations are

$$\sigma_1 + 2\sigma_2 + \sigma_3 = \frac{2Q}{A} \qquad \sigma_3 = \sigma_1 \qquad (4)$$

Similarly if ε_1, ε_2 and ε_3 are the strains in the bars,
with q the vertical displacement of the common joint, then the
strain displacement relations are

$$\varepsilon_1 = \varepsilon_3 = \frac{q}{4L} \qquad \varepsilon_2 = \frac{q}{L} \qquad (5)$$

The constitutive relations for steady isothermal creep
are

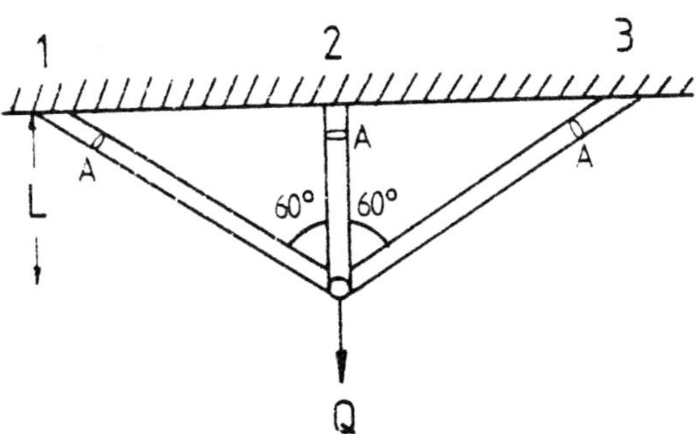

Fig 3 :   Geometry of a three bar truss

$$\dot{\varepsilon}_1 = B\sigma_1{}^n \qquad \dot{\varepsilon}_2 = B\sigma_2{}^n \qquad \dot{\varepsilon}_3 = B\sigma_3{}^n \tag{6}$$

These equations may be solved on inverting the constitutive relations, eqns (6), expressing stress in terms of strain-rate, substituting into equilibrium, eqn (4), and using the strain-displacements relations, eqn (5) in rate form to give finally

$$\frac{\dot{q}}{L} = \frac{B(Q/A)^n}{(1 + (\tfrac{1}{2})^{2/n})^n} \tag{7}$$

$$\sigma_1 = \frac{(Q(A)\,(\tfrac{1}{2})^{2/n}}{1 + (\tfrac{1}{2})^{2/n}} \qquad \sigma_2 = \frac{(Q/A)}{1 + (\tfrac{1}{2})^{2/n}}$$

These expressions can now be evaluated for the three sets of material parameters described above.   If we look at a particular example, say Q/A = 10,000 psi, then the deformation rate and maximum stress are

METHOD	$\dot{q}/L$	σ_1 (psi)
1	4.75×10^{-6}	0.568×10^4
2	2.22×10^{-8}	0.543×10^4
3	3.03×10^{-8}	0.544×10^4

Two observations can be made. Firstly, the variation in predicted stress is small; this is related to the observation that, as a good approximation, the maximum stress varies linearly with the inverse exponent (1/n). Secondly, the variation in predicted deformations is unacceptably high. The crucial question is, if we had looked at a more complex structure which could only be analysed with considerable expense, would the variation in predicted deformation (displacements or strain) have been acceptable for the price. The simple answer is, probably not.

The Reference Stress Method is used to control this uncertainty. In effect there are <u>two</u> basic approaches which we call here the "limit-type (or global) method" and the "local method". These are distinct and embody two different interpretations of the method, although the fundamental idea - to reduce the effect of uncertainty in the material model on calculated deformation rates in a structure - is the same. This distinction has not been emphasised in previous reviews of the Reference Stress Method [2, 3] and should be borne in mind when referring to these. Both approaches are based on one fundamental relation. That is, <u>structural deformations are related to a uniaxial creep test held at a particular stress, called the reference stress, in such a way that the constant of proportionality, called the scaling factor, is insensitive to variations in the material parameters.</u>

So, if $\dot{q}$ is some characteristic displacement rate

$$\dot{q} = \delta \times \dot{\varepsilon}(\sigma_R) \tag{8}$$

where σ_R is the REFERENCE STRESS and δ the SCALING FACTOR.

We will examine how this is done, and attempt to emphasise the two distinct approaches for the three-bar truss.

2.3 The Reference Stress Method - Limit approach

Let $\dot{q}$ be some characteristic displacement rate in a structure undergoing steady creep. As in eqn (8) we can write

$$\dot{q} = \delta \times \dot{\varepsilon}_o$$

where $\dot{\varepsilon}_o = B\sigma_o^n$ and σ_o is arbitrary. Then from eqn (7) the scaling factor for the three-bar truss is by definition

$$\delta(\sigma_0,n) = \frac{L(Q/A\sigma_0)^n}{(1 + (\tfrac{1}{2})^{2/n})^n}$$

For simplicity we may write $\sigma_0 = \alpha Q/A$ for some arbitrary scalar α.   In Fig 4

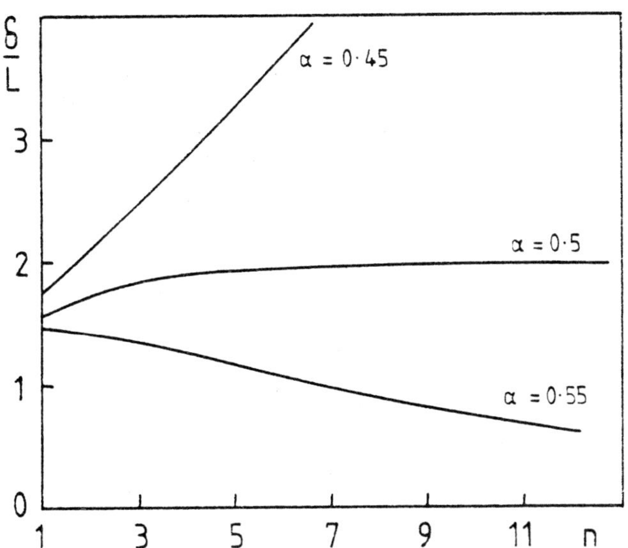

Fig 4 :   Variation of scaling factor with n for
         different values of parameter α

the variation of δ with n is examined for various values of α. It can be seen that if $\alpha < \tfrac{1}{2}$ the scaling factor $\delta \to \infty$ as $n \to \infty$, whereas if $\alpha > \tfrac{1}{2}$ the scaling factor $\delta \to 0$ as $n \to \infty$. But, perhaps surprisingly, if $\alpha = \tfrac{1}{2}$, the scaling factor tends to a limit as $n \to \infty$ (in fact $\delta \to 2$).   Furthermore, to a good approximation

$$\delta(\sigma_R,n) \simeq \delta(\sigma_R,\infty) \tag{9}$$

provided we choose the reference stress $\sigma_R = \frac{1}{2}\frac{Q}{A}$. Therefore we have approximately

$$\dot{q} \simeq \delta(\sigma_R,\infty) \times \dot{\varepsilon}(\sigma_R) \tag{10}$$

and now the scaling factor is independent of either B or n!

As it stands, this expression for the displacement rate remains subject to the uncertainty in the material parameters through the term $\dot{\varepsilon}(\sigma_R)$. But, and this is the most crucial part of the Reference Stress Method, we can interpret $\dot{\varepsilon}(\sigma_R)$ as <u>the secondary creep rate from a uniaxial creep test held at</u> <u>the reference stress</u> σ_R. That is, we do not need to know the material parameters — at all, we just perform a creep test at the stress indicated!

There is a subtle assumption here of course: the approximation inherent in eqn (10) has been derived on the basis of a power law. We really also need to establish that the approximation is valid for other constitutive models.   That is, not only should the scaling factor and reference stress be indepenent of the parameters B and n for a power law, but also of the constitutive model used. Furthermore, there is the tacit assumption that the required test at the reference stress can be carried out. If not, one would presumably put scatter bands on the original creep data and consequently on the minimum creep rate at the reference stress level required. But, there is no guarantee that we will not still get an un- acceptable variation in the structural deformations as a result.   The important point is, we can get a good answer if necessary.   Furthermore, <u>the decision about the effect of</u> <u>scatter reduces to the consideration of the uniaxial data,</u> <u>independent of the structure under consideration!</u> This has useful implications for design, as we will see.

Thus, in summary, the limit-type reference stress method is as follows.   Some characteristic deformation rate is expressed in the reference stress form, eqn (8); the reference stress, σ_R, is chosen such that the scaling factor has a limit as $n \to \infty$.   The scaling factor is subsequently approximated by this limiting value, if this can be allowed.   This is essent- ially the procedure adopted by Anderson et al [4] in the first paper dealing with reference stress techniques (although the procedure was used previously by Soderberg [5]).   This may seem a rather complex procedure, if it were not for a rather simple, but powerful, result observed by Sim [6].   The reference stress is given by

$$\sigma_R = \frac{Q}{Q_L}\, \sigma_y \tag{11}$$

where Q_L is the limit load for an equivalent structure composed of a perfectly plastic material with yield σ_y.   As we will see this is a quite general result, <u>but</u> there are two obvious drawbacks.   Firstly, the limit of the scaling factor may be zero; secondly, the limit may not be a good enough approximation for a large enough range of the exponent n. We will give two examples of these later.

2.4   The reference stress method - local approach

In the limit type approach an attempt is made to render

the scaling factor independent of the material parameters;
as a consequence the reference stress itself is also independ
-ent of these parameters. As mentioned above, sometimes
this approach will not work. In the alternative local
approach it is assumed that we have a reasonable estimate of
the exponent n. For example, for the data of Fig 2, a
reasonable estimate would be $n \simeq 8$ (discounting the erroneous
linear unweighted fit to the log transformed data). The
procedure used in the local approach is as follows: again
write

$$\dot{q} = \delta \times \dot{\varepsilon}_o$$

for some arbitrary σ_o. The scaling factor is now expanded
in a Taylor series about the estimated value of n, which
we denote by $\bar{n}$

$$\delta(\sigma_o,n) = \delta(\sigma_o,\bar{n}) + (n - \bar{n}) \left.\frac{d\delta}{dn}\right|_{\bar{n}} + 0(n - \bar{n})^2$$

It is assumed that the estimate $\bar{n}$ is reasonable so
that terms of order $(n - \bar{n})^2$ are negligible. Then we can
write

$$\delta(\sigma_R,n) \simeq \delta(\sigma_R,\bar{n}) \tag{12}$$

provided we choose $\sigma_o = \sigma_R$ in such a way that

$$\left.\frac{d\delta}{dn}\right|_{\bar{n}} = 0 \tag{13}$$

The local reference stress approximation is then

$$\dot{q} \simeq \delta(\sigma_R,\bar{n}) \times \dot{\varepsilon}_R \tag{14}$$

In the case of the three-bar structure, evaluating the
derivative in eqn (13) and solving for σ_R we find

$$\log\frac{\sigma_R}{Q/A} = \frac{\frac{2}{\bar{n}}(\frac{1}{2})^{2/\bar{n}}}{1 + (\frac{1}{2})^{2/\bar{n}}} \log(\tfrac{1}{2}) - \log(1 + (\tfrac{1}{2})^{2/\bar{n}}) \tag{15}$$

The variation of reference stress with $\bar{n}$ is shown in Fig 5; as
$\bar{n} \to \infty$, σ_R tends to the limit value, eqn (11). Also in
Fig 6 we show the variation in scaling factor for this value
of reference stress with the exponent n, assuming $\bar{n} = 8$ as an
example. The scaling factor is sensibly constant in the
range $5 < n < 11$.

This form of reference stress method was suggested by
Johnsson [7] and in the present form by the writer [8].
One immediate drawback is the necessity of evaluating the

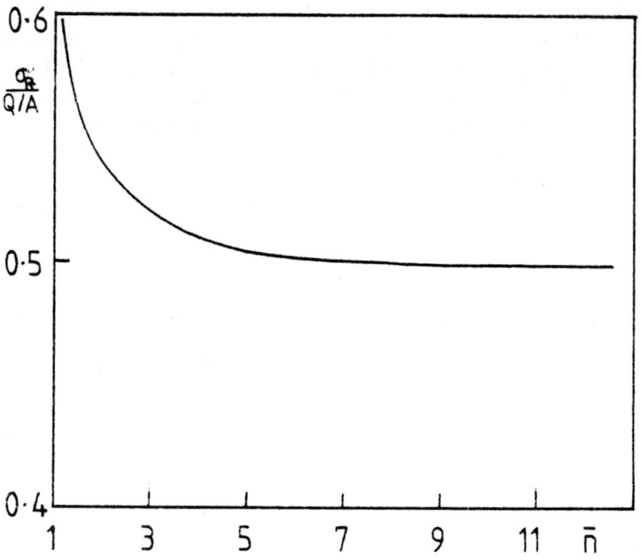

Fig 5 :   Dependence of local reference stress on
estimated value of n

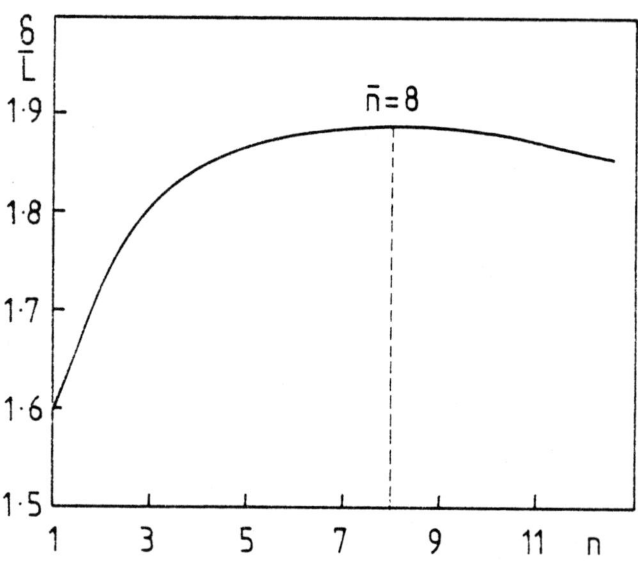

Fig 6 :   Variation of scaling factor with n
for estimated value, $\bar{n}$ = 8

derivative in eqn (13), which of course is only possible if there is an expression for the scaling factor which can be differentiated.   An obvious approximation is to replace the derivative by a finite difference

$$\frac{d\delta}{dn}\bigg|_{\bar{n}} \approx \frac{\delta+ - \delta-}{2\Delta n}$$

where $\delta+ = \delta(n+)$, $\delta- = \delta(n-)$ and $n+ = \bar{n} + \Delta n$, $n- = \bar{n} - \Delta n$ for some small Δn.   Thus the reference stress is chosen such that

$$\delta+ = \delta- \tag{16}$$

In the case of the three-bar truss it can be found that

$$\sigma_R = \left[\frac{(1 + (\tfrac{1}{2})^{2/n-})^{n-}}{(1 + (\tfrac{1}{2})^{2/n+})^{n+}}\right]^{\frac{1}{n+ - n-}} \times \frac{Q}{A} \tag{17}$$

This is compared in Fig 7 for the case $\bar{n} = 8$, assuming various values of Δn with the exact local reference stress, eqn (15). It can be seen that the expression given by eqn (17) closely approximates that of eqn (15) for quite a wide range of values of Δn.

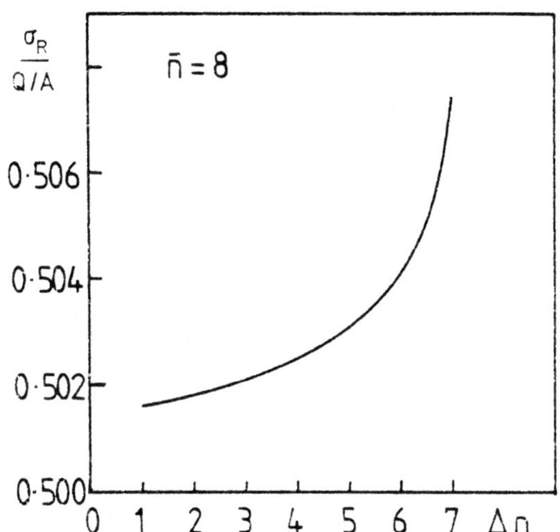

Fig 7 :   Dependence of approximate reference stress
on Δn for estimated value, $\bar{n} = 8$

The definition of a reference stress given by eqn (17) is essentially that proposed by Sim [9 - 12] which predated the suggestion by Johnsson.   It is also similar to that originally suggested by Mackenzie [13] on identifying $n- = 1$ and $n+ = \bar{n}$. In fact Mackenzie's basic approach is different and it is worth commenting further here.   In Mackenzie's method the scaling factors corresponding to $n = 1$ and $n = \bar{n}$ are equated

$$\delta(\sigma_R, 1) = \delta(\sigma_R, \bar{n})$$

and the reference stress is identified accordingly, i.e. for the three-bar truss from eqn (17)

$$\sigma_R = \left[\frac{5/4}{(1 + (\tfrac{1}{2})^{2/\bar{n}})^{\bar{n}}} \right]^{\frac{1}{\bar{n}-1}} \times \frac{Q}{A} \tag{18}$$

If we plot the variation of σ_R thus derived with $\bar{n}$, Fig 8, it can be seen that σ_R is very nearly constant.   An <u>average</u> value is selected for σ_R and then

$$\dot{q} \simeq \delta(\sigma_R^{AV}, 1) \times \dot{\varepsilon}(\sigma_R^{AV}) \tag{19}$$

This procedure is often inaccurate, Fig 9, but has the advantage that the scaling factor can be found from a simple <u>linear</u> analysis corresponding to $n = 1$ (analogous to linear incompressible elasticity).

2.5 A general local reference stress method

So far we have only considered the material constitutive relation described by a simple power law, eqn (3).   The question is - can the reference stress method be extended to other constitutive relations?   It turns out that the local approach discussed above can also be applied to different constitutive relations with little change, provided the scaling factor depends on a single material parameter as, for example, in the case of the Prandtl and Dorn Laws.   The reference stress is chosen to satisfy the basic condition given by eqn (13) replacing the exponent n in the power law by the required material parameter.

For example, in the case of the three bar truss, assuming Prandtl's law in the form

$$\dot{\varepsilon}_o = B \sinh (\alpha\sigma)$$

the vertical displacement rate is

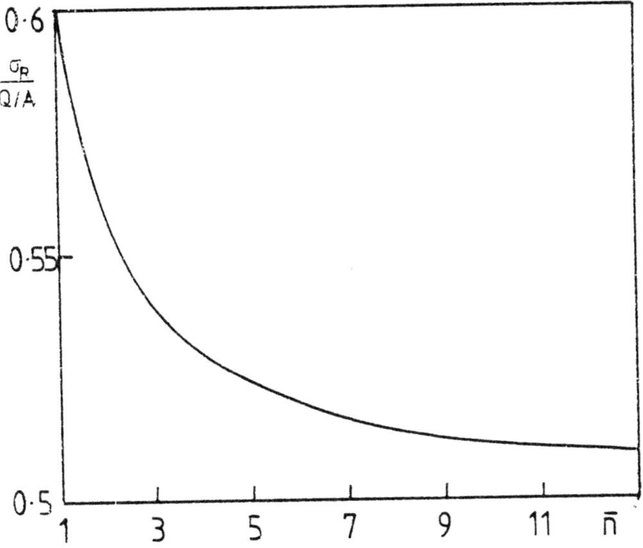

Fig 8 :   Dependence of reference stress calculated using Mackenzie's method on $\bar{n}$

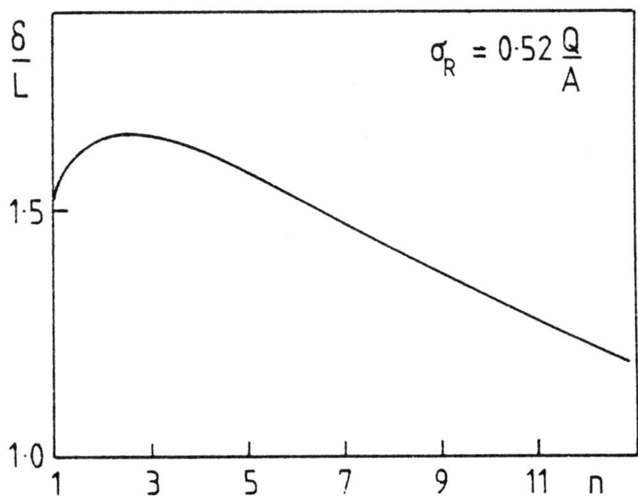

Fig 9 :   Variation of scaling factor calculated using Mackenzie's method with n

$$\frac{\dot{q}}{L} = \frac{B\sinh(\alpha Q/A)}{(\frac{17}{16} + \frac{1}{2}\cosh(\frac{\alpha Q}{A}))^{\frac{1}{2}}} \tag{20}$$

The reference stress could be found by forming the scaling factor and setting the derivative with respect to α equal to zero. This should be done numerically, so it is better to use the approximate method given by eqn (16). The variation of the reference stress σ_R thus derived with the parameter $(\alpha Q/A)$ is shown in Fig 10.

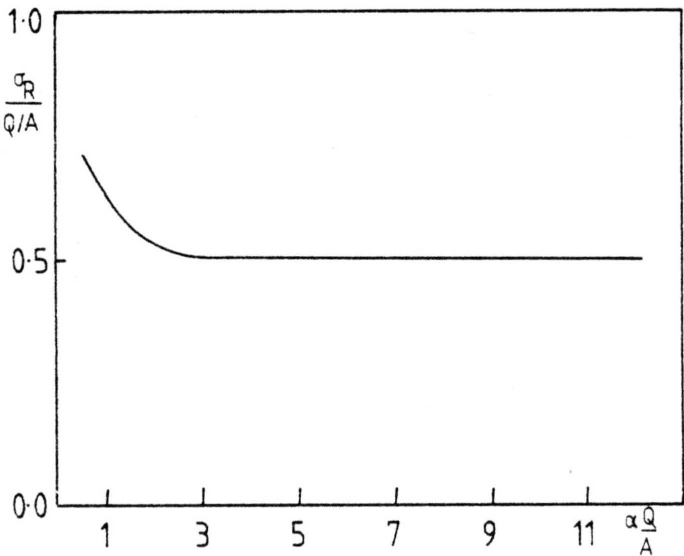

Fig 10 : Variation of reference stress for Prandtl's
Law with parameter $\alpha Q/A$

We can see that the reference stress is nearly insensitive to this parameter, with a limit $\sigma_R = \frac{1}{2} Q/A$ as for a power law!

This extension of the reference stress method was suggested by Johnsson [8] and in the present form by the writer. In fact we have seen that the reference stress for the three bar truss closely approximates the limit reference stress for the power law. Assuming that the reference stress was the same for all constitutive relations, Fairbairn [14] demonstrated that the appropriate scaling factor was sensibly insensitive to the constitutive relation also. This conclusion can also be made from the writer's work [15, 16]. In this a method was given for extending the local approach to constitutive relations which lead to scaling factors

dependent on more than one material parameter, e.g. as for
Garofalo's law, based on __statistical__ considerations. Numerical studies on several simple components demonstrated that
the reference stress and scaling factor were both independent
of the creep law which was used.   Thus, not only does this
remove one of the criticisms of the reference stress method,
but also it can be inferred that, provided we adopt a reference
stress approach (and this is crucial) it is sufficient to use
the simple and familiar Norton power law!

2.6 Bounding theorems and a simplified method for design

In both the limit-type and local interpretations of the
reference stress method, it is necessary to carry out the
necessary creep stress analysis to determine the appropriate
value of the scaling factor.   If the limit approach is
successful then no creep analysis is required to determine the
reference stress, eqn (11).   A creep analysis can therefore
be avoided completely if we can estimate the scaling factor.
For a structure under the action of a single load Q, the
scaling factor δ corresponding to the associated displacement rate $\dot{q}$ can be bounded above by

$$\delta \leqslant \frac{V\sigma_R}{Q} \tag{21}$$

where V is the volume of the structure, and σ_R is the limit
type reference stress [17].   The interesting fact is that
__this bound is valid no matter what the constitutive relation__
[18].   By way of example the scaling factors for both
Norton's and Prandtl's law for the three bar truss are shown
in Fig 11, together with the bound, which takes the value
$\delta < 5L/2$.

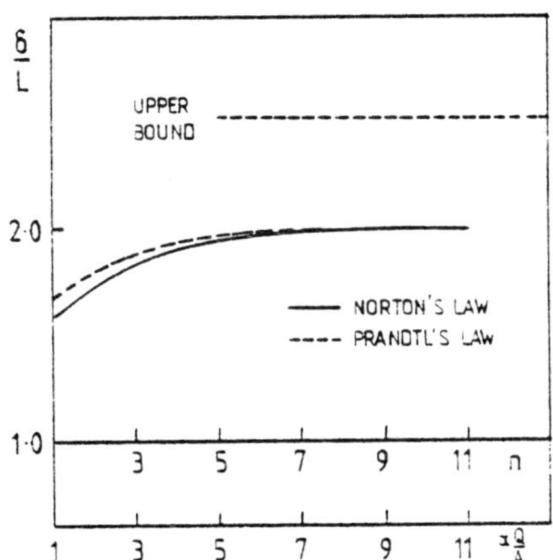

Fig 11 : Comparison of scaling factor for Norton
and Prandtl's laws with upper bound

330

     This simple result can be extended to multiple loading in
the case of the power law [19]; the reference stress, eqn(11)
is replaced by the plastic limit surface in load space for an
equivalent perfectly plastic material.

     It is worth summarising here the results we have discover-
ed.   So far the aim of the reference stress method has been
to minimise the effect of the uncertainty in the constitutive
relation - we have described two methods of doing this. For
the simple example studied here, the reference stress is the
same for both approaches and independent of the creep law
used.   This appears to be true for many (but, as we will see,
not all) structures.   The reference stress may then be
evaluated according to eqn (11) from an equivalent limit
analysis. The simple bound on the scaling factor, eqn (21)
is given in terms of this limit reference stress, alternatively we could use Mackenzie's approximation to the scaling
factor based on an equivalent linear elastic calculation. The
possibility then exists of obtaining estimates and bounds on
the deformation rates in a steadily creeping structure without
recourse to any creep analysis. This suggests a simplified
approach to creep design based on plastic limit analysis; we
will examine this further in a later section.   The success
of this simplified method depends on the applicability of the
limit type approach.   Unfortunately the limit approach can
fail - surprisingly for rather simple examples [20, 21].   In
the case of a cantilevered beam under end load, the appropriate
caling factor is zero (although it can still be bounded):
also, many examples of non-uniformly heated structures [1] do
possess a valid non-zero limit to the scaling factor, but
this is a poor approximation for a wide range of values of the
exponent n for a power law.   But it is important to realise
that the local approach will always be successful and therefore viewed as a means of merely minimising the uncertainty in
the constitutive relations, the reference stress method
remains valid.

2.7 Reference temperatures

It was mentioned above that the limit type reference stress
method gives poor results for non-uniformly heated structures.
If the local method is used this poses no great problems
[1, 7] but it would be attractive to have some simple result
useful in design.   One possibility which has received some
investigation is to introduce a reference temperature, viz.
the temperature at which the reference stress test is carried
out can be varied and it is possible that this could lead to a
simple result.   In some cases a meaningful reference temperature can be found [22], but the concept is not generally
successful [23].   One way round this [24] which involves
modifying the reference stress such that the reference
temperature is nearly independent of the material parmeters
can also often lead to difficulties [25].

It is possible that the above mentioned investigations have varied success because the concept of a reference temperature is not needed.   Let us reconsider the three-bar truss, but now suppose that the outer bars are at a temperature T_1 with the inner bar at a temperature T_2.   The relative constitutive relation now takes the form

$$\dot{\varepsilon}_0 = C \exp(-\gamma/T)\sigma_0^{\,n}$$

where we have written $\gamma = \Delta H/R$.   The displacement rate is now given by

$$\frac{\dot{q}}{L} = \frac{C(Q/A)^n \exp(-\gamma/T_1)}{[1 + (\tfrac{1}{2})^{2/n}e^\beta]^n}$$

where $\beta = \dfrac{\gamma}{n}(1/T_2 - 1/T_1)$ assuming $T_1 \geqslant T_2$.   It can be verified that if we define $\sigma_R = \tfrac{1}{2}Q/A$ as in the isothermal case ($\beta = 0$) then the corresponding scaling factor tends to a limit which exhibits considerable variation with n [21].   However let us write

$$\dot{q} = \delta(\sigma_R,n) \times \dot{\varepsilon}(\sigma_R^\beta,T_1) \tag{22}$$

where $\delta(\sigma_R, n)$ is the scaling factor for isothermal conditions and σ_R^β is a modified reference stress given by

$$\sigma_R^\beta = e^{-\tau(\beta,n)}\sigma_R \tag{23}$$

where σ_R is the usual limit reference stress, eqn (11).   The reference test is now held at the stress σ_R^β and the maximum temperature T_1.   The quantity $\tau(\beta,n)$ is defined by

$$\tau(\beta,n) = \ln\left[\frac{(1 + (\tfrac{1}{2})^{2/n}e^\beta)}{(1 + (\tfrac{1}{2})^{2/n})}\right]$$

which is shown in Fig 12, keeping the quantity β fixed.   It can be seen that this is virtually insensitive to n for large n and tends to a limit as $n \to \infty$.

Therefore we can replace the scaling factor in eqn (22) by the limiting value $\delta(\sigma_R,\infty)$ and the modified reference stress in eqn (23) by $\sigma_R^\beta = e^{-\tau(\beta,\infty)}\sigma_R$, both of which are independent of the exponent n.   Apparently the quantity β, which does depend on n remains;   however it can be verified that β can be obtained directly from the uniaxial stress/strain rate curves for steady creep [25].   It is the horizontal separation of the lines in a log/log plot corresponding to temperatures T_1 and T_2 - that is, to find the constant β, we do not need to know the exponent n!

The important point about the above approximate procedure

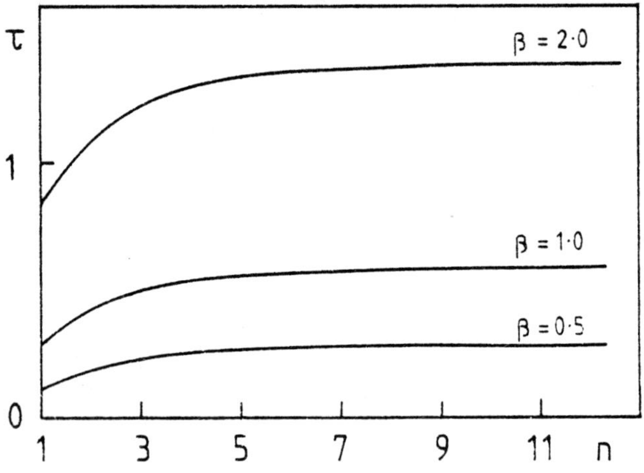

Fig 12 : Variation of factor $\tau(\beta, n)$ with n

which is related to that of Walter & Ponter [24], does <u>not</u> require a special reference temperature. Perhaps it is generally unnecessary, we just use the maximum temperature.

3   EXTENSIONS OF THE REFERENCE STRESS CONCEPT

We have discussed the reference stress method as it has been formulated for steady creep. In effect it has only really been demonstrated theoretically for steady creep. Nevertheless the fundamental concept - relating component behaviour to that of a uniaxial creep test held at the reference stress in such a way that the reference stress and approximate scaling factor are independent of the material constitutive relation used - can be applied to other aspects of material creep behaviour. We will describe some of these extensions here. Throughout it should be borne in mind that the aim of these extensions is largely to produce simple estimates of structural creep behaviour without recourse to complex stress analysis. Mostly they are extensions of the limit approach.

3.1 Forward creep

Forward creep is usually defined as the behaviour of a creeping structure under constantly applied loading [1].

For example, a typical displacement q in this structure will
vary in time in a manner similar to Fig 13.   After an

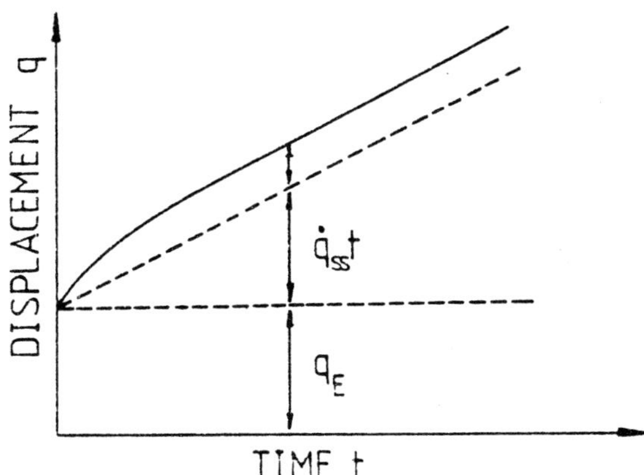

Fig 13 : Typical time variation of displacement
in forward creep

initial elastic response, q_E, a period of stress redistribution will follow causing the rate of change of the displacement to decrease to a constant steady state $\dot{q}_{ss}$.   It is
always possible to construct a lower bound on   the total
displacement by simply adding the steady state solution to the
initial elastic solution, i.e.

$$q \geqslant q_E + \dot{q}_{ss}t \tag{24}$$

Alternatively if the standard creep curves are given by a
time-hardening model of the form $\varepsilon_c = G(t)f(\sigma)$ for some time
function $G(t)$, the total displacement may be approximated by
[26]

$$q \simeq q_E + G(t)\dot{q}_{ss} \tag{25}$$

In both of the above we can introduce a reference stress
approximation for the steady displacement rate.   It is also
possible to relate the elastic deformation to the result of
the reference stress test by writing [9]

$$q_E = \delta_E \times \varepsilon_E \, (\sigma_R) \tag{26}$$

334

where $\varepsilon_E(\sigma_e)$ is the initial elastic strain from the reference stress test and, by definition the elastic scaling factor is

$$\sigma_E = \frac{Eq_E}{\sigma_R} \tag{27}$$

Using this approach two approximations to the forward creep behaviour, eqn (24) and (25) can be constructed which rely on the elastic and steady state strains from the reference stress test. However, another possibility exists. Why not, if the reference stress and scaling factor are independent of the creep law for steady state creep, simply assume this will hold true for the whole forward creep response? That is

$$\dot{q} \simeq \delta \times \dot{\varepsilon}(\sigma_R, t) \tag{28}$$

where $\dot{\varepsilon}(\sigma_R, t)$ is the measured strain rate from the reference stress test and δ is the scaling factor calculated from the steady state. This is, of course, only a proposition, but it can be tested by comparison with experiment.

Penny and Marriott [26] went one step further and suggested the scaling factor derived from an <u>elastic</u> calculation, eqn (27), as an approximation to the steady state scaling factor - this is known as the 'method of similarity', being based on the similarity of the displacement response, Fig 13, to the standard creep curve. The reference stress is thus obtained from a plastic limit analysis, and the scaling factor from an elastic analysis. Naturally this is an approximation.

Finally, it is worth noting that a reference stress has been evaluated for the time to the steady state [27]. However this is based on an estimate of the redistribution time [28] which is often wholly inaccurate.

3.2 Relaxation

Relaxation occurs in a creeping structure which has imposed upon it boundary displacements which are fixed in time [1]. Eventually the internal stresses will relax to zero and the strains will achieve a constant value. Of interest here is the relaxation of the resultant loads on the boundary restraints.

Spence and Hult [29] described a simple method of estimating the relaxation of a load due to creep which can be related to a uniaxial relaxation test. Consider the three-bar truss and suppose a vertical displacement q_o is imposed upon the common joint. We are interested in the relaxation of the load $Q(t)$; in fact this can be simply evaluated for this problem. Spence and Hult's approximation is based on decomposing the total displacement q_o into an elastic part found purely from an elastic analysis

$$q_E(t) = \frac{L}{E} \frac{4}{5} \frac{Q(t)}{A}$$

and a creep part, found from a creep analysis ignoring elastic effects. For a time hardening power law of creep $\dot{\varepsilon}_c = g(t)\sigma^n$ this is (c.f eqn (7))

$$\dot{q}_c(t) = \frac{Lg(t) (Q(t)/A)^n}{(1 + (\frac{1}{2})^{2/n})^n}$$

Since $\dot{q}_o = 0$ we have approximately

$$\dot{q}_E(t) + \dot{q}_c(t) \simeq 0$$

which gives a simple differential equation for $Q(t)$ which can be solved to give

$$\frac{Q(t)}{Q(0)} \simeq \{1 + (n - 1) \, E\sigma_o^{n-1} \int g(t)dt\}^{\frac{-1}{n-1}} \qquad (29)$$

where

$$\sigma_o = \alpha(n)\frac{Q(o)}{A}, \qquad \alpha(n) = \left[\frac{5/4}{(1 + (\frac{1}{2})^{2/n})^n}\right]^{\frac{1}{n-1}}$$

The importance of this result lies in the fact that if we look at the relaxation of stress in a uniaxial relaxation test then

$$\frac{\sigma(t)}{\sigma(o)} = \{1 + (n-1)E\sigma(0)^{n-1} \int g(t) \, dt\}^{\frac{-1}{n-1}} \qquad (30)$$

Thus the <u>relaxation of the load $Q(t)$ can be equated to the relaxation of stress in a uniaxial creep relaxation test started at the stress $\sigma(o) = \sigma_o$</u> from eqn (29) and (30). In this the initial stress σ_o is dependent on the exponent n. But it is easily verified that the scalar $\alpha(n)$ decreases with n to a limit and varies by no more than 5% for $n > 3$. Moreover

$$\sigma_o \to \frac{1}{2} \frac{Q(0)}{A}$$

as $n \to \infty$, i.e. the limit reference stress, eqn (11)!

Again use of the limit reference stress for relaxation will be an approximation since we have started from eqn (29). This concept of a relaxation reference stress has been found useful for estimating the behaviour of more complex problems, e.g. curved pipes [30] and as a means of determining the creep properties of sea ice [31].

3.3/

336

3.3 Cyclic loading

The behaviour of creeping structures under cyclic loading
is inherently complex, and the stress analysis expensive.   In
addition the material constitutive relation is not well
established and the necessary material parameters even more
difficult to determine.   For these reasons the reference
stress approach is attractive.

An heuristic extention to arbitrary proportional time
variable loading was made by Sim [9].   Essentially, if the
loading admits a time variation $\lambda(t)$ and if σ_R is the
appropriate reference stress for steady creep, then the
creep deformation due to variable loading could be predicted
by a reference stress test held at the stress $\lambda(t)\sigma_R$
multiplied by the scaling factor for steady creep.   As we
will see, this procedure can be remarkably accurate.

There have been numerous investigations over the past
decade on the behaviour of cyclically loaded structures.
Below we will generalise these and select those which have
particular application to reference stresses.   The reader is
advised to refer to the original papers in these cases.   In
a structure under cyclic load it is well established that the
internal stresses will eventually reach a cyclic state(usually
called the steady or stationary cyclic state).   The deformations do not reach a cyclic state;   however, once the
stationary cyclic state is reached, the accumulated deformation over a cycle will be constant.   If the stationary cyclic
state can be calculated based on some material constitutive
relation, then it is possible to relate the accumulated
deformation over a cycle to the result of a uniaxial creep
test at a constant stress using the local reference stress
method.   Alternatively, due to the expense of creep stress
analysis, approximations to the stationary cyclic stress
distribution have been suggested [32 - 34].   Ainsworth [34]
has used the local reference stress method in Johnsson's form
[7] on the approximated accumulated creep deformation over a
cycle corresponding to these approximate stress distributions,
again relating to a reference stress test at constant load.
This sort of application of the reference stress method is
fairly straight forward, but some additional results can be
obtained.   The approximate stationary cycle stress histories
can be shown to provide <u>bounds</u> on the inelastic work per cycle
[32 - 34], thus indicating a rough means of assessing the
accuracy of the assumed stresses.   Interestingly, it is
possible to relate the inelastic work over a cycle to the
results of reference stress tests.

Consider a structure under the action of a cyclic load
$Q(t) = \lambda(t)Q$ with period T which results in an associated
deformation rate $\dot{q}$.   Assuming a time hardening law of creep

337

the average accumulated inelastic work over a cycle can be
bounded by

$$\delta \times \dot{\varepsilon}_C^T (\lambda(t)\sigma_R) \leqslant \frac{1}{T} \int_o^T Q(t)\dot{q}\ dt \leqslant V\sigma_R^S\ \dot{\varepsilon}_C\ (\sigma_R^S) \qquad (31)$$

where

$$\dot{\varepsilon}_C^T (\lambda(t)\sigma_R) = \frac{1}{T} \int_o^T \dot{\varepsilon}_C(\lambda(t)\sigma_R)dt$$

is the average creep strain from a uniaxial reference stress
test at the cyclic stress $\lambda(t)\sigma_R$, where σ_R is the steady
state referen ce stress, given by eqn (11) with associated
scaling factor δ and V is the volume of the structure. A new
reference stress σ_R^S has been introduced, defined by

$$\sigma_R^S = \frac{Q}{Q_s}\ \sigma_y \qquad (32)$$

where Q_s is the shakedown load for an equivalent structure
composed of a perfectly plastic material with yield σ_y.
Although these bounds were originally derived for the power
law [32, 33] the upper bound can be established for other
constitutive relations [18].

Finally, a wholly different application of the reference
stress method to proportional cyclic loading has been
described by Williams & Leckie [35, 36] (and further
discussed for non-proportional loading by Williams & Dimmer
[37]).   The bounds described above, eqn (31), are based on
the time-hardening theory of creep - it is well established
that this does not represent well cyclic creep behaviour.
Williams and Leckie [35] describe a technique whereby the
results of time-hardening calculations (which can be related
to reference stress tests as described above) could be used
together with specific additional tests (at constant and
cyclic reference stress levels) to better represent the
cyclic behaviour.   Effectively the time scale in the time-
hardening theory is distorted according to the results of
these additional tests.   Unfortunately the technique does
 not appear to have been validated experimentally: indeed,
Radhakrishan [38] carried out cyclic creep tests on pure
aluminium which cast some doubt on the basic assumptions.
Nevertheless, Williams & Leckie's technique is novel and
attractive and represents a real attempt to deal with uncertainty in the form of the constitutive model for time-
varying stress.

3.4 Creep buckling

In the classical theory of the buckling of elastic
structures design is based on the critical load at which
instability occurs.   This is insufficient if material creep
is present, since buckling may occur at a load less than the

338

critical elastic buckling load.   Then design should be based
on the expected lifetime of structure.   Even in elasticity,
an assessment of the stability of a real structure or component.
poses a formidable problem to the stress analyst. Buckling is
usually associated with large deformations and is highly
dependent on the nature of imperfections, for example, in
shape due to manufacturing tolerances or in loading.   This
difficulty is amplified if material creep is present since now
the imperfection can relate to material inhomogeneity or
residual stresses, unevenness of heating, local yielding or
deterioration.   Buckling calculations are normally based on
highly idealised structures and are consequently often
unreliable when applied to real components.

The nature of the buckling phenomenon leads to further
uncertainties in the material constitutive model.   Firstly,
the inherent large deformations naturally are also associated
with local plasticity.   Secondly, there is no accepted
mathematical representation of material inelasticity with large
strains.   This additional uncertainty again brings into
question the usefulness of expensive finite element calculations.   It would be attractive if a simple technique, like a
reference stress approach, could be applied to creep buckling.
There is one fundamental difficulty.   For creep under constant
or cyclic loading we would relate component deformations to
those in a uniaxial reference stress test by suitable scaling.
For relaxation we were able to approximately represent the
relaxation of a resultant load in a component to the relaxation of stress in a uniaxial reference relaxation test.   For
creep rupture, as we will see, we can relate the rupture time
of a component to that of a uniaxial reference rupture test.
How cam we relate buckling of a component to the behaviour of
a simple uniaxial creep test?   In fact this cannot be done
directly.

Penny and Marriott [26] suggested relating the deflection
of the component to the deformation in a reference stress test
in the usual way, but recognised that it would be necessary to
account in some way for the large deformations.   They describe
a simple ad hoc procedure (which has no real theoretical
justification) for estimating the variation of component
deflection with time as large deformations occur.

We start from the method of similarity, Sec 3.1, eqn (28)
using an elastically calculated scaling factor, δ_E which is
subsequently modified.   For a single load Q resulting in an
elastic deflection q_E in some component, we may write

$$Q = k_E q_E$$

where k_E is the (linear) elastic stiffness from a load-
deflection diagram Fig 14.   Then the elastic scaling factor,
eqn (27), can be expressed as

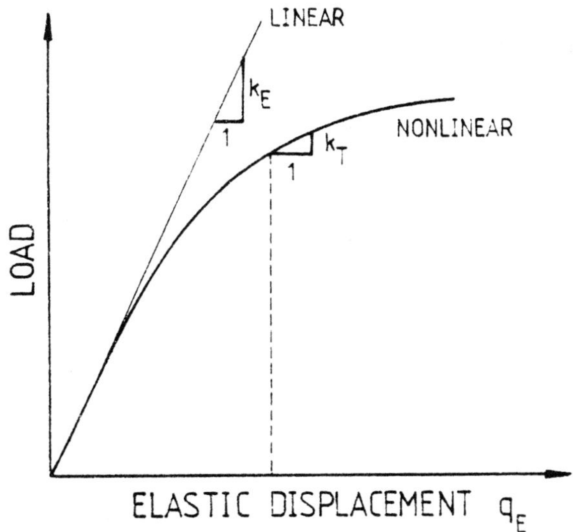

Fig 14 :   Elastic load-deflection diagram

$$\delta_E = \frac{q_E}{\sigma_R/E} = \left(\frac{E}{k_E}\right)\left(\frac{Q_L}{\sigma_y}\right)$$

using the limit type reference stress, eqn (11).   To account
for large deformations, Penny & Marriott suggested replacing
k_E by the instantaneous (tangential) stiffness for the current
deflection.   We denote this by $k_T(q)$, Fig 14, and modify the
scaling factor

$$\delta_T(q) = \left(\frac{E}{k_e}\right)\left(\frac{k_E}{k_T(q)}\right)\left(\frac{Q_L}{\sigma_y}\right)$$

and the reference stress approximation is

$$\dot{q} \simeq \delta_T(q) \times \dot{\varepsilon}_C(\sigma_R) \tag{33}$$

Thus, given the result of the reference stress test,
together with the load/deflection curve, both $\dot{\varepsilon}_C(\sigma_R)$ and
$\delta_T(q)$ can be determined.   Then eqn (33) is a simple differential equation for the displacement q;   this can be solved to
give the variation of displacement with time.   When this
grows without bound, the component is considered to have
buckled.

340

Finally, it should be noted that Gerdeen & Sazawal [39 - 41] have also described a "reference stress" approach to buckline. However, it is not really a reference stress method in the conventionally accepted sense, i.e. relating component behaviour to that of uniaxial creep tests. The method of Gerdeen & Sazawal assumes power law creep and is wholly influenced by any variation in the material parameters which the conventional reference stress approach aims at reducing. It is best not thought of as a reference stress method for buckling and is better interpreted as a simple procedure for assessing the buckling time of simple shell structures, which can be quite successful [42].

3.5 Creep rupture

A structure under constant load and subject to material creep will not be able to sustain that load indefinitely, eventually it will fail. This phenomenon is known as "creep rupture". The failure may be of a ductile nature accompanied by high local flow deformations, or of a brittle nature from material deterioration at high temperature. In the latter, failure occurs through the initiation and propagation of a region of failed material through the body of a structure. The study of such problems form the basis of the relatively new field of "continuum damage mechanics". This is treated in more detail elsewhere in this book.

The usefulness of a detailed stress analysis of a structure or component in the stages preceding failure is again brought into question due to the additional uncertainty which is introduced in attempts to mathematically model tertiary creep. Considerable scatter can often be observed in plots of applied stress and associated rupture time of uniaxial tests. In addition the criteria chosen for failure under multiaxial stress states can have a profound effect on the calculated rupture time. One of the best examples of this has been given by Henderson & Sneddon [43] who compared experimental and calculated rupture times for solid bars in torsion; they found that the incorrect multiaxial rupture criterion can lead to the latter being several orders of magnitude in error.

Considerable interest has focussed on simplified methods of assessing the lifetime of a creeping structure. Some success has been obtained for structures under constant load using the reference stress method. These will be discussed below: of course, once the rupture time has been computed it will depend on the estimated values of the material parameters. Then it is possible to use the local reference stress approach. For example, if t_R is the computed rupture time of a component, then it may be related to the rupture time of a uniaxial test, t_{RO}, by

$$t_R = \delta \times t_{RO} \tag{34}$$

in terms of some scaling factor, δ. As before, the reference stress for rupture may be chosen so that δ shows little variation with the material parameters. This approach has been investigated for a simple problem elsewhere by the writer [1, 44].

In the above, some expression for the rupture time is required either in the form of numerical values or as an estimate. Several investigators have examined the possibility of developing a simple estimate of the lifetime of a component based on the limit type reference stress concept. The aim is to choose a "reference rupture stress" which gives the <u>same</u> rupture time in the associated uniaxial test and in the component (thus in eqn (34) we have the restriction $\delta = 1$). Initially it had been thought that perhaps the reference stress for steady creep might coincide with that for rupture [2, 14]. But this takes no account of the known dependence on the multiaxial failure criterion. To remedy this Goodall & Cockroft [45] suggested a modification to the limit type reference stress for deformation, eqn (11), to allow for this. The reference rupture stress should be

$$\sigma_R = \frac{Q}{Q_L^R} \, \sigma_y \tag{35}$$

where Q_L^R is the limit load for an equivalent structure composed of a perfectly plastic material with yield σ_y <u>calculated using the appropriate multiaxial rupture criterion as the multiaxial yield criterion.</u> This would theoretically provide an upper bound, that is an unconservative estimate, of the final failure of the structure. In practice it would be preferable to have a lower estimate and in addition a structure may also be judged to have failed not only when it breaks but also, for example, in the case of a pressure vessel, when it leaks. There is thus also a need to consider local effects. Goodall et al [46, 47] derived an empirical formula which should provide both a lower estimate and take into account local failure. The reference rupture stress from eqn (35) would be modified as

$$\sigma_R^{mod} = \sigma_R \{1 + 0.13(\chi - 1)\} \tag{36}$$

where χ is the ratio of the limit load of the structure to the load for first yield. (And it should be noted that Ainsworth & Goodall [47] also considered proportional cyclic loading, but we will not discuss this further here).

An alternative to eqn (36) for local failure was suggested by Leckie & Hayhurst [48]. They recommended modifying the limit type reference stress, eqn (35) as

$$\sigma_R^{mod} = \sigma_R\{1 - \{1 - (1 + (K_e-K_p)/\nu K_p)^{-\nu}\}^{1-\nu}\}^{1/\nu} \qquad (37)$$

where K_e is the elastic local stress concentration factor, K_p the equivalent plastic stress concentration factor and ν the slope of the log stress to log rupture time graph (a material parameter). In this form the modified reference stress is not independent of all the material parameters.

Finally, it should be noted that Leckie & Hayhurst [48] derived bounds on the rupture time for kinematically determinate structures. The investigation of Goodall & Cockroft [45] was also aimed at developing a bound on the life -time (for a simple material model). There have been numer- ous attempts at bounding rupture time (a review and discussion can be found in the paper by Ponter [49]). Some of these can be expressed in the reference stress form; the reader is referred to these for further consideration.

3.6 Creep fracture

In our discussion of creep rupture above it was implicitly assumed that the failure would not originate from an existing defect (in particular a crack) which was able to grow by creep mechanisms. The study of the creep behaviour of structural components containing cracks is fairly recent and present knowledge is limited. Previous design philosophy presumed the component to be free of defects and adopted the criterion that no crack should appear during the design life. Current thinking accepts the presence of defects caused by manufacture and suggests basing the design on a consideration of the effects of crack initiation and propagation. Some work has been done on introducing the reference stress concept in creep fracture. This will be described briefly below.

The creep rupture of defect free structures results from the growth of a region of damaged material. A similar pheno- menon occurs in the presence of cracks. Two distinct situat- ions can occur. In the first, the growth of the crack is slower than the growth of the damaged region which moves ahead of the crack. In the second, the rate of crack growth exceeds the propagation rate of the damaged region. These two cases are treated separately.

If the growth of the damaged material dominates, then it has been argued by Goodall & Chubb [50] that the propagation of the crack will have little effect on the rupture life. Consequently the approach used for defect free structures, wherein the lifetime can be estimated from the limit-type reference stress (eqn (11), (35)) should remain applicable, provided the limit load is evaluated for the component with the initial defect. Of course this applies only to structures containing an initial defect and not to situations where a crack appears during service.

If the growth of the crack dominates, then the propagation of the crack will affect the eventual failure time. It is necessary to examine creep crack growth initiation and propagation rate. Attempts have been made to correlate creep crack growth rate with reference stress - this has been reviewed by Haigh [51], and Ellison & Harper [52]. In some situations crack growth rate can indeed be correlated with a reference stress such that the latter still governs the failure time[53]. In addition to knowing how the crack propagates, it is necessary to know the time at which it starts to grow (the initiation time). Ainsworth [54], using an idealised model, has given a relationship between initiation time t_i and the critical crack opening displacement at initiation, d_i, using the rupture time of a uniaxial test at the limit reference rupture stress, $t_R(\sigma_R)$, eqn (35). This takes the form for power law creep

$$\frac{t_i}{t_R(\sigma_R)} = \frac{\alpha(n)}{\epsilon_m} \left(\frac{d_i}{R}\right)^{\frac{n}{n+1}} \tag{38}$$

where

$$\alpha(n) = \frac{2(n + 1)}{n\sqrt{3}} \left(\frac{2n}{3n + 4}\right)^{\frac{n}{n+1}}$$

and R is a characteristic macroscopic distance which can be estimated either from tables for some geometries, or approximated from plane strain calculations. The material constant ϵ_M is the so-called Monkman-Grant constant which relates minimum creep rate $\dot{\epsilon}_{min}$ in a uniaxial test at stress σ_o to rupture time, t_R

$$\dot{\epsilon}_{min} \, t_R = \epsilon_M$$

This is an approximation [1].

The right hand side of eqn (38) varies with the exponent n and a reference stress interpretation for the initiation time t_i is not appropriate. Nevertheless, Ainsworth argues that eqn (38) could be used to give a direct assessment of the need to take creep crack growth effects into consideration. When the right hand side of eqn (38) is close to, or greater than, unity, then failure is governed by the reference stress approximation of Goodall & Chubb [50].

3.7 Summary

In this section we have described the extension of the reference stress concept, developed originally for steady creep, to other aspects of creep behaviour - forward creep,

relaxation, cyclic creep, buckling and failure.
It is apparent that, to some extent at least, the simple limit
type reference stress could be used in each case.   It is not
clear how this can be applied to the situations where we know
the simplified approach is inapplicable.

4   VERIFICATION OF THE REFERENCE STRESS METHOD

There have been several experimental studies of the
validity of the reference stress method.   With a few
exceptions, these have been concerned with rather simple
components and have been aimed at verifying the simplified
limit-type reference stress method.   Most of these will be
summarised below: again in each case the reader is referred
to the original papers.

4.1   Constant and cyclic bending tests on beams

Sim and Penny [55] tested a number of rectangular cross
section beams made from a commercially pure aluminium which
creeps at room temperature.   The beams were tested both in
forward creep and under cyclic loading.   The measured total
strains at the outside fibre of the beams are shown in Fig 15,

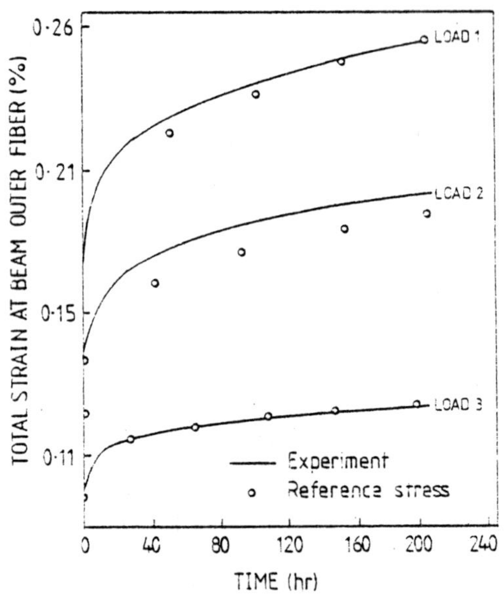

Fig 15 : Constant bending test on a rectangular
cross section beam [55]

together with the results of reference stress predictions.
The limit type reference stress, i.e. $\sigma_R = M/bd^2$, where M is

the applied moment, b the beam breadth and d the semi-
depth, was not used.   Rather a slightly different value, σ_R =
1.014 M/bd² was derived using the local approach, eqn (16).
It is not clear how the reported reference stress predictions
for the beam strains were related to the result of the
reference stress test, although the comparison is very good.
The results for cyclic loading are equally impressive, Fig 16,
where the measured outer fibre strain at the upper load level
is compared.

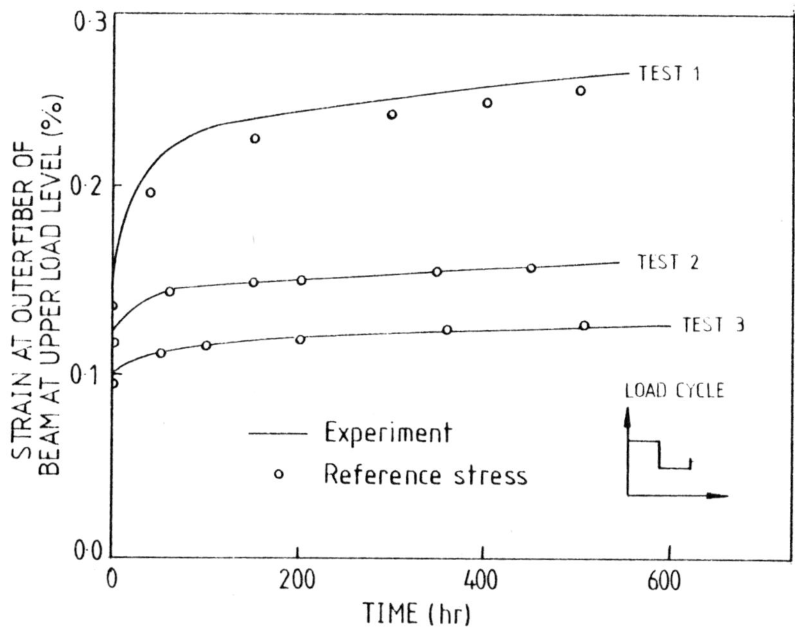

Fig 16 :   Cyclic bending test on a rectangular
cross section beam [55]

4.2 Constant and cyclic bending tests on thick tubes

Fairbairn [56] tested a number of thick tubes made from
aluminium alloy RR58 at 180°C under both constant and cyclic
bending.   He measured the maximum creep strain at the outside
surface of the tube, and compared these to the results of
appropriate reference stress tests.   It was found that
reference stress predictions underestimate the measured surface
strains in both cases.   Fairbairn used a reference stress
slightly different from that derived from the limit load.   In
fact if the reference stress was increased by only 5% a far
better comparison could be obtained, Fig 17.

346

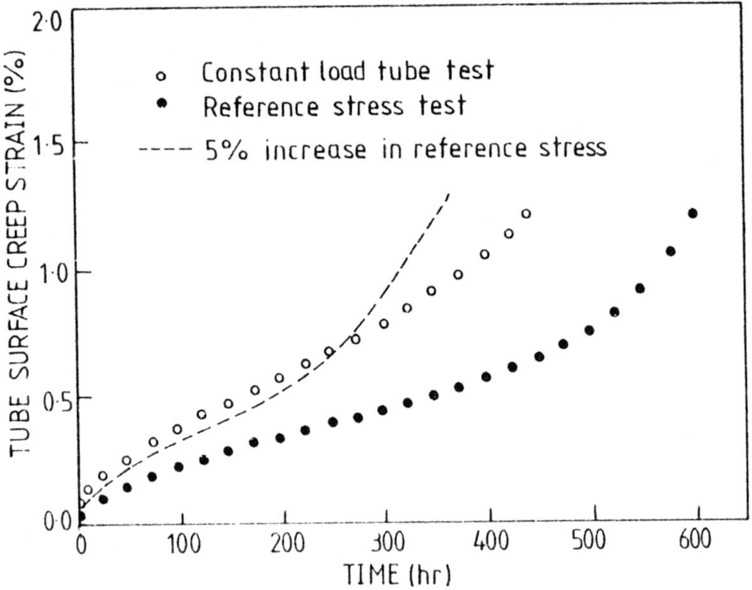

Fig 17 :   Constant bending test on a thick tube [56]

4.3   Pressure tests on model nozzle-sphere intersections

Penny and Marriott [57] carried out a series of tests on model aluminium pressure vessels at 180°C and compared the vertical nozzle deflection in forward creep with a reference stress test, Fig 18.   The limit type reference stress was used and the required limit load found from collapse tests on the models themselves; the scaling factor was found from elastic calculations, eqn (27).   It can be seen that the reference stress predictions diverge from the measured result; to account for this Penny & Marriott argued that large deflections were present and modified the reference stress to account for this (in a manner similar to that for buckling, Sec 3.4). This modified result is also shown.

4.4   Sway tests on portal frames

Leckie & Ponter [58] tested several portal frames made from both commercially pure aluminium at room temperature and from aluminium RR58 at 200°C under point loading.   In Fig 19 the comparison between the measured steady state sway rate and a reference stress test is given for RR58 frames.   The reference stress is based on the limit load solution, the required limit

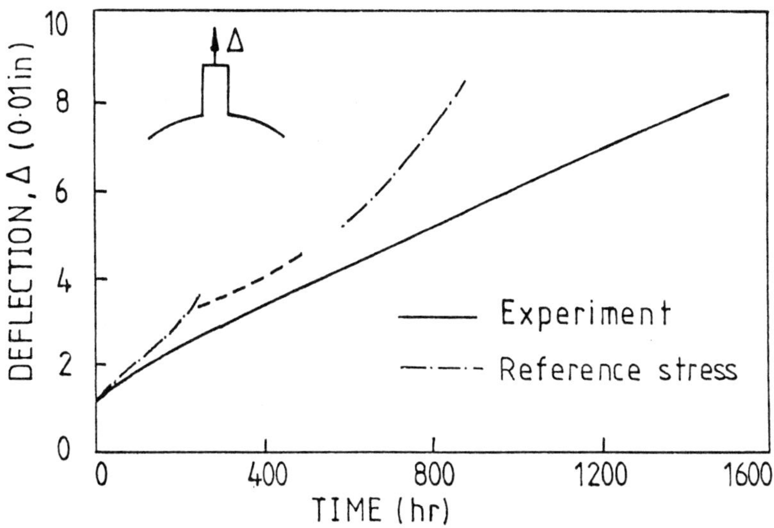

Fig 18 : Pressure test on a nozzle-sphere intersection [57]

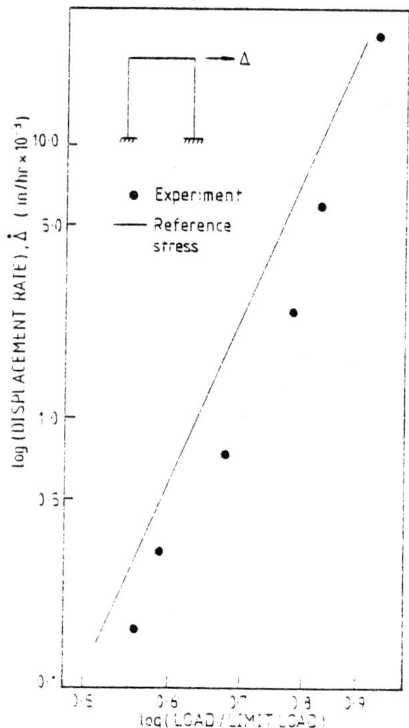

Fig 19 :   Sway test on a portal frame [58]

load being measured from experiment.   The scaling factor is
derived from an upper bound (similar to eqn (21)) and is
clearly quite unconservative.

4.5   Three-bar structure under variable temperature

Ainsworth [59] conducted a series of experiments on a
simple three-bar structure made from 316 stainless steel
subject to a constant mechanical load and to a variable
temperature between 600°C and 700°C.   Under cyclic thermal
loading the structure eventually exhibits a stationary cyclic
state and a repeated increment in displacement per cycle.
Reference stress calculations are also reported based on
approximate stationary cyclic stresses [34] in the manner
discussed in Sec 3.3.   The comparison is given between the
theoretical local reference stress σ_R and an "experimentally
determined reference stress" σ^{ε}_R which allows the measured
displacement increment to   "exactly"match the uniaxial
steady state data.   Results for a number of different tests
subject to different cycles are given below.

σ^{ε}_R	89.1	97.7	167.1	171.4	158.5	167.3	MN/m^2
σ_R	90.0	95.2	181.4	180.4	175.3	179.5	MN/m^2

The error in the theoretical reference stress σ_R is small;
however if this is translated to the measured average displacement increment over a cycle, the theoretical predictions
can greatly overestimate the experimental results.

4.6   Non-uniformly heated propped cantilever

Walter and Ponter [24] performed a series of tests on
commercially pure aluminium propped cantilevers subject to a
central vertical load and a temperature gradient.   They
showed by trial and error that the displacement rate in steady
creep at the centre of the span could be expressed in the form

$$\dot{q} \simeq \lambda \, \frac{L^2}{d} \, \dot{\varepsilon}(\mu\sigma_R, T_R)$$

where L is the length of the cantilever and d its semi-depth.
The reference stress they used approximated the limit-type,
$\sigma_R = M/bd^2$ and the factors μ, λ and "reference temperature" T_R
were chosen to be insensitive to the exponent n in a power
law in such a way that the above approximation was theoretically
valid.   A comparison between the measured steady deflection
rates and the reference stress prediction is given in Fig 20;
in view of the known difficulties with the reference
temperature approach, Sec 2.7, they are remarkably close.

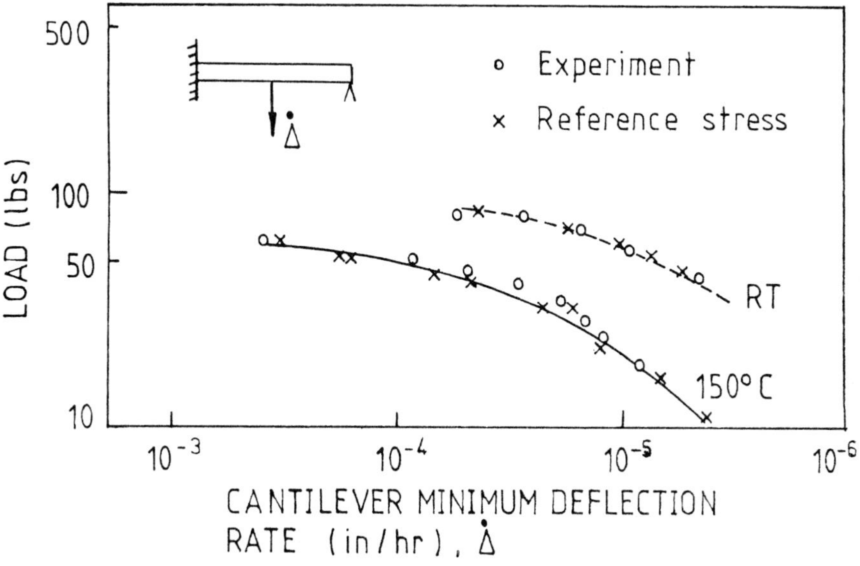

Fig 20 : Non-uniformly heated propped
cantilever [24]

4.7   Lead model component tests

Fessler, Hyde and Webster have carried out a number of val
-idations of the reference stress method on lead model beams
on cantilevers under constant load [60], and cyclic load [61]
and on hemispherically-ended cylindrical pressure vessels under
cyclic and constant load [62] and to rupture [63] as part of an
admirable independent investigation of the range of validity of
a simplified methods for creep. Here we will examine one set
of these tests in more detail - the tip loaded cantilever for
which we know the limit-type reference stress is not applicable [21].

Fessler et al tested three nominally identical cantilevers
made from a lead-antimony-arsenic alloy which creeps at room
temperature. The loads were chosen so that the anticipated
reference stress would be (i) within the elastic range of the
material, (ii) near the yield point and (iii) in the plastic
range. The measured tip creep deflection is shown in Fig 21
for these load levels. Also shown are the results of
reference stress predictions based on Mackenzie's method,
eqn (19) and using the limit type reference stress, eqn (11).

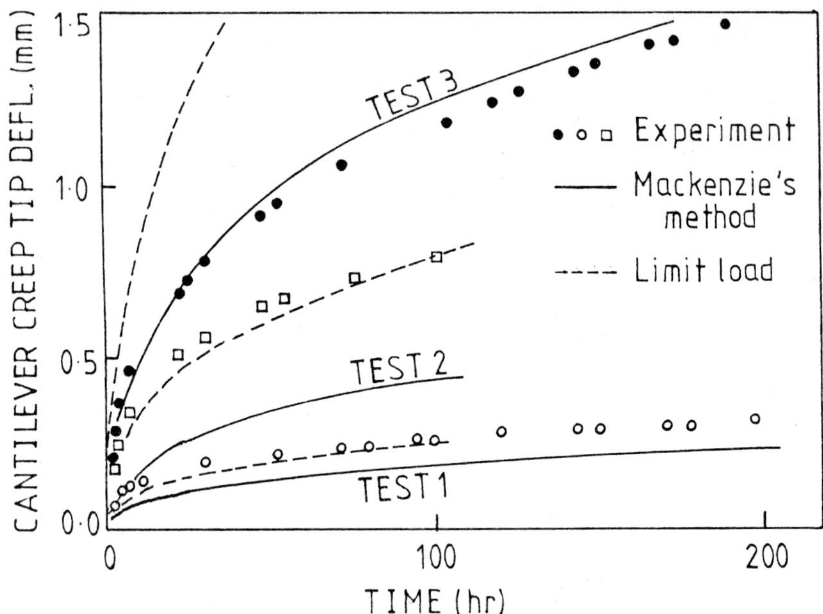

Fig 21 : Lead model cantilever under constant
load [60]

In the former

$$\sigma_R = 0.83 \frac{QL}{bd^2}$$

and in the latter

$$\sigma_R = \frac{QL}{bd^2}$$

The scaling factors on the creep deformations using eqn (25)
are in the former

$$\delta = 0.6 \frac{L^2}{d}$$

and in the latter, using the elastic solution eqn (27)

$$\delta = 0.5 \frac{L^2}{d}$$

It can be seen that these do not compare consistently with the
experiments. The predictions based on the limit type
reference stress are significantly greater than those based on
Mackenzie's method. The experiments lie close to one or the
other and, although the limit-type reference stress is better
for lower loads, it is much worse for higher loads.

These comparisons demonstrate well the potential dangers

in using the reference stress method.   We know [21] that the limit-type reference stress does not guarantee a scaling factor which is insensitive to the exponent n in a power law.   In the above components particular values of the scaling factor have been chosen.   Also Mackenzie's approach is based on choosing an average reference stress, which can lead to gross errors. The reference stresses used differ greatly and lead to dissimilar estimates of deflections.   With hindsight, it would have been better perhaps to use the local approach for the cantilever beam.   For this lead alloy the exponent in a power law has the estimated value, n = 4.7;   then using the local approach

$$\sigma_R = 0.866 \frac{QL}{bd^2} \qquad \delta = 0.473 \frac{L^2}{d}$$

It is probable that these values would give a more consistent comparison with the experiment.

The reported reference stress validations on beams in bending and in hemispherically ended cylindrical pressure vessels on the whole gave better comparisons.   But it is worth noting that the creep rupture predictions on the pressure vessels based on the simple approximations of Sec 3.5 differed from the measured rupture times by more than three orders of magnitude.

4.8   Estimation of creep deformations in pressure vessels

Matsuda has described how the reference stress approach can be used to estimate the creep deformations in flat plate pressure vessel heads [64] and in tapered cylindrical vessels [65].   Suitable reference stresses were evaluated so as to predict the strains at several locations on these vessels.   In the case of flat pressure vessel heads, the experimentally measured strains at these locations were compared to reference stress predictions using calculated reference stresses from other locations, Fig 22.   These compare favourably.   It would have been interesting to make a more conventional reference stress comparison using the limit-type reference stress; Matsuda evaluated his reference stresses using a method similar to Mackenzie's.

4.9   Sphere-cylinder intersections under internal pressure

Leckie et al [66] gave some detailed comparisons between approximate methods and experimental results on a pressurised sphere/cylinder intersection made from aluminium RR58 and tested at 200°C.   They obtained a theoretical upper bound on the vertical nozzle displacement in the steady state (similar to eqn (21)) as

$$\dot{q} \leq 12.5 \frac{\sigma_R}{p} \dot{\varepsilon} (\sigma_R)$$

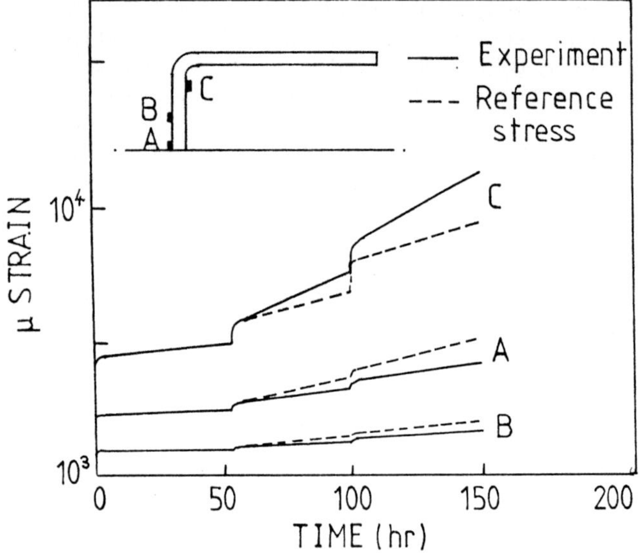

Fig 22 :   Flat pressure vessel head [64]

where $\sigma_R = \dfrac{p}{p_L}\,\sigma_y$ is the limit reference stress and p_L the

limit pressure on the vessel.   In Fig 23 this upper bound is

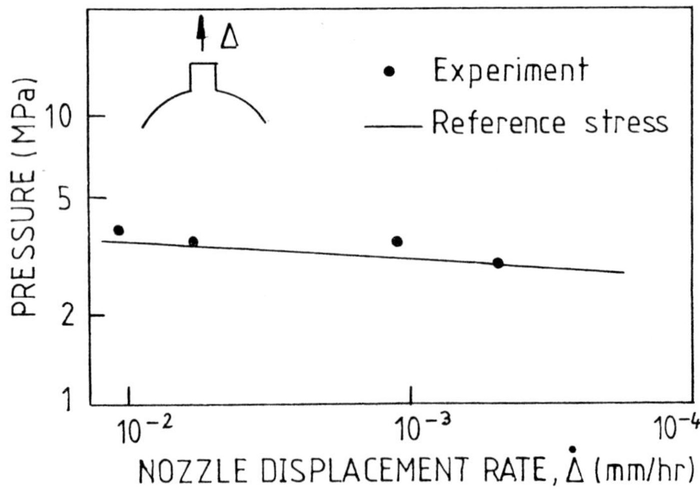

Fig 23 : Pressurised sphere-cylinder intersection [66]

compared to the measured nozzle displacement rate for constant
load.   The comparison is extremely good.   An equally good
comparison was obtained for the average nozzle displacement
rate over a cycle in the cyclic stationary state assuming
cyclic loading.

Leckie et al also examined the rupture of this vessel and
concluded, as pointed out by Penny & Marriott (Sec 4.3), that
it was very conservative to calculate deformation near rupture
based on the undeformed geometry.   Leckie [67] has discussed
this point further and argues that the lifetime of this
component could be underestimated by a factor of four using
the limit load approximation.

4.10 Buckling of boss loaded spherical shells

Penny & Marriott [68] studied the creep in stability of an
aluminium boss loaded spherical shell at 200°C.   Using the
procedure outlined in Sec 3.4, they were able to compare the
measured deflection of the boss to an approximate analysis
which included the effects of large deformations based on a
modification to the limit-type reference stress.   These
comparisons are shown in Fig 24 for two buckling tests at
different load levels.   In both of these the estimated failure
time is conservative and considerably so at the higher load.

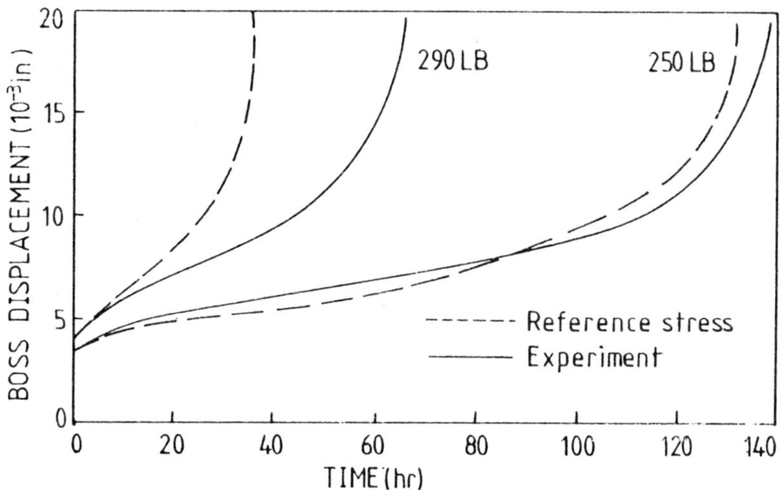

Fig 24 : Buckling of a boss-loaded
spherical shell [68]

354

4.11 <u>Complex torsional creep tests on solid bars and thick tubes</u>

Henderson and Ferguson [69] used complex torsional creep
tests on solid bars and thick tubes made from three different
steels and an aluminium alloy to attempt a severe test of the
capabilities of the reference stress method.   In each test a
complex load history was followed consisting of (i) constant
load followed by (ii) removal of load to allow creep recovery
until (iii) the angle of twist is held constant to maintain
relaxation conditions and, finally, (iv) a constant torque was
imposed leading to rupture.   The measured history of twist in
these torque tests on one of the steels is shown in Fig 25.
Also shown are results from a uniaxial test following a
similar programme of load determined by the reference stress.
In general it was found that the reference stress prediction
was acceptable in the first three stages of loading, but
greatly underestimated the final rupture time.

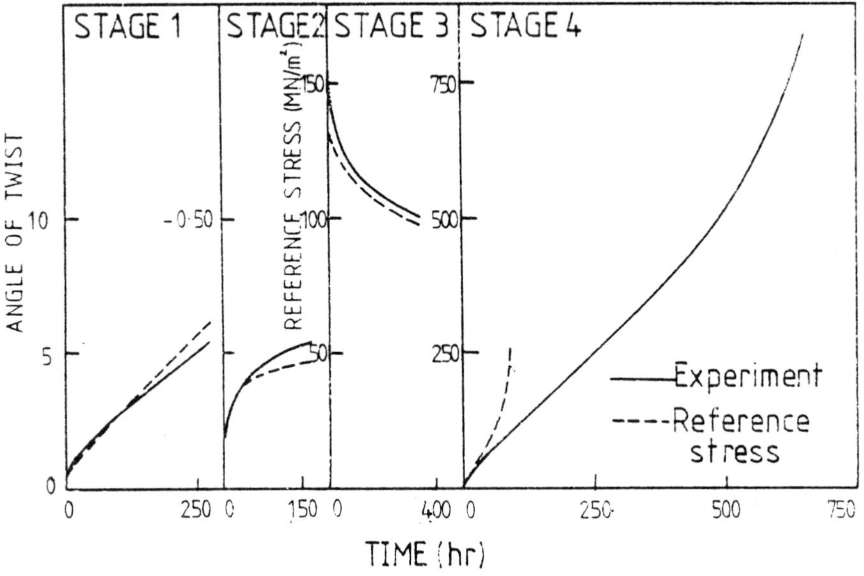

Fig 25 : Torsion of a thick tube [69]

The interpretation of the reference stress method adopted
by Henderson and Ferguson is essentially different from, but
related to, the conventional one we have described here.   In a
torsion test the angle of twist, φ, is related to the effective
strain at some radius r by

$$\varepsilon_e = \frac{r}{\sqrt{3}} \frac{\varphi}{\ell} \tag{39}$$

where ℓ is the length of the tube.   From experimental observation there exists some radius at which the effective stress does not alter significantly during creep - this is known as the "skeletal point" [26].   Let σ^*_e be the effective stress at the skeletal point r^* then from eqn (39)

$$\varphi = \frac{\sqrt{3}\ell}{r^*} \; \varepsilon_e(\sigma^*_e) \tag{40}$$

Since σ^*_e does not depend on the creep constitutive relation (it is unaltered during creep) we may identify it as a reference stress and consequently $\varepsilon_e(\sigma^*_e)$ is the strain from a uniaxial test held at the reference stress σ^*_e;   from eqn (40) the scaling factor takes a simple known form.

The success of this procedure depends on the existence of a skeletal point and on the assumption of kinematic determinancy, eqn (39).   In general, even if a skeletal point exists, it is not possible to determine deflections if the strain is known at only one point.   At most the strain at the skeletal point can be obtained.   There is also some question as to whether the skeletal point concept is applicable to rupture.   It is known that a region of failed material spreads through the component;   the position of the skeletal point would have to change and probably also the magnitude of the stress.   It is then not surprising that in the above the comparisons for rupture time are poor.

4.12 Failure of pressurised pipes and tubes

Cane and Browne [70] also used the skeletal point concept to compare analytically derived "reference stresses" and experimentally derived reference stresses on a series of pressurised steel tube tests conducted $565^{\circ}C$.   From the tests the measured steady state strains on the outer surface of the tube were converted to strains at the skeletal point.   These are plotted against applied pressure scaled to the limit-type reference stress, Fig 26.   To provide a comparison the associated uniaxial minimum creep rates are superimposed on this plot.   It can be seen that the result of a reference stress test would overestimate the measured strain at the skeletal point.

4.13 Verification of simplified estimates of component lifetime

The simple estimates of lifetime based on reference stress methods using the limit-type reference stress derive from the work of Goodall and Cockroft [45].   There has been considerable effort by this group aimed at experimentally assessing this simple approach for design.   These will be summarised below.

Goodall [71], using the results of [45] obtained bounds on the pressure carrying capacity of a cylinder-cylinder

356

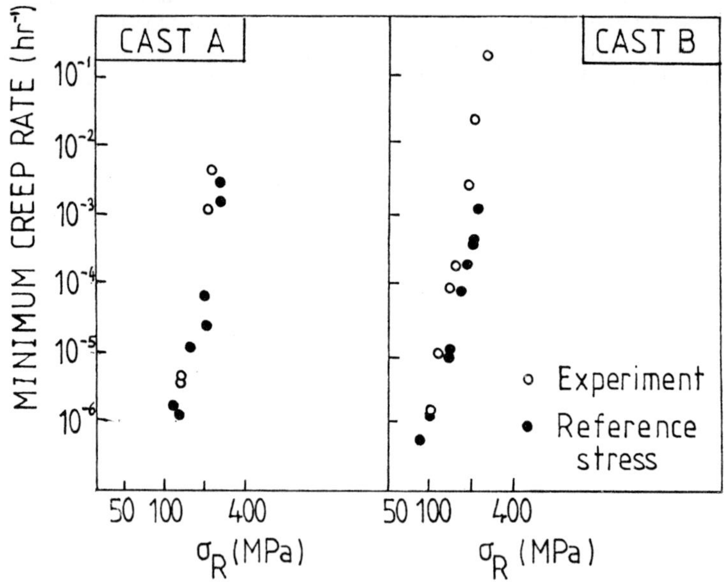

Fig 26 : Pressurised tube test [70]

intersection in terms of the reference stress. These bounds
take the form

$$0.86 \, \sigma_R \; \leqslant \; \frac{p_R r}{h} \; \leqslant \; \sigma_R$$

where h is the thickness of the main cylinder, r its mean
radius and p_R the pressure required to give rupture in a time
t_R.   The reference stress σ_R is the uniaxial stress required
to give rupture in the same time t_R.   From reported experiments it was found that $p_R r/h \simeq \sigma_R$;   thus rupture coincided
with the upper bound (i.e. the time predicted by the reference
stress test).

A number of experimental investigations have been reported
by Goodall and Ainsworth [41].   These indicated that the
modified reference rupture stress (eqn (36)) was conservative,
but not unduly so.   For example, for a uniaxial constant load
rupture test on an aluminium alloy plate with a slot, tested at
150°C, the rupture time was plotted against the net section
uniaxial stress obtained from experiment.   Superimposed on
this was the rupture time obtained from a uniaxial test at the
modified reference rupture stress.   The comparison
was   greatly conservative in this case, but better results
were obtained for plates with holes.

Goodall and Chubb [50] and Ainsworth [54] examined the
use of the limit-type reference stress for fracture (Sec 3.6)

by comparison with tests on various sized double-edge notched
tensile specimens of 1% Cr MoV steel [72] and compact tension
specimens of several Cr MoV steels [73].   Some results for
the latter are reproduced below for two different materials.

Material	Experimental $t_i/t_R(\sigma_R)$	Predicted (eqn (38)) $t_i/t_R(\sigma_R)$
1	0.32	0.75
1	0.26	0.96
2	0.29	0.32
2	0.31	0.38
2	0.52	0.36

The experimental and predicted values of the ratio of
crack initiation time to lifetime at the reference stress
compare favourably for only one of the materials.   In each
case it can be seen that $t_i < t_R(\sigma_R)$ so that as expected the
reference stress prediction is unconservative.

4.14 Summary

The verification of the reference stress method has
largely been aimed at assessing its use as a simplified method
for estimating component creep behaviour.   It appears that,
when applied to constant or cyclic loading in the stages well
before failure, the method can be, perhaps surprisingly, quite
successful.   Its application to rupture remains an open
question.   Unfortunately there has been no real study of the
validity of the fundamental premise of the reference stress
method.   That is, that it enables the known variability in the
estimation of component creep behaviour, which results from the
uncertainty in the material creep law, to be reduced by
relating to further, more specific and relevent tests.   It
would be most interesting to examine this.

5   THE REFERENCE STRESS METHOD IN CREEP DESIGN

There are two related, but fundamentally different, interpretations of the reference stress method.   The local approach
which follows the original aims of the method in dealing with
uncertainties in the material constitutive equations alone,
and the simplified approach based on plastic limit analysis.
The latter is perhaps more relevant to the designer since it
avoids the need for any detailed creep stress analysis and leads
to simple yet meaningful results which are similar to the

358

design methods adopted below the creep range.   Not surprising
-ly a comprehensive design methodology based on the reference
stress concept has been proposed.   This will be outlined
below, but firstly some background to current design practice
should be given.

5.1 Design criteria

The design process is concerned with the identification
and prevention of several modes of failure.   In the first
instance it will be useful to list and briefly explain the
principal structural failure modes which can occur.   For
steady loading these are:

(i)   Extensive short term yielding which can lead to ductile
      rupture.

(ii) Creep rupture from long-term loading.

(iii) Loss of function due to excessive deformation in the form
      of large creep distortion which can obstruct moving parts
      or in the form of inelastic buckling.

For cyclic loading the additional failure modes are:

(iv) Gross distortion due to incremental collapse or ratchett-
      ing in the plastic regime which can be enhanced by creep.
(v)   Combined creep and low-cycle fatigue.

Other failure modes can be envisaged (for example, high cycle
fatigue, material deterioration due to environmental effects)
but adequate design rules have not yet been formulated for
these.

5.2 ASME Boiler and Pressure Vessel Code, Case N-47-14: Class 1 components in elevated temperature service

Most existing design codes are aimed at preventing the
occurrence of the above failure modes, but only one, the
American Society of Mechanical Engineers (ASME) Code Case N-47
represents any formal attempt to regulate design practice for
both steady and cyclic loading at elevated temperature taking
into account real component behaviour [74].   The Code Case
is an offshoot of Section III of the ASME Code which applies to
water-cooled nuclear plant.   It is intended to provide rules
for materials, design and certification of components whose
service temperatures exceed 800°F for austenitic and high alloy
steels and 700°F for ferritics.   The Code Case is complex and
its contents, of necessity extensive; here we can only outline the general features.

The ASME Code recognises that improvements are possible to
design if the results of detailed stress analysis are available.   Thus for steady loading the calculated maximum stress
in the steady state must be limited by an appropriate "time-

dependent material stress", S_t which is defined as

(i)   the minimum stress to rupture in time t, divided by 1.5,

(ii) the minimum stress to 1.0% creep strain in time, t,

(iii) 80% of the minimum stress to reach tertiary creep in time t.

These limits are intended to prevent creep rupture; additional limits on the maximum calculated accumulated inelastic strain are also imposed to avoid excessive deformation; these are:

(a)   Strains averaged through the wall thickness of a vessel, 1%.

(b)   Linear strain distributions, 2% at the surface of a vessel.

(c)   Local strain concentrations, 5%.

This presumes that the detailed stress analyses can be carried out; if not, Appendix T of the Code Case provides simple procedures based on <u>elastic</u> analysis to estimate total inelastic strain.

A similar approach is adopted for cyclic loading: it is possible to use the results of inelastic analysis, but alternative routes are allowed using the results of elastic analysis.   However, except for certain simple well documented situations, the ASME Code only allows the use of elastic analysis if the elastically calculated stresses are below yield at all times during the service life.

5.3   An approach to design based on reference stress concepts

There is an obvious need to relax the criteria adopted in the ASME Code Case.   The inelastic route based on detailed stress analysis is expensive, particularly so for cyclic loading, and of course this must be based on appropriate constitutive models which can be unavoidably inexact.   The elastic route is conservative and its derived rules are based on gross generalisations to the real mechanics of the problem.

Several attempts have been made outside the U.S. to provide improvements to the ASME Code Case.   The most interesting in the present context is based on reference stress concepts proposed by Ainsworth and Goodall [75] for discussion by the Design Codes Working Group for the Fast Reactor in the U.K.   Summaries of these proposals are available elsewhere in the published literature [76, 77, 78].

The application of the reference stress concept is straightforward for <u>constant</u> loading.   Three design criteria are satisfied in the following form:

(i) Excessive deformation is considered using the limit type reference stress, eqn (11) and an elastic scaling factor, eqn (27).

(ii) Failure due to ductile rupture is considered using the modified reference rupture stress, eqn (36) for defect-free (uncracked) structures.

The treatment of failure due to brittle rupture is slightly different; the maximum steady state stress in a structure composed of a power law material is given approx-imately by

$$\sigma_{max}^{ss} \simeq \sigma_{max}^{\infty} + \frac{1}{n} (\sigma_{max}^{e} - \sigma_{max}^{\infty})$$

where σ_{max}^{e} is the maximum elastic stress (corresponding to $n = 1$ and Poisson's ratio $\nu = \frac{1}{2}$) and σ_{max}^{∞} is the maximum stress in the limit state for a perfectly plastic material (corres-ponding to $n \rightarrow \infty$). The modified reference stress for brittle rupture is given approximately by the above expression replacing σ_{max}^{∞} by the limit-type reference stress, eqn (11). The Ainswoth and Goodall proposals also extend these concepts to cyclic loading. In this, "modified shakedown limits" are suggested which allow a relaxation of the restrictions of the simplified elastic route of the ASME Code Case to permit its application beyond yield to a modified shakedown boundary. A full discussion of these is given in Ref [78].

5.4 Discussion

The use of reference stress concepts in design is attract-ive for many reasons. Not only does it avoid full inelastic analysis and provide simple design procedures but also, by the very nature of the method some consideration is given to real component creep behaviour and the accepted variability of the basic creep material data. Of course the above described proposals are only a first attempt for discussion. There are as many caveats. We know the limit type reference stress can fail for simple problems and that there is some difficulty in its use for thermal loading (Sec 2.7). Some proposals for dealing with the latter are given by Ainsworth and Goodall using simplifications developed in an unpublished report by Ainsworth [79]. Nevertheless, some difficulties still remain unresolved. It is not clear how the design methods proposed deal with multiple loading on a component. In addition the applicability of the so called creep modified shakedown limit to strongly strain hardening alloys such as austenitic steels has been brought into question [18].

Again it must be stressed that only an initial attempt has been made at formulating a design methodology based on

reference stress concepts and naturally problems will occur.
The results of the method are both attractive and powerful,
but numerous details still need to be addressed.

6   CONCLUSIONS

"...if a man will begin with certainties, he will end with
doubts, but if he will be content with doubts he shall end in
certainties..." Francis Bacon

There is perhaps an impression that modern engineering is
an exact science, probably due to its use of the scientific
method, complex mathematics, computers, and so on.   It is
often forgotten that we are dealing with real physical
structures and their vagaries.   Indeed, it is very easy to
become involved in a detailed stress analysis of some component
and to lose sight of the fact that the final result only
approximates reality.   This is particularly true in high temperature design since the material constitutive relation
required by an analysis is in itself only a relatively crude
representation for real loading histories.   We have seen in
Sec 2 that uncertainties in the constitutive relation can lead
to an un acceptably wide range of predicted deformations in
steady creep of a simple structure.   Applying this uncertainty
to design could lead to unnecessary conservatism, or worse -
the design may not in fact be safe.   One possibility would be
to introduce the fact of the basic uncertainty at the beginning of the design process and formulate the rules accordingly
[80].   Another possibility would be to minimise the effects
of the uncertainty on the stress analysis.   By its very nature
the reference stress approach to design takes into account any
doubts regarding the actual material behaviour by simply
referring the component response back to the uniaxial creep
data.   In principle then any uncertainty also reduces to a
consideration of the available creep data alone.

The crucial question must now be - how successful has the
method proved itself as a genuine tool for design?   In this
review we have identified two different but related interpretations of the reference stress concept.   One, which we have
called the "local approach" deals only with the effects of any
uncertainty on the results of a stress analysis; theoretically
it should always be successful.   This has not found application in design so far.   The other, which we have called the
"limit approach", and which has been suggested as a suitable
design procedure, leads to simple, powerful results requiring
only elastic and plastic limit analysis.   It is this approach
which is normally referred to as the reference stress method,
and which we will discuss further.   The limit approach has
been most successful when dealing with steadily loaded
structures, although we have seen for example in the case of a
cantilevered beam, it is sometimes not applicable.   Further,
with the evidence we have so far, it is unproven for varying

loads and in the estimation of failure times.   There is
considerable work yet to be done on validating the <u>details</u> of
the reference stress method for design.   Yet, as we have seen
in the experimental comparisons of Sec 4, it can lead to some
very striking results.   For this reason alone it cannot be
dismissed.   If the reference stress method is not yet in a
useable form, then neither should it be unduly criticised.
More research should be undertaken into discovering the nature
of the method and why it should allow the time dependent creep
response of a component to be expressed in a particularly
simple and meaningful form.   The emphasis has certainly been
on the reference stress itself in the understanding of how it
relates to the plastic limit load.   Little work has been done
on the nature of the equally important scaling factor.   For
example, can the limiting value of the scaling factor be
expressed in an identifiable form?   When applied to the estimation of failure times in structures, the method is less
acceptable.   The basis for this is the upper bound on rupture
time established by Goodall & Cockcroft [45].   This restricts
the method since no scaling factor is required; in fact
modifications to the reference rupture stress have also failed
to introduce a scaling factor.

The key to the reference stress method is perhaps not in
its simplicity, but in its treatment of the uncertainty in
modelling the real material behaviour.   The conclusion which
must be drawn here is that, viewed as a simplified method for
design, it is far from complete.   But the alternative, a full
inelastic analysis should not be judged to be any better given
the uncertainties in the basic material model for complex,
multiaxial combined mechanical and thermal loading histories.
An abundance of line printer output and a heavy bill for
computer usage can lead to a false sense of security.

7   ACKNOWLEDGEMENT

Much of the research carried out by the writer on the
reference stress method was funded by the Science and Engineer
-ing Research Council through an Advanced Fellowship.

8   REFERENCES

GENERAL:

1. BOYLE, J. T. and SPENCE, J. - _Stress Analysis for Creep_ Butterworths, 1983.

2. MARRIOTT, D. L. - 'A review of reference stress methods for estimating creep deformation', IUTAM Symposium on _Creep in Structures_ 1970, Ed. J. Hult, Springer-Verlag, 1972, p.137.

3. KRAUS, H. - 'Reference stress concepts for creep analysis', Welding Research Council Bull. No.227, 1977.

DEVELOPMENT OF REFERENCE STRESS CONCEPTS FOR STEADY CREEP:

4. ANDERSON, R. G., GARDNER, L. R. T. and HODGKINS, W. R. 'Deformation of uniformly loaded beams obeying complex creep laws', J.Mech.Eng.Sci. 1963, $\underline{5}$, 238.

5. SODERBERG, C. R. - 'Interpretation of creep tests on tubes', Trans. A.S.M.E., 1941, $\underline{63}$, 737.

6. SIM, R. G. - _Creep of structures_, PhD Dissertation, University of Cambridge, 1968.

7. JOHNSSON, A. - 'An alternative definition of reference stress for creep'. Int. Conf. on Creep and Fatigue in elevated temperature applications, I. Mech.E., 1973, Paper C205/73.

8. JOHNSSON, A. - 'Reference stress for structures obeying the Prandtl and Dorn creep laws', J.Mech.Eng.Sci., 1974, $\underline{16}$, 298. Also Discussion, J.T. Boyle, ibid, 1975, $\underline{17}$, 178.

9. SIM, R. G. - 'Reference stress concepts in the analysis of structures during creep', Int. J. Mech.Sci., 1970, $\underline{12}$, 561.

10. SIM, R. G. - 'Evaluation of reference parameters for structures subject to creep', J. Mech. Eng. Sci., 1971, $\underline{13}$, 47. Also Discussion, C. R. Calladine, ibid, 1972, $\underline{14}$, 159.

11. SIM, R. G. - 'Reference results for plane stress creep behaviour', J. Mech. Eng. Sci., 1972, $\underline{14}$, 404.

12. SIM, R. G. and PENNY, R. K. - 'Plane strain creep behaviour of thick-walled cylinders', Int. J. Mech. Sci., 1971, $\underline{13}$, 987.

13. MACKENZIE, A. C. - 'On the use of a single uniaxial test to estimate deformation rates in some structures undergoing creep', Int. J. Mech. Sci., 1968, $\underline{10}$, 441.

14. FAIRBAIRN, J - 'Applications of the reference stress concept', Natl. Eng. Lab. Repl No.436, 1969.

15. BOYLE, J. T. - 'A consistent interpretation of the reference stress method in creep design', Mech. Res. Comm., 1977, $\underline{4}$, 41.

16. BOYLE, J. T. - 'Approximations in the reference stress method for creep design', Int. J. Mech. Sci., 1980, $\underline{22}$,73.

17. PONTER, A. R. S. and LECKIE, F. A. - 'Application of energy theorems to bodies which creep in the plastic range', J. Appl. Mech., 1970, $\underline{37}$, 753.

18. PONTER, A. R. S. - 'On the creep modified shakedown limit', 3rd IUTAM Symposium on <u>Creep in Structures</u>, 1980. Ed. A. R. S. Ponter and D. R. Hayhurst, Springer, 1981.

19. BOYLE, J. T. - 'The theorem of nesting surfaces in steady creep and its application to generalised models and limit reference stresses', Res. Mech., 1982, $\underline{4}$, 275.

20. ANDERSON, R. G. - 'Some observations on reference stresses, skeletal points, limit loads and finite elements', 3rd IUTAM Symposium on <u>Creep in Structures</u> 1980. Ed. A. R. S. Ponter and D. R. Hayhurst, Springer, 1981.

21. BOYLE, J. T. and SPENCE, J. - 'The use of the reference stress method in creep analysis - a caution to designers', <u>Material Behaviour at Elevated Temperatures and Components Analysis</u>, ASME Special Publ. PVP-60, 1982.

REFERENCE TEMPERATURES:

22. SIM, R. G. - 'Reference stresses and temperatures for cylinders and spheres under internal pressure with a steady heat flow in the radial direction', Int. J. Mech. Sci., 1973, $\underline{15}$, 211.

23. WILLIAMS, J. J. and COCKS, A. C. F. - 'Reference stress and temperature for non-isothermal creep of structures' J. Appl. Mech., 1979, $\underline{46}$, 795.

24. WALTER, M. H. and PONTER, A. R. S. - 'Some properties
     of creep of structures subjected to non uniform
     temperatures', Int. J. Mech. Sci., 1976, 18, 305.

25. PONTER, A. R. S. - 'The creep of structures subjected
     to temperature gradients : the evaluation of a refer-
     ence temperature', Univ. of Leicester, Dept of Engin-
     eering Rep. 79-10, 1979.

FORWARD CREEP:

26. PENNY, R. K. and MARRIOTT, D. L. - Design for Creep.
     McGraw-Hill, 1971.

27. BILL, I. M. and MACKENZIE, A. C. - 'Reference stress
     for redistribution time in creep of structures',
     J. Mech. Eng. Sci., 1969, 11, 429.

28. CALLADINE, C. R. - 'Time scales for redistribution of
     stress in creep of structures', Proc. Roy. Soc.
     London, Ser A, 1969, 309, 363.

RELAXATION:

29. SPENCE, J. and HULT, J. - 'Simple approximations for
     creep relaxation', Int. J. Mech. Sci., 1973, 15, 741.

30. BOYLE, J. T. and SPENCE, J. - 'A comparison of
     approximate methods for estimation of the creep
     relaxation of a curved pipe', Int. J. Press. Vess. &
     Piping, 1976, 4, 221.

31. LADANYI, B. - 'Borehole relaxation test as a means
     for determining the creep properties of ice covers'.
     Proc. 5th Port and Ocean Engineering under Arctic
     conditions conf., Trondheim, 1979, p.757.

CYCLIC LOADING:

32. PONTER, A. R. S. - 'Deformation, displacement and
     work bounds for structures in a state of creep and
     subject to variable loading', J. Appl. Mech., 1972,
     39, 953.

33. PONTER, A. R. S. - 'On the stress analysis of creep-
     ing structures subject to variable loading', J. Appl.
     Mech., 1973, 40, 589.

34. AINSWORTH, R. A. - 'Approximate solutions for creep-
     ing structures subjected to periodic loading',
     Int. J. Mech. Sci., 1976, 18, 149.

366

35. WILLIAMS, J. J. and LECKIE, F. A. - 'A method for
     estimating creep deformation of structures subjected
     to cyclic loading', J. Appl. Mech., 1973, 40, 928.

36. WILLIAMS, J. J. and LECKIE, F. A. - 'A method for
     quantifying creep strain due to cyclic stress',
     J. Appl. Mech., 1974, 41, 953.

37. WILLIAMS, J. J. and DIMMER, P. R. - 'An approximate
     method for computing structural creep deformation
     due to non-proportional cyclic loading', Int. J.
     Mech. Sci., 1976, 18, 453.

38. RADHAKRISHNAN, V. M. - 'A parametric approach to
     cyclic creep', J. Appl. Mech., 1976, 43, 28.

CREEP BUCKLING:

39. GERDEEN, J. C. and SAZAWAL, V. K. - 'A review of
     creep instability in high temperature piping and
     pressure vessels', Welding Research Council Bulletin,
     1974, 195, 33.

40. SAZAWAL, V. K. and GERDEEN, J. C. - 'Creep buckling
     of cylindrical shells using a reference stress
     concept', ASME Paper No.76-PVP-50, 1976.

41. MAYVILLE, R. and GERDEEN, J. C. - 'Creep buckling of
     spherical shells', ASME Paper No.76-PVP-45, 1976.

42. CHERN, J. M. - 'A simplified approach to the
     prediction of creep buckling time in structures' in
     Simplified Methods in Pressure Vessel Analysis,
     ASME Special Publ. PVP-PB-029, 1978, p.99.

CREEP RUPTURE:

43. HENDERSON, J. and SNEDDON, J. D. - 'Creep fracture
     of copper cylindrical bars under pure torque',
     Natl. Eng. Lab. Rep.No.509, 1972.

44. BOYLE, J. T. - 'The use of statistical reference
     stress techniques in nonlinear structural analysis',
     IUTAM Symp. on Physical Non-linearities in
     Structural Analysis, Ed. J. Hult, J. Lemaitre,
     Springer-Verlag, 1981, p.29.

45. GOODALL, I. W. and COCKROFT, R. D. H. - 'On bounding the life of structures subjected to steady load and operating in the creep range', Int. J. Mech. Sci., 1973, 15, 251.

46. GOODALL, I. W., COCKROFT, R. D. H. and CHUBB, E. J.- 'An approximate description of the creep rupture of structures', Int. J. Mech. Sci., 1975, 17, 351.

47. GOODALL, I. W. and AINSWORTH, R. A. - 'Failure of structures by creep', Proc. 3rd Int. Conf. on Pressure Vessel Technology, Vol II, 1977, p.871.

48. LECKIE, F. A. and HAYHURST, D. R. - 'Creep rupture of structures', Proc. Royal Soc. London, Ser A, 1974, 340, 323.

49. PONTER, A. R. S. - 'Upper bounds on creep rupture life of structures subjected to variable load and temperature', Int. J. Mech. Sci., 1977, 19, 79.

CREEP FRACTURE:

50. GOODALL, I. W. and CHUBB, E. J. - 'The creep ductile response of cracked structures', Int. J. Fracture, 1976, 12, 289.

51. HAIGH, J. R. - 'The mechanisms of macroscopic high temperature crack growth. Part II - Review and reanalysis of previous work', Mater. Sci. Engng, 1975, 20, 225.

52. ELLISON, E. G. and HARPER, M. P. - 'Creep behaviour of components containing cracks - a critical review' J. Strain Anal., 1978, 13, 35.

53. AINSWORTH, R. A. - 'Some observations on creep crack growth', Int. J. Fracture, 1982, 20, 147.

54. AINSWORTH, R. A. - 'The initiation of creep crack growth', Int. J. Solids Structures, 1982, 18, 873.

EXPERIMENTAL VERIFICATION OF THE REFERENCE STRESS METHOD:

55. SIM, R. G. and PENNY, R. K. - 'Some results of testing simple structures under constant and variable loading during creep', Exptl. Mech., 1970, 152.

56. FAIRBAIRN, J. - 'A reference stress approach to creep bending of straight tubes', J. Mech. Eng. Sci., 1974, 16, 125.

57. PENNY, R. K. and MARRIOTT, D. L. - 'Creep of pressure
vessels', Int.Conf. on Creep and Fatigue in elevated
temperature applications', I.Mech.E., 1973,
Paper C204/73.

58. LECKIE, F. A. and PONTER, A. R. S. - 'Theoretical and
experimental investigation of the relationship between
plastic and creep deformation of structures',
Arch. Mech., 1972, $\underline{24}$, 419.

59. AINSWORTH, R. A. - 'An experimental study of a three-
bar structure subjected to variable temperature',
Int. J. Mech. Sci., 1977, $\underline{19}$, 247.

60. FESSLER, H., HYDE, T. H. and WEBSTER, J. J. -
'Stationary creep predictions from model tests using
reference stresses', J. Strain Anal., 1977, $\underline{12}$, 271.

61. HYDE, T. H. and WEBSTER, J. J. - 'Assessment of some
methods for predicting creep deformations and life of
components', Conf. on Engineering Aspects of Creep,
I.Mech.E., 1980, Paper C192/80.

62. FESSLER, H., HYDE, T. H. and WEBSTER, J. J., -
'Experimentally determined reference stresses for the
prediction of creep deformations and life of
components using models', Conf. on Recent Developments
in High Temperature Design Methods. I.Mech.E., 1977,
Paper C272/77.

63. FESSLER, H., HYDE, T. H. and WEBSTER, J. J., -
'Prediction of creep rupture of pressure vessels'
ASME Energy Technology Conf., Houston, 1977,
Paper No.77-PVP-54.

64. MATSUDA, F. - 'Reference stress approach for
estimating creep strain of flat plate pressure
vessels heads', Proc. 6th Int. Conf. on Structural
Mechanics in Reactor Technology, Paris, 1981,
Paper No.M5/4.

65. MATSUDA, F. and FUJIKAWA, T. - 'Reference stress
approach for estimating creep deformation of tapered
cylindrical vessels', Conf. on Engineering Aspects of
Creep, I.Mech.E, 1980, Paper C241/80.

66. LECKIE, F. A., HAYHURST, D. R. and MORRISON, C. J. -
'The creep behaviour of sphere-cylinder shell inter-
sections subjected to internal pressure', Proc. Roy.
Soc. London, Ser A, 1976, $\underline{349}$, 9.

67. LECKIE, F. A. - 'The role of geometry change in high temperature design', Int. J. Mech. Sci. Special S. S. Gill Conference Issue, 1982.

68. PENNY, R. K. and MARRIOTT, D. L. - 'Creep buckling of boss loaded spherical shells', Proc. 1st Int. Conf. on Pressure Vessel Technology, Delft, 1969, Paper II-68.

69. HENDERSON, J. and FERGUSON, F. R. - 'Assessment of the reference stress method by complex torsional creep tests', Conf. Engineering Aspects of Creep, I.Mech.E. 1980 Paper C245/80.

70. CANE, B. J. and BROWNE, R. J. - 'Representative stresses for creep deformation and failure of pressurised tubes and pipes', Int. J. Press. Vess. & Piping, 1982, 10, 119.

71. GOODALL, I. W. - 'The creep of branch connections' Conf. on Creep Behaviour of Piping, I.Mech.E., 1974, Paper C45/74.

72. TAYLOR, E. and BATTE, A. D. - 'Creep crack formation in a 1% Cr MoV rotor forging steel' Conf. on Engineering Aspects of Creep, I.Mech.E, 1980, Paper C216/80.

73. HAIGH, J. R. - 'The mechanisms of macroscopic high temperature crack growth - Part I: Experiments on tempered Cr MoV steels' Mat. Sci. Engng., 1975, 20, 213.

DESIGN FOR CREEP:

74. —, ASME Code Case N47 of the ASME Boiler and Pressure Vessel Code, 1977.

75. AINSWORTH, R.A. and GOODALL, I.W. - 'Proposals for primary design above the creep threshold temperature- homogeneous and defect free structures' C.E.G.B Rep RD/B/N4394, 1978.

76. TOWNLEY, C.H.A. - 'Design Criteria of High-Temperature Plant' Chap 10 of Creep of Engineering Materials and Structures, Ed. Bernasconi, A and Piatti, G, Applied Science, 1979.

77. GOODMAN, A.M - 'Design related aspects of creep and fatigue' Chap 8 of Creep and Fatigue in High Temperature Alloys, Ed. Bressers, J, Applied Science, 1981.

78. GOODALL, I.W. et al - 'The development of high
    temperature design methods based on reference stresses
    and bounding theorems'. Trans ASME, 1979, 101, J.Eng.
    Mat & Techn, 349.

79. AINSWORTH, R.A. - 'A rapid method for the evaluation
    of reference parameters for creeping structures under
    non-uniform temperature conditions' CEGB Rep.
    RD/B/N3985, 1977.

80. BLOCKLEY, D.I. - The Nature of Structural Design and
    Safety. Ellis Horwood, 1980